Incarnation and Physics

Recent titles in
AMERICAN ACADEMY OF RELIGION
ACADEMY SERIES

SERIES EDITOR
Carole Myscofski, Illinois Wesleyan University

A Publication Series of
The American Academy of Religion
and
Oxford University Press

Profile of the Last Puritan: Jonathan Edwards, Self-Love and the Dawn of the Beatific
David C. Brand

Victor Turner Revisited: Ritual as Social Change
Bobby C. Alexander

Chinese Women and Christianity, 1860–1927
Kwok Pui-Lan

The Man in the Yellow Hat: Theology and Psychoanalysis in Child Therapy
Dorothy W. Martyn

The Grace of Difference: A Canadian Feminist Theological Ethic
Marilyn J. Legge

The Intersubjectivity of the Mystic: A Study of Theresa of Avila's *Interior Castle*
Mary Frolich

Narrating History, Developing Doctrine: Friedrich Schleiermacher and Johann Sebastian Drey
Bradford E. Hinze

Analogical Possibilities: How Words Refer to God
Philip A. Rolnick

Womanist Justice, Womanist Hope
Emilie M. Townes

Women Don't Count: The Challenge of Women's Poverty to Christian Ethics
Pamela K. Brubaker

The Exploration of the Inner Wounds—*Han*
Jae Hoon Lee

Comprehending Power in Christian Social Ethics
Christine Firer Hinze

The Greening of Theology: The Ecological Models of Rosemary Radford Ruether, Joseph Stiller, and Jürger Moltmann
Steven Bouma-Prediger

The Spirit and the Vision: The Influence of Christian Romanticism on Development of 19th-Century American Art
Diane Apostolos-Cappadona

The Freedom of the Spirit: African Indigenous Churches in Kenya
Francis Kimani Githieya

Bridge-Makers and Cross-Bearers: Korean-American Women and the Church
Jung Ha Kim

God Bless the Child That's Got Its Own: The Economic Rights Debate
Darryl M. Trimiew

Energies of the Spirit: Trinitarian Models in Eastern Orthodox and Western Theology
Duncan Reid

The Goddess Laksmi: The Divine Consort in South Indian Vaisnava Tradition
P. Pratap Kumar

Creative Dwelling: Empathy and Clarity in God and Self
Lucinda A. Stark Huffaker

Hospitality to Strangers: Empathy and the Physician-Patient Relationship
Dorothy M. Owens

The Bonds of Freedom: Feminist Theology and Christian Realism
Rebekah L. Miles

The Specter of Speciesism: Buddhist and Christian Views of Animals
Paul Waldau

Incarnation and Physics: Natural Science in the Theology of Thomas F. Torrance
Tapio Luoma

AAR
AMERICAN ACADEMY OF RELIGION

Incarnation and Physics

Natural Science in the Theology of Thomas F. Torrance

TAPIO LUOMA

OXFORD
UNIVERSITY PRESS

2002

OXFORD
UNIVERSITY PRESS

Oxford New York
Auckland Bangkok Buenos Aires Cape Town Chennai
Dar es Salaam Delhi Hong Kong Istanbul Karachi Kolkata
Kuala Lumpur Madrid Melbourne Mexico City Mumbai Nairobi
São Paulo Singapore Taipei Tokyo Toronto

and an associated company in Berlin

Published by Oxford University Press, Inc.,
198 Madison Avenue, New York, New York 10016

www.oup.com

Library of Congress Cataloging-in-Publication Data
Luoma, Tapio, 1962–
Incarnation and physics : natural science in the theology
of Thomas F. Torrance/Tapio Luoma.
p. cm. — (American Academy of Religion academy series)
Includes bibliographical references (p.) and index.
ISBN 0-19-515189-5
1. Incarnation. 2. Religion and science. 3. Torrance, Thomas Forsyth, 1913–
I. Title. II. Series.
BT220 .L86 2002
261.5'5'092—dc21 2001051379

1 3 5 7 9 8 6 4 2

Printed in the United States of America
on acid-free paper

To Pirjo, Antti, Juho, and Anni

Preface

As with any genuine effort, academic work requires commitment. This means that researchers cannot simply apply a half-hearted attitude toward their projects but must surrender to them with their whole essence. My commitment has led me, as it were, on a long journey during which time a variety of feelings, ranging from hilarious joy to inconvenient depression, became familiar to me. The process of academic work is indeed a journey, not only to the landscape of the subject under investigation but also to the innermost self of the researcher. For me this process has been extremely rewarding, and what is most important to me is that I have not had to make this journey alone. Many people, theologians, physicists, philosophers, and others, have generously helped me with their time and knowledge.

My journey into the frontier area between theology and science actually began when I won a book in a boys' club competition at the Lutheran parish of Vaasa. The book, George Gamow's *Mr Tompkins in Wonderland*, first introduced me to the enchanting world of Einstein's theories of relativity. I could not imagine then how useful it would eventually prove to be some twenty years later, when Professor Eeva Martikainen suggested that I could take the thinking of the Scottish professor Thomas F. Torrance into consideration in my postgraduate studies. I am deeply grateful to her for this suggestion and for her constant support during this process. Indeed, Thomas Torrance's thinking has been an inspiring companion to me throughout. Membership in the Research Project on Modern Theology, led by Professor Martikainen, has been immensely helpful.

This work would probably have been delayed for an indefinite amount of time without the help of a bursary granted to me by the Board of World Mission at the Church of Scotland, which enabled me to study at the University of Edinburgh in 1993–94. My supervisor, Dr. Gary Badcock, after having read my first primitive papers, gave me a most useful piece of advice: "You should start from the Incarnation!" To him, to the people in the Church of Scotland, and to the Faculty of Divinity at the University of Edinburgh I wish to express my deep gratitude. I realise that the continuous support and encouragement of the staff in the Office for Theological Affairs (Department for International Relations) at the Evangelical Lutheran Church of Finland was of indispensable help during my bursar months in Scotland, and for this I am also truly grateful.

Numerous other people in Finland and abroad have significantly contributed to the completion of this dissertation. In my efforts to put Torrance's thinking in a wider intellectual and historical context, I owe many thanks Professors Reijo Työrinoja, Simo Knuuttila, and Miikka Ruokanen (Department of Systematic Theology of the University of Helsinki), who gave me valuable points of view; without their input my work would certainly have lacked something essential. Docents Kari Enqvist (Department of Physics) and Jaakko Rusama (Academy of Finland) have kindly read my work and offered many useful suggestions. Rev. Dr. John Polkinghorne and Professor William R. Stoeger, S.J., both of whom are authoritative experts in this field, have not hesitated in giving freely their time in providing me with encouragement and important points of view. To these scholars, and to many other people with whom I have had the opportunity to discuss issues relating to this work, I would also like to express my deep gratitude.

I have been priviledged to have done this research in Peräseinäjoki and Ilmajoki, rural townships in Southern Ostrobothnia, which have provided an excellent and inspiring environment for reflection and writing and where I have been constantly encouraged by my fellow workers and by the members of the Lutheran parishes in Peräseinäjoki and Ilmajoki.

In addition to the Church of Scotland, my work has been financially supported by the Christian Foundation for the Support of Science and Arts, the Finnish Cultural Foundation (South Ostrobothnia), the Research Institute of the Evangelical Lutheran Church of Finland, the Annala Foundation, Peräseinäjoki Community and the Lions Club of Peräseinäjoki. I sincerely thank them all for their help and contributions.

Lic. Theol. Michael Cox has carefully read my work as a language consultant and helped me to make it more comprehensible.

It is a great honor to have this book published jointly by the American Academy of Religion and Oxford University Press. I cordially thank Dr. Carole Myscoski of the AAR and Mr. Theodore Calderara of the Oxford, as well as other persons in these organizations with whom I have been privileged to work, for their relentless help and advice in the final stages of the manuscript preparation.

It certainly goes without saying that I could not have succeeded in this journey without the encouragement of my friends and relatives and, in particular, of my mother, Sirkka, and father, Veikko, who have always showed their confidence in my ability to accomplish difficult tasks. There is, however, a very special source of strength and inspiration, constantly sharing with me the joys and strains of life and of my academic work. There are simply no appropriate words to describe my gratitude to my wife, Pirjo, and our children, Antti, Juho, and Anni. I dedicate this book to my family with love.

Ilmajoki
Feast of the Incarnation 2001 — Tapio Luoma

Contents

Incarnation and Physics

1

Introduction

> [A]s our science presses its inquiries to the very boundaries of being, in macrophysical and microphysical dimensions alike, there is being brought to light a hidden traffic between theological and scientific ideas of the most far-reaching significance for both theology and science.
>
> From the General Foreword to the series Theology and Science at the Frontiers of Knowledge

In recent decades the advance in the natural sciences and especially in quantum physics has raised a number of questions with ontological and metaphysical insights. The ultimate nature of microcosmic events and the vastness of the universe with the problem of its beginnings have challenged scientists critically to reevaluate modes of thought that were applied during the era of Newtonian mechanics and its philosophical implications. Since the seventeenth and eighteenth centuries, the view of theology and science as two totally contradictory and exclusive manners of approaching reality has come to attract not only scientific circles but also the popular mind. This tradition of the idea of conflict has now, however, to a great extent lost its actual point of strict controversy or, at least, its ultimate mode detectable in the so-called *warfare theory*, expounded in the last decade of the nineteenth century, according to which the relationship can be explained in military terms.[1] This approach has been replaced by one in which a new understanding of the relationship is searched for by posing the modest question, "Is it possible to take religion seriously in an age of science?"[2] In spite of this new shift, problems cannot be easily removed, a constant reminder of which fact is the popular view of the fate of Galileo Galilei (1564–1642) when confronted with papal authority, although historically his case is a very human tale of ambition, obstinacy, and unfortunate coincidences rather than one of sheer conflict between emerging science and the Christian Church and theology.[3]

The history of the relationship between theology and science is a long and diverse one, and their interrelation has been seen as a distinctive feature of humankind.[4] Not always has this relationship been an explicit one, but especially in ancient times it was a subtle and conceptually imprecise intermingling of both religious and scientific motives, as in ancient Greece, where a view of nature was inevitably connected with a conception of deities interfering arbitrarily in mundane affairs. Only insights generated by the rise of philosophical rigor helped to draw a sharper demarcation line, resulting partly in a form of atheism and denial of any divine action, but also in the great philosophies of nature in the thought of Plato and Aristotle, to the structure of which a certain conception of a god essentially belongs.[5] Christianity, having emerged within the immensely complex religiosity of Antiquity, interpreted the interaction of God and the created universe on its own terms, which were occasionally sensed as threatening by the Imperial authorities, especially with regard to the cult of the Emperor.[6] A special case that has been uncritically taken to represent the hostile relationship between Christianity and the philosophical thinking of Antiquity is the acclamation of Tertullian of Carthage (c. 155–c. 230), who asked, "What indeed has Athens to do with Jerusalem?"[7] There is, however, weighty evidence for concluding that Tertullian's words were not directed against intellectual philosophical enterprise as such but rather against such philosophies as tended to favor speculations resulting in heresies like Gnosticism.[8]

A shift to another long period of relations between Christian theology, a term preferable to that of *religion*, began with St. Augustine (354–430), who reflected carefully on the connection between faith and reason. On the one hand, he maintained that there can be no understanding without faith and that revelation is the ultimate authority.[9] But on the other hand, the relationship between "faith and reason is therefore that between the precondition and the ultimate objective, the means and the end; in that sense we find in Augustine the subordination not of reason to faith but of faith to reason."[10] Augustine's thought with its Neoplatonic context was assumed and applied by the Christian West for almost a millennium, at the end of which it met a serious challenge in the philosophy of Aristotle, which was revived after a number of his works had been translated from Arabic into Latin in the twelfth century. Aristotle's philosophy appeared to attract widely both natural philosophers and theologians, and one of its exponents, the Englishman Roger Bacon (c. 1220–c. 1292) held that theology "does not oppress the sciences . . . but puts them to work, directing them to their proper end."[11] The harmonious coexistence of theology and science, so typical of the Middle Ages in general, reached its highest peak in the impressive synthesis of Christian theology and Aristotelian philosophy achieved by St. Thomas of Aquinas (c. 1225–1274). In his solution, however, one can hardly detect any independent natural philosophy or natural science but one serving the ends of theology, "the noblest of all sciences."[12]

Quite soon after the discovery of Aristotle's philosophy, its problematical nature from the standpoint of Christian theology became apparent. The conflict culminated in 1277 when the bishop of Paris condemned more than two hundred articles of the Aristotelian natural philosophy. On the one hand, this act certainly was an intrusion of the ecclesiastical authority into the territory of science, but, on the other, a novel road for the development of scientific thinking was thereby opened. In the rejection of Aristotelian ideas, such as the eternity of the world or the view that there are several truths, others accessible to sciences and others to faith, the church tried to safeguard its understanding of the universe. At the same time, however, this effort to defend the omnipotence of God, which was seriously threatened by the Aristotelian view of natural impossibilities (e.g., the existence of other worlds than ours is impossible), forcefully confirmed belief in the radical contingency of the universe and rejection of necessitarian and deterministic interpretations.[13]

The fifteenth century launched enterprises that anticipated the beginning of a new era in the history of Western Christendom, including the Age of Discovery with the acquisition of new information about remote countries and nations and the increasing dissatisfaction with the Roman Catholic Church with its scholastic theology.[14] The

Reformation did not necessarily mean a thoroughly radical revolution in the relationship between science and theology because the main issues at stake were purely theological efforts to give rationale for the justification of sinners as well as the absolute sovereignty of God. But, indirectly, the consequent idea of the passivity of human beings before God has been described as having paved the way for the untroubled reception of a mechanistic understanding of the universe which, in its turn, was a "reassertion of a Platonic view of mathematics over the view of Aristotle, which had dominated natural philosophy since the thirteenth century."[15] It should be remembered that the Reformation was involved more directly in the development of science. The Lutheran "Wittenberg Circle," led by Philipp Melanchthon, played a crucial role in conveying Copernicus's major work *De Revolutionibus* to the West. Within this circle, particularly the mathematician Georg Joachim von Lauchen, known also as Rheticus, and the theologian Andreas Osiander shared a common interest in Copernicus's astronomical work and in its propagation.[16]

The relationship between science and theology entered a new phase in the seventeenth century through the work of Sir Isaac Newton (1642–1727), who has generally been accredited with launching what has come to be known as the *Scientific Revolution*.[17] Although his own efforts in the scientific field, in which the publication in 1687 of his work *The Mathematical Principles of Natural Philosophy* was the most eminent contribution to physical science for the centuries to come, were "apparently entirely subordinate to his religious beliefs,"[18] his way of using God as "an explanatory and regulative principle within the system of efficient causes"[19] later led, with the advance of scientific explanation, to the exclusion of this *God of the gaps* from rigorously pursued science.[20] One of the representatives of the Newtonian mechanistic conception of the universe, Pierre Simon Laplace (1749–1827), drew the ultimate conclusions from this, reporting, according to the well-known story, to the Emperor Napoleon that he had no use for "the hypothesis" of God.[21] This reaffirms the fact that the mechanistic view of reality promoted by Newtonian science could not but affect religious views as well, resulting, for instance, in a considerable deviation from traditional orthodox Christianity and a turn toward such beliefs as deistic theologies.[22] Accordingly, this understanding proved to be an effective tool in the hands of the materialists of the French Enlightenment in their attacks upon Christianity and in their constant appeals to the sole authority of reason.[23]

This materialism found further support in the biological theory of evolution and natural selection as formulated by Charles Darwin (1809–1882) in his groundbreaking work *On the Origins of Species*. Although it was not his intention to attack Christianity with his conception of the emergence of life through chance natural selection, it nevertheless offered another proof of a supposed scientific nature to those who preferred to see a radical separation between theology and science. The tension between the disciplines grew not only because of the shocking implications of Darwin's theory for theology and their ready use in the hands of materialists but also because of the rigorous conservative positions taken in defense of theology.[24] But conservatism was not the only option available to theologians. Broadly speaking, the nineteenth century testified to a single theological shift of emphasis in response to the continuing advance of the Enlightenment program with its materialism, rationalism, and, finally, positivism, supported by alleged but highly controversial evidence from the natural sciences. Religion in general and theology in particular were gradually viewed as making claims, not so much with regard to an objective, truly existent reality but rather with regard to the subjective inner religious states and feelings of a human being. Especially in Germany, efforts of a relatively liberal spirit were made to remove the tension between science understood in a positivist way and theology with its supernatural basis.[25] This took place, however, at the expense of the traditional orthodox interpretation of the central dogmas of the Christian faith, including the doctrines of creation, Incarnation, and redemption. Beginning from the foremost German theologian of the continental Romantic era, Friedrich Schleiermacher (1768–1834), this resulted in a situation where Bible scholars and theologians saw Christianity "as an ethical and social teaching born from and acting upon human feeling, not knowledge."[26] Theology and science were thus radically separated, even with the consent of notable theologians, for decades to come, reserving objective reality for the inquiries of the natural sciences alone and leaving moral and emotional issues to theology.[27] In a truly positivistic spirit, references to metaphysics were excluded.[28]

The development in physics, however, provided hints at another approach to reality with a less positivistic ideology. In particular, *the theories of relativity* of Albert Einstein (1879–1955) and *the quantum theory* had a remarkable influence on the traditional mechanistic conception of nature upon which naturalistic positivism had been constructed. The theories of special and general relativity pay attention to the macrocosmos with the implication that space and time cannot be conceived as absolute entities but as integral parts of observable phenomena. This led to the question of the origins of the universe, which provided a new point of contact for theological interests, for example in the efforts to accept the so-called Big Bang theory as a scientifically expressed version of the biblical understanding of creation.[29] Quantum theory, on the other hand, concentrating on the elementary particles of matter, has strongly underlined not only the inadequacy of a mechanistic, Newtonian view of the universe and the element of surprise inherent in scientific research but also the existence of limits to rational inquiry.[30] Together these two physical theories have intimated the existence of what lies *beyond* observable phenomena or historical traces and raised critical questions concerning the ability of science to give mankind a lasting and enduring means to prosperity and happiness.[31] However, it would be naive to assume that all physicists agree upon these features or their implications for theology and interdisciplinary dialogue.[32] Suggestions for such a general agreement undoubtedly provoke feelings not only in physicists but also among theologians who have found it unnecessary to become involved with issues other than those belonging strictly to their own discipline.

Those who have found mutual discussions fertile refer not only to the theories of relativity or quantum mechanics but, following the advance of physics, have discerned new insightful challenges which seem to emerge continuously. Among them can be named a variety of ideas relating to dynamical systems, chaos theory, thermodynamics as well as the physics of self-organization and of complex systems constellation.[33] The interaction is considered so crucial that it has been suggested that the issue should be entitled *the discipline of science and religion*. The importance of this interaction lies in its ability to provide theology with new insights, based on contemporary scientific understanding, with which to explicate classical theological affirmations. Its beneficial educative influence in both academic and religious communities has been emphasized as well.[34]

Many of the theological implications of the recent development in the natural sciences have been raised by scientists. There are, however, notable theologians who have taken the challenge seriously and shown profound interest in the relations between science and theology. Above all, their involvement indicates in an articulate manner the deeply ecumenical nature of the theological response to the challenge of the modern natural sciences. The Lutheran scholar Wolfhart Pannenberg, for instance, has reflected on the epistemological question of the relation between faith and reason as well as the role of natural theology and, most interesting, physical space and time as aspects of the activity of the Holy Spirit.[35] In America, Philip Hefner, professor of systematic theology at the Lutheran School of Theology in Chicago, has contributed to the discussion not only with his theology of creation but also by editing an eminent magazine in the field, *Zygon*. The Roman Catholic theologian Karl Rahner has researched frontier questions between theology and the natural sciences, including methodology, suggesting that a strong metaphysical orientation guides scientific enterprise, even unconsciously, and demanding the appreciation of theology as a complementary partner.[36] Representatives of the Jesuit order have made noteworthy contributions to the present discussion through their relentless interest in questions of natural science and authoritative argumentation.[37] John Polkinghorne, an Anglican theologian as well as a scientist, has combined his vast scientific knowledge with theological motives, resulting in challenging, insightful and sophisticated arguments in which the relationship between theology and science is carefully reflected upon. The number of theologians and scientists from various Christian denominations who have contributed to the interdisciplinary dialogue is impressively high, so that the persons named above are indicated only with the purpose of underlining the ecumenical nature of the enterprise.

Perhaps the most widely known contemporary theologian in the Reformed tradition who has participated in the discussion is the former professor of Christian dogmatics in the University of Edinburgh, Thomas Forsyth Torrance (1913–). Torrance's contribution to the discussion in question, as well as its uniqueness, is based on his Barthian background and the centrality of Christological themes, both of which elements are examined in the course of this work. What makes Torrance's point interesting is that, unlike many other theologians, he does not attach himself primarily to the doctrine of creation as the theological starting-point for interdisciplinary interaction but to Christology and especially to the doctrine of the Incarnation, involving the concept of *homoousion*, the Greek term with which the Early Church Fathers expressed their belief in the consubstantiality of the Son with the Father. For Torrance, the *homoousion* explicates the overlap in the common reality of God and human beings and the universe in which each of them is open toward others. Indeed, the *homoousion* is the substantial and structural principle of Torrance's theological thinking.[38]

The aim of this work is to examine Torrance's view of the significance of the doctrine of the Incarnation, involving the patristic term *homoousion*, for both theology and the natural sciences and, accordingly, to their mutual relations.[39] The main problem to be investigated lies in Torrance's conviction that in physics, too, and, implicitly, in the other natural sciences the substance of the concept of *homoousion* has weighty relevance.[40] On what basis Torrance can claim this and what in his view is the actual role of the *homoousion* for the two disciplines and their relationship are questions to be answered in the course of the work. The method to be used is that of systematic analysis, with the aid of which a model for the solution of the problem is sought on the basis of Torrance's literary production.

Torrance's literary activity has been enormous, including essays, articles, and monographs. In his newspaper writings he has expressed his own stance on acute issues in Christianity, in his own Church of Scotland and in Western society in general. The principal sources of this work are Torrance's books, often collections of articles published previously, in which the relationship between science and theology and his view of the *homoousion* are articulated. His other writings serve as secondary sources. Torrance's interest in the theology-science relations grew in the 1960s, resulting in the publication of *Theology in Reconstruction* (1965), his first book with articulate reflections on matters of science. The two books published in 1969, *Theological Science* and *Space, Time and Incarnation*, represent a culmination in Torrance's thinking, as his subsequent works have more or less repeated views and insights originally presented in these two earlier books. Torrance's interest in the issue continued during the following decades, and he published more books on the subject. *God and Rationality* (1971) concentrates on a reflection on the intelligibility of theology and the created universe and thus bears an emphatically philosophical slant. *Space, Time and Resurrection* (1975) applies the problem of space and time to the questions of mortality and eternity; in *Divine and Contingent Order* (1981), Torrance discusses the role of the idea of indeterminacy and contingency in both theology and the natural sciences, while *Ground and Grammar of Theology* (1980) and *Reality and Scientific Theology* (1985) continue the treatise of the view of theology as science but now with more links to the natural sciences. *Transformation and Convergence in the Frame of Knowledge* (1984) includes the most detailed discussions and therefore offers an authoritative picture of Torrance's view on the history of science and Western thought with regard to theology.

The books named above occasionally contain very sophisticated argumentation, and therefore Torrance has consciously made efforts to present his thoughts in a more comprehensible way in publications with a more pastoral intent. To these belong works such as *Christian Theology and Scientific Culture* (1980), in which the close relationship between the faith of the Church and the prevailing Western context is emphasized, and *The Mediation of Christ* (1992) where Christological themes are developed to give a consistent view of the role of Israel, Christ, and our scientific society, as well as the place of the humans and their response to the Word of God. In *Reality and Evangelical Theology* (1982), Torrance presents his view of an adequate interpretation of the Bible, rejecting both liberal and fundamentalist positions as untenable.

In his recent works *The Trinitarian Faith* (1988), *Trinitarian Perspectives* (1994), and *Divine Meaning* (1995), Torrance combines his authoritative knowledge of patristic theology with his understanding of the philosophy and history of science as well as with Christological themes from his Barthian background. These books also have an ecumenical intention which Torrance has articulated with regard to contemporary interchurch relations in *Theology in Reconstruction* (1975). Perhaps his most significant book, ultimately, is *The Christian Doctrine of God: One Being, Three Persons* (1996), in which he draws together his theological understanding in concise form, repeating themes and viewpoints familiar to his readers from earlier writings and summing up the very heart of his mature theology.[41] The insights it contains are considered against their historical background in Torrance's most recent book, *Scottish Theology* (1996), in which he examines the development of theological thinking in his own tradition, thus offering an enlightening account of his own position as a Reformed theologian.

Partly due to the fact that quite a number of the books mentioned are actually collections of articles published earlier independently, Torrance's reader cannot avoid the feeling that he makes continuous repetition of his central themes using a mode of argumentation that has prevented him from presenting his views in a strictly systematic form. The chapters of his books are long and without any easily conceivable structure. However, one can sense the fluency of his writing, and, accordingly, his reluctance to use subheadings may be best understood as an indication of the fact that he considers them as being a hindrance to his argumentation rather than a helpful way of arranging the material under discussion. On the one hand, his style of writing makes it rather easy to gain a grasp of his arguments, but, on the other hand, finding minor but significant details can prove quite an arduous task.

Torrance's thoughts have aroused a significant response from all over the Western world, and the number of studies on different aspects of his theology is likely to rise in the future. Quite recently Alister McGrath published an authoritative and comprehensive work on Torrance's life and theological thought, entitled *T. F. Torrance. An Intellectual Biography* (1999). In the biographical section of this book McGrath highlights the overall line of Torrance's life from his childhood until his retirement. The second part of the book concentrates on Torrance's theological thought and its development, including the role of the concept of *homoousion* and the relationship between theology and science, themes that are also central to the discussion in this book. McGrath's work is the first serious effort to offer a clear overview of Torrance's life and thought as well as of their mutual interdependence. As to the theme of theology-science dialogue, McGrath does not penetrate very deeply into the specific problems that are dealt with in this book. This could be due to the introductory nature of his treatise which determined its structure. Nevertheless, McGrath's book is an invaluable companion for every Torrance scholar, undoubtedly, for many years to come.

A remarkable number of other studies have been written on Thomas F. Torrance and the various aspects of his theology. Most of them have been produced in British or American universities or seminaries, a fact that reveals the principal areas of the influence of his thinking. Many of the studies more or less touch upon the relation between natural science and theology in Torrance's thought from the perspective of a theological doctrine. One of the earliest dissertations, underlining the role of the Incarnation and the relation between revelation and reconciliation, was prepared by Joannes Guthridge, S.J., whose study of Torrance's Christology is an interesting indication of how Roman Catholic theology has responded to Torrance's Christocentrism.[42] Worthy of note is the fact that also in the second remarkable study on Torrance, Catholic theology formed the context within which the first examination of Torrance's philosophy of science and, respectively, of his view of theology-science relations was authoritatively carried out. In an outstanding and exemplary manner, Brian J. A. Gray's dissertation concentrates on the intellectual and historical background to Torrance's thought as well as on an analysis of the adequacy of Torrance's efforts to view theology as a special science.[43] However, neither Guthridge's nor Gray's work, despite their sharp argumentation and indubitable advantages, can answer questions exposed by Torrance's later thinking, in which continuous development has taken place up to the present decade.

There are a few more recent studies made of Torrance's theology relating it explicitly to the natural sciences. Douglas Alan Trook's dissertation, "The Unified Christocentric

Field" (1986), is informative and most insightful. It is a serious attempt to see the profound interconnection that unites theology and the natural sciences, in Torrance's opinion. The main problem in Trook's work is that his interpretation of Torrance's thought is taken so far that his theology is given "scientific" features which are actually alien to it. The imprecise mixture of theological and scientific concepts and patterns of understanding so typical of his discussion are discernible even in the title of the work. Another attempt to relate Torrance's thought to the natural sciences is Wolfgang Achtner's study of Torrance's natural theology.[44] Its merits include an appreciation of the fact that, unlike Barth, Torrance has not totally abandoned natural theology but wants to *integrate* it into what is understood as revealed theology. Achtner's work is evidence of broad learning and precise thinking, and it cannot be blamed for being unsystematic. As far as the subject of the present study is concerned, Hing Kau Yeung's work comes perhaps closest to it, although the starting points and modes of approach differ considerably from each other.[45] Yeung takes seriously Torrance's insistence upon the centrality of the Incarnation and the concept of *homoousion*, but, in spite of his efforts, he fails to detect the real profundity of this insight and its relevance for theology and the natural sciences.

While the studies of Trook, Achtner, and Yeung relate theology to the natural sciences in general and physics in particular, certain scholars have chosen a more articulate approach from the standpoint of the philosophy of science and especially Torrance's conception of realism. Kurt A. Richardson has been able convincingly to interrelate Torrance's Barthian context, his Trinitarian theology, realism, and the challenges of the philosophy of modern physics, as it appears in Einstein's theories of relativity, to such an extent that his work could be described as a convenient introduction to Torrance's thinking.[46] The Swede Roland Spjuth gives an outstanding presentation of Torrance's Christological orientation with regard to his doctrine of creation and provides an illuminating discussion of contingency in Torrance's thought.[47] His argument is subordinated to his analysis of some aspects of Torrance's philosophy of science, but he can give an insightful account of the possibility of a Barthian theology for a consistent doctrine of creation and the divine presence within it. An emphatically epistemological approach has been taken by John D. Morrison, who has raised the problem of Torrance's realism by showing his close adherence not only to Barth but also to Sören Kierkegaard.[48] Greater clarity characterizes Stephen A. Simmons's effort to underline Torrance's discussion with the natural sciences through semantic relations.[49] His work is another example of the conviction that Christological themes, the person of Christ in particular, determine Torrance's thought, also in the semiotic sense.

A challenging attempt to set Torrance's thinking in the wider context of the contemporary intellectual world (this period is sometimes called *postmodernism*) is made by James B. Miller, who has chosen Torrance as a representative of a theological enterprise that attempts to reject any dualistic view of the world. Miller's sharp analysis, with an appeal to process theology, brings to the fore several important points in Torrance's theology.[50] From the standpoint of the present work, some relevance must be attributed also to the studies by Robert J. Stamps[51] and Brian H. Meardon.[52] Stamps wrote his thesis on Torrance's eucharistic theology, and what makes it interesting is its way of dealing with the question of the Incarnation, although because of this principal subject he does not take further the implications of the Incarnation for the problems of theology and science. In this respect a more helpful work is that by Meardon, whose study reflects on Torrance's understanding of space and time, an area in which he has succeeded in bringing to light the close relation between patristic theology and recent developments in physical science.

Although each of these works illuminates the issue of the relation between science and theology in Torrance's thinking from their specific point of view, what they fail to provide is a coherent general picture. The following points especially should be emphasized. First, studies on Torrance's thought concerning theology-science relations have not critically evaluated Torrance's claim that modern natural science was anticipated by the incarnational theology of the Early Church to the extent that the former may actually be indebted to the latter for its existence. Second, although Torrance's realism has been discussed in a number of studies, it has usually been done purely philosophically so that none of them has sufficiently reflected on its relation to one or more specific theological doctrines. Third, Torrance's interpretation of some historical figures, especially Newton, and his use of the concept of *dualism* have been generally taken for granted without further critical analysis. Fourth, the relation between Torrance's view of the achievements of the natural sciences and his programmatic suggestions for a renewal in theology has not been given adequate consideration. And finally, in spite of the appreciation that the *homoousion* is indeed the most central concept in Torrance's thought, the implications of these insights have not been taken far enough.

The present work attempts to take these points into account by analyzing Torrance's writings in which a "bilateral traffic" between theology and science is clearly discernible. Torrance makes three weighty claims concerning this interaction. The first one is that theology has had an immense impact on the development of the natural sciences. Secondly, the modern natural sciences are currently having an impact on present-day theology. The third point is that, in order to be truly scientific, both theology and the natural sciences have to accept the impact of objective reality.[53] These influences can be further described with three specific and structuring ideas detectable in Torrance's thinking. The *idea of contribution* expresses the impact of theology on the natural sciences. Torrance's conviction is that Christian theology, as it was practiced in the Early Church during the first centuries, actually "injected" such ingredients into Western thinking that the rise of modern empirical science became possible in the first place. The *idea of compulsion* indicates his view of realism, according to which genuine scientific thinking never dictates what reality ought to be like but humbly and attentively obeys its self-testimony and commands. And, finally, through his *idea of reminder* Torrance points out that nowadays it is the natural sciences, and physics in particular, which have something important to tell theologians, who should replace their dualistic modes of thinking by a renewed realistic attitude toward their object of inquiry. As will be shown, all the three ideas are essentially linked with incarnational theology and the concept of *homoousion*.

The outline of the present work follows these ideas. Before beginning the actual analysis of Torrance's thought, a short account of Torrance's theological position and a preliminary description of his view of the relation between theology and science is given. Even at the risk of simplification, the second chapter is intended to give as succinct a picture of his thought as possible, concentrating on details necessary for conceptualizing the issues in the subsequent chapters. Introductions to the entirety of his thinking are available elsewhere,[54] which is a valid reason for restricting the presentation only to what is conceived as essential for the argumentation of the present work.

The third chapter begins by featuring the background to Torrance's idea of contribution. The actual analysis deals with epistemological questions, the problem of natural theology and the role of faith included, and moves on to Torrance's idea of the ontological contribution of the incarnational theology of the Early Church. A special discussion of space and time is offered. The main attention is directed to critically weighing the evidence Torrance presents in support of his contribution argument due to the sole fact that his claim concerning the crucial contributive role of Early Church theology is both bold and provocative.

After describing the embarrassing situation in the discussion of realism and its content, the third chapter points to the idea of compulsion as the most important characteristic of Torrance's realism. The idea of compulsion, in turn, is claimed to have a subtle but firm relation to the Reformed doctrine of election in a genuinely Barthian manner. The second part of the chapter is dedicated to the problem of dualism, and a lengthy analysis of the concept is provided because of the lack of such a procedure in earlier studies. Torrance's view of Isaac Newton's role in the rise of modern science, as well as of the scientist's religious position, is critically studied and Torrance's argument is evaluated in the light of Newton's own writings and contemporary Newton scholarship.

The final chapter concentrates on analyzing the factors in modern physics which, in Torrance's opinion, should also be taken seriously in theology. The question to be answered is, what exactly are the elements in contemporary natural science which remind theology of its foundation basis. A discussion from both ontological and epistemological points of view will be provided. Thereafter Torrance's conception of the theological implications of the novelties in natural science is examined, for example, with regard to the fundamentals of theology, the question of the Trinity and biblical scholarship. The final part of the chapter is dedicated to the implications of Torrance's ideas for ecumenical theology because of the centrality of ecumenical concerns in Torrance's thought, as an indication of the possibilities for a practical application of Torrance's suggestions in the life of the churches.

The structure of this work is intended to stress the "bilateral traffic" between theology and the natural sciences, as intimated by Torrance. The idea of contribution explicates the direction from theology to the natural sciences while the idea of reminder refers to that from the natural sciences to theology. The actual issue and content of the two ideas can be seen as rooted in the idea of compulsion, its realism and antidualism. If anyone wishes to search for the nucleus of the work, I personally would find it in the treatment of the doctrine of election in chapter 4.

The reader will note that the terms *physics* and the *natural sciences* or simply *sciences* are commonly used interchangeably. Although the scope of the natural sciences is not limited merely to physics, Torrance himself is of the opinion that it is precisely the pioneering development of this discipline that points the way forward not only to the other natural sciences but to all other disciplines as well. For reasons of consistency, the same usage is followed in this study. Also, Torrance's preference for the term *theology* rather than *religion* is archetypal for this study. Thereby attention is directed to a disciplined and methodologically well-reasoned scholarly enterprise rather than to the wider concept of religion. This procedure is based on the presupposition that intellectually both theology and science belong in the same category.

As an academic work this is a critical study dealing with issues related to one of the implications of quantum physics, namely the limitations of human knowledge. Broadly speaking, I detect two diverse types of criticism in scholarly approaches. The first is firmly aware of its superiority to the object of inquiry, claiming to possess final solutions to the problems it poses. The second admits its restrictions from which the critical questions arise. The first mode of criticism affirms: "I know better!" The second form of criticism concedes: "I wish I knew better!" The spirit of the latter approach, I believe, underlies every genuinely academic study, and it is in this spirit that I would like to proceed in the subsequent discussion.

2

Torrance's Thought in Outline

The source of all our knowledge of God is his active revelation of himself.
Thomas Torrance, *Divine Meaning*

The discussion and analysis in this work starts from the conviction that there are two determinate elements in Torrance's thought which characterize, in essential points, his view of the significance of the doctrine of the Incarnation and the concept of *homoousion* for theology and the natural sciences. Therefore it is necessary to pay proper attention to them in order to obtain a balanced idea of his thinking in this respect. While almost every study made of Torrance's thought ponders these two characteristics, here an effort is made to view them from an angle that would illuminate the subsequent discussion.[1] First of all, Torrance's position within his own Reformed tradition is to be considered, and thereafter attention will be paid to his quite original conception of the history of ideas in Western civilization and his general conception of the relations between science and theology.

Reformed Theology

Thomas Forsyth Torrance was born in 1913 in China, where his parents served as missionaries of the Church of Scotland. In his own words, he states that Bible stories and single verses from the Scriptures left an imprint upon him in his early childhood. The tension between a conservative and a more liberal interpretation of the Bible was deeply felt on the mission field, where missionaries from various strands of Christianity found themselves in disagreement. Torrance's family belonged to the more conservative group of missionaries, a feature that reveals something essential in Torrance's own position.[2] After having returned to Scotland with his family, Torrance commenced his studies in philosophy at the University of Edinburgh in 1931. Three years later he entered the Faculty of Divinity, and a scholarship enabled him to travel to the Continent, first to Berlin and then to Basel, where he came into contact with Karl Barth, who encouraged him to write a doctoral thesis entitled *The Doctrine of Grace in the Apostolic Fathers* (published in 1946 and 1948). Quite shortly after having returned to Scotland in 1938, Torrance moved to North America for further studies. He came back to Scotland in 1940 in order to work as a parish minister and later as an army chaplain, assigned to

Southern Europe and North Africa.[3] In 1948 he founded the *Scottish Journal of Theology* anticipating his future academic career in his native country. Torrance took up the office of professor of church history in Edinburgh in 1950 and in 1952 Torrance accepted the professorship of Christian Dogmatics, a post he held until his retirement in 1979.[4] His close relationship with his own church is clearly discernible in the responsible ecclesiastical posts he has held during the past decades. Participation in ecumenical activities and interchurch dialogues, as well as his moderatorship of the Church of Scotland in 1976–77, are among the most notable elements of his career.

These main lines of Torrance's life reveal his most intimate relation to the Presbyterian tradition in which he grew up and in which he has been active. He has been described as truly Reformed with respect to the central doctrines of his tradition, including the primacy of God's grace, election, justification by Christ alone, and the supremacy of God's Word.[5] Most significant, however, is Torrance's appreciation of John Calvin and his theology, which is evident throughout his works, providing a challenging interpretation of the Reformer's views. Occasionally Torrance can name him as the foremost example of an intellectual disposition that has essentially contributed to the rise of empirical science.[6] From the standpoint of strict, "scientific" theological research, Calvin's contribution has been comparable to none, according to Torrance, primarily because he sharpened the view of theology as the proper knowledge of God, based in its essentials on Christology.[7] Torrance also endeavours to show that Calvin, in fact, stood firmly on the ground of patristic theology not only with regard to the knowledge of God but also with the doctrine of the Trinity as the actual basis of all theology.[8] One implication of this interpretation is inevitably that Calvin is seen as attaching himself closely to the patristic tradition in the understanding of the doctrine of the Trinity, an approach that is further confirmed by Torrance's insistence that Calvin consciously followed Athanasius's formulation of the doctrine of the Trinity, seen through the concept of *homoousion*.[9]

This appreciation of Calvin's thought and its anchoring in patristic theology is not sufficient, however, to elucidate Torrance's position within his own Reformed tradition. It is important to note that the view of the Reformer and his relation to Athanasius described above is an *interpretation*, which alone does not provide a complete picture of Torrance's Reformed theology. He consciously distances himself from the kind of Calvinism that found its doctrinal formulation in the so-called *Westminster Catechisms* (1648), accepted widely as a normative credal formula in seventeenth-century British Puritanism and articulating, for instance, that religion in general and theology in particular were gradually viewed as making claims, not so much with regard to an objective, truly existent reality but rather with regard to the subjective inner religious states and feelings of a human being.[10] He contrasts these catechisms sharply with some earlier catechisms of the Reformed churches and points out that there are four major differences between them.[11] First, the Westminster Catechisms lack the universality and Catholicity that are characteristic of the older ones. Instead, they bear the characteristic inclinations of their authors.[12] Second, the Westminster Catechisms are less Christocentric than their predecessors, and accordingly, in third place, they are also more rationalistic. Finally, objectivity is explained as appearing quite differently in the older catechisms since in them the object is the Incarnate Word, not we as human beings and our capabilities, as is the case in the Westminster Catechisms.[13]

Torrance's way of drawing distinctions between the two types of catechisms is certainly highly revealing. From his list of differences it is now possible to conclude what are the most important elements in his interpretation of the Calvinist tradition and his reading of the Reformer himself. First, Torrance would like to see his own tradition as firmly established on the foundation of the undivided heritage of the early Catholic Church, the foundation laid by the Early Church Fathers. This discloses a markedly ecumenical orientation in Torrance's thought, deviating from any conception of Reformed theology as an independent and self-explaining or even sectarian branch of Christianity. Secondly, the heritage of Calvin, according to Torrance, must be seen as crucially Christological in character so that legalistic moralism can find no central place in it. Related to this detail is one of the most original insights proposed by Torrance, namely the notion of the *vicarious humanity of Christ*. In this view, Christ has completed our belief, conversion, repentance, worship and everything in our Christian life in his own human life so that "I do not need to look over my shoulder all the time to see whether I have given myself personally to him, whether I really believe and trust him, whether my faith is at all adequate, for in faith it is not upon my faith, my believing or my personal commitment that I rely, but solely upon what Jesus Christ has done for me, in my place and on my behalf, and what he is and always will be as he stands in for me before the face of the Father."[14] Salvation means participation in this fulfilled work of Christ so that nothing is left for us to fulfill. Everything depends on God's grace in Jesus Christ, who alone should be the focus of Christian attention. Any thought of the meritorious works of men and women, including full obedience to God, is dismissed. In this interpretation of the Reformed tradition, the notion of *objectivity* retains a crucial role, one that is ultimately and entirely centered on Christ.

The third important feature of Calvin's theological heritage, in Torrance's opinion, is that it avoids formulating itself in terms of some alien patterns of thought but tries to find its expression in thought forms derived from the Scriptures and the theology of the Early Church. While this approach is designed to abandon the overtly "rationalistic" tendency of the Westminster Catechisms, it is easy to conclude that Torrance himself would like to stress a more "flexible" and "fluent" mode of theology, with a conscious effort to avoid strict mechanical or causal explanations within theology. Torrance is especially critical of so-called *Federal theology*, in which a particular school of thought, in this case one that considered the relation of God and human beings strictly in terms of a pact or covenant, injected an ossifying tendency into theology.[15]

These features in Torrance's interpretation of the Reformed tradition sufficiently disclose his deviation from the kind of Calvinist orthodoxy that began to prevail shortly after the period of the Reformation and that found its doctrinal expression in the Confession and Catechisms of Westminster. These details in mind, it is not difficult to discern the direction from which a major influence upon Torrance's thinking has come. With regard to Federal theology within the Reformed tradition, Karl Barth evidently paved the way for Torrance's interpretation. Offering a long discussion of this special and influential branch of Calvinism, Barth states that its defectiveness resulted from its advancement of medieval scholasticism and its being surrounded by Protestant scholasticism.[16] The interesting point is that Barth explains the difficulties of Federal theology in a very similar way to Torrance, by referring to an alien conceptuality brought into the corpus of theological thinking. As a result, says Barth, particularly the doctrine of the

Atonement in Jesus Christ suffered because now it came to be viewed through the concept of covenant so that its extraordinary character faded away.[17]

The question of proper objectivity is no less revealing. The conviction that only in Christ can a true object of theological study be found is a well-known central theme in Barth's thinking.[18] Everything depends on Christ with regard to both correct knowledge of God and our salvation.[19] In the former case, any hint of natural theology is abandoned, and in the latter, our capability of contributing anything to our own salvation is denied. This objectivity is an *anathema* to all other efforts to replace this object with something else, such as a Christian *view of faith*[20] or *what concerns us ultimately*.[21] The object is Jesus Christ. Torrance's view of natural theology and his notion of the vicarious humanity of Christ have to be seen as direct correlates of Barth's conception of the objectivity of theology both in epistemology and soteriology.

However, Torrance's willingness to see his own tradition as an heir of patristic theology and the Early Church itself can be viewed as having its roots, at least partly, in Barth's thinking. One example of Barth's efforts to place the Reformed tradition firmly in an ecumenical context on the ground of patristic theology is, first, that at the beginning of his *Evangelical Theology*, a concise treatment of his conception of what theology is or should be, he defines the name by stating that "'evangelical' refers primarily and decisively to the Bible, which is in some way respected by all confessions. Not all so-called 'Protestant' theology is evangelical theology; moreover, there is also evangelical theology in the Roman Catholic and Eastern Orthodox worlds, as well as in the many later variations, including deteriorations, of the Reformation departure."[22] Accordingly, he concludes his work by quoting a section from the liturgy of the Early Church, namely the *Gloria* to the Triune God. Throughout his book Barth speaks emphatically of the Church as a totality, not as a certain denomination, a feature that also characterizes his massive *Church Dogmatics*.

On this ground Torrance constructs his own strongly ecumenical thinking. In fact, one can discern a certain ecumenically oriented shift of emphasis in the development of his thought from a narrower view of the Reformed tradition, and an almost exclusive appreciation of Calvin, to a broader horizon, based more on the theological heritage of the undivided Church. Even a short glance at the list of his major works suffices to tell us that an interest in patristic Trinitarian theology has increasingly gained his attention. Studies of Calvin and his relevance for interchurch relations are subjects of his writings in the 1950s, whereas the 1960s reveal another strand of emphasis in his thinking through his efforts to build bridges in the direction of the natural sciences and to use this knowledge for ecumenical purposes. Strongly Christological themes are enriched with a reinforcing Trinitarian understanding, culminating in his books *The Trinitarian Faith* and *The Christian Doctrine of God*. I do not want to say that patristic and Trinitarian theology were earlier alien to Torrance's thought but that their implicit status has been gradually changed into an explicit and more normative one. When considering reasons for this shift of emphasis, one cannot bypass his practical involvement in the ecumenical movement from the first half of the 1950s, marking the beginning of a period in his life that helped broaden the horizon toward an ecumenical approach and particularly toward the Eastern Orthodox theology.[23]

In all this development and involvement, Torrance has never ceased to be a theologian in the Barthian context.[24] However, Torrance cannot be seen as an uncritical representative of the Barthian theological paradigm without any deeper insights of his own. In fact, he has been called a "modified" Barthian.[25] Particularly in one respect this is most apparent. It has been pointed out that, while Torrance focuses on the doctrine of creation, it proves to be quite controversial in the Barthian Christocentric context, which is generally supposed to involve a kind of polite indifference toward questions of creation and, accordingly, toward the natural sciences.[26] However, Torrance's approach is to be seen as a remarkable contribution in that it attempts to give adequate consideration to the doctrine of creation, thus proving that claims about the inability of Barthian theology to give this doctrine full and proper attention are highly exaggerated.[27] Naturally, the doctrine of creation is not the only case in which Torrance has developed Barth's themes further, and therefore it is justified to call him a "modified" Barthian, a theme that will be given further attention in the course of this work. For the moment it is sufficient to point to the significance of Torrance's approach in developing further certain aspects of his former teacher's theology, a direct implication of which is his effort to become involved in dialogue with the natural sciences. Before engaging in a wider discussion of this aspect it is useful to consider briefly Torrance's position within the Scottish Reformed tradition.

Torrance's position as a Barthian Calvinist has proved to be quite controversial within his own nation. On the one hand, from the standpoint of that branch of Calvinism which wishes to preserve the doctrinal tones of the Westminster Confession and Catechisms, his Barthian context alone must appear rather liberal. For instance, his fierce criticism of double individual predestination, as described above, means a definite deviation from traditional Calvinist orthodoxy, not to mention his continual attacks upon the deterministic and mechanistic overtones of the Westminster Calvinism, and his notion of vicarious humanity. Torrance has recorded an illuminating memory that succinctly sheds light on this controversy.

> During my first week of office as Moderator of the General Assembly of the Church of Scotland when I presided at the Assembly Gaelic Service, a highlander asked me whether I was born again, and when I replied in the affirmative he asked when I had been born again. I still recall his face when I told him that I had been born again when Jesus Christ was born of the Virgin Mary and rose again from the virgin tomb, the first-born from the dead. When he asked me to explain I said: "This Tom Torrance you see is full of corruption, but the real Tom Torrance is hid with Christ in God and will be revealed only when Jesus Christ comes again. He took my corrupt humanity in his Incarnation, sanctified, cleansed and redeemed it, giving it new birth, in his death and resurrection."[28]

Another indication of Torrance's "liberalism" from the standpoint of Westminster Calvinism and Federal theology is his appreciation of John McLeod Campbell (1800–1872), a minister of the Church of Scotland who was discharged from his office for preaching ideas that were found to be contrary to the formulations of the Westminster Confession. The heart of the incident was Campbell's conviction that Christ died for all men, believers as well as nonbelievers, resulting in a view of *universal atonement.* The General Assembly of the Church of Scotland could not but consider his teaching heterodox since it denied the commonly accepted view, taught by the Westminster Confession, that Christ actually died only for the elect and that in this sense the Atonement is merely *limited.*[29] Torrance determinedly defends Campbell's position, interpreting it as based essentially on the Incarnation and a profound understanding of the filial relations between the Father and the Son, and rejecting, respectively, any legalistic view of a Christian's relation to his Lord.[30]

If Torrance is considered a liberal from the perspective of legalistic Calvinist orthodoxy, then, on the other hand, he must be viewed as a conservative from the standpoint of the modernist branch of Scottish Christianity which makes efforts to reformulate Christian doctrines in new terms. A lively illustration is offered by the vivid discussion in the Scottish media shortly after Christmas 1994 initiated by the then moderator of the General Assembly of the Church of Scotland, Dr. James Weatherhead, who explained in one of his sermons, "We are not necessarily saying anything derogatory at all about the virgin birth if we say it is a symbol."[31] Torrance did not hesitate in responding that "Any minister who denies the Deity of Christ should have the moral courage and honesty to resign from the Holy Ministry."[32] Accordingly, he claimed that "To speak of the Virgin Birth as a 'symbol' is theologically misleading and linguistically inept."[33] The discussion aptly reveals Torrance's position with regard to traditional Christian dogma and its interpretation. His approach is characterized by a conscious and determinate adherence to the classical formulations of Christian understanding and constant reference to the incarnational theology of the Early Church. As is clear from the above examples, for Torrance, the doctrine of the Incarnation provides the main argument against both conservatives and liberals. This controversial position between the two fronts also reveals the difficulty inherent in any effort to define Torrance and his stance within his own tradition. This feature results in a somewhat confusing situation in which his views seem to be more warmly accepted abroad than in his own country.

Theology and Science

It was pointed out above that Torrance's attempts to forge dialogue with the natural sciences is a proof of the capability of Barthian theology to focus its attention properly on the questions of the doctrine of creation, contrary to widely held assumptions. Before entering into the actual issues of this dialogue, we should consider the rough outlines of Torrance's views on science. A good overall picture of the different modes of relating theology and the natural sciences to each other is provided by an eminent researcher in the field, Ian G. Barbour, who holds that there are basically four main ways of reflecting the relationship. The first one is that of *conflict*, whereby science and religion are seen as hostile opponents, as in scientific materialism and biblical literalism. The second mode of relationship is that of *independence*, where the total originality and distinctiveness of different modes of human investigation are emphasized, either by means of the idea of contrasting methods or by means of the view of differing languages. The third alternative is that of *dialogue*, seeking discussion on the basis of what is common to various disciplines, such as the so-called boundary questions or the search for methodological parallels. In the fourth option, *integration*, efforts are made to mold together science and religion so that either science is used to provide support for theological ideas or vice versa.[34]

In this classification of interscientific relations Barbour sets Karl Barth and his followers, Torrance explicitly, in the category of *independence*, on the basis that Barth's adherence, on the one hand, to God's exclusive revelation and, on the other, his hostility toward any form of natural theology had the consequence that he considered both theology and natural sciences as being so different from each other that their absolute independence is confirmed.[35] However, Barbour also sets Torrance in the class of *dialogue* because Torrance is interested in boundary questions raised by quantum physics.[36] Although Barbour offers valuable insights with his categories, it would not be quite appropriate to consider Torrance merely as a representative of the *independence* view or, to a minor degree, of *dialogue* on the boundary questions. If Barth's own alleged disinterest in the relations between theology and science is underscored as with Barbour, the more significant fact at this point is, as described earlier, that Torrance's contribution means a further application of Barthian theology to this discussion, revealing its possibilities for such a dialogue.[37] This fact alone makes it hard to place Torrance primarily in the *independence* category and only secondarily in that of *dialogue*. The converse is nearer the truth, for, despite his insistence upon the diverse fields of inquiry of both theology and the natural sciences, Torrance makes continuous efforts to bring the disciplines to contact in conversation and interchange of opinions, views, and insights.[38] And, on the other hand, blaming Barth for indifference toward the natural sciences is not, after all, wholly justified. Theology and science, in Barth's opinion, differ from each other primarily with regard to their diverse objects.[39] In several respects they have much in common, for instance, in that they both reject a special worldview as the basis of scientific enterprise.[40]

This discussion proves how difficult it can be to attempt to categorize Torrance according to the kind of manifold scheme suggested by Barbour. Innovative thinking seldom bends to fit tight straitjackets of categorization, although this approach can prove its helpfulness in disclosing some essential characteristics in its structure and the motivation behind it. However, much more relevant for the purposes of this work is the classification or categorization made by Torrance himself regarding the history of Western science. The theological points of the Incarnation and the context of Barthianism demand another category, a more scientific and historic one, by which Torrance arranges his thoughts and which he uses as a means to forward his arguments concerning the relationship between theology and the natural sciences. Torrance's original *periodization* of the Western history of ideas is a significant and ambitious effort to distill the essential from the nonessential, the profound core from what is seen as mere superficial dregs. In short, the essential that Torrance would like to underscore through his periodization is the correct understanding of the *Incarnation*, and the dregs he would like to wipe out are the phenomenon he calls *dualism*.

Torrance divides the history of ideas into three periods of cosmological views, each of which has, according to him, made an immeasurable contribution to the development of Western scientific thinking. Each period has begun with a scientific revolution which has meant a turning point in a profoundly paradigmatic way. Torrance maintains that the first mutation took place in the early Christian centuries as a result of Platonic philosophy giving rise to *Ptolemaic cosmology*, "which envisaged a deep split in the cosmos, between the intelligible realm of celestial realities and the sensible realm of terrestrial phenomena."[41] Greek science was thus affected with what Torrance calls *dualism*, against which the Christian Church felt compelled to struggle in its strife to defend the reality of the Incarnation of the Son of God. Furthermore, Torrance argues that the heresies, against which the ancient Church had to fight, were deeply based on this dualistic mode of thinking, while the Greek Fathers adhered to a different mindset, namely to that of Jewish and Christian faith.[42]

The second scientific revolution giving rise to another period in the history of science, as described by Torrance, took place in the days of the Renaissance and the Enlightenment when new cosmological ideas "initiated by Copernicus, Kepler and Galileo were developed by Sir Isaac Newton and mathematically elaborated in his magnificent 'system of the world.'"[43] This *Newtonian cosmology*, according to Torrance, had the effect of building a deep-seated bifurcation of a dualistic nature, the distinction between absolute space and time and relative, apparent, and contingent space and time, into the heart of Western science. The present era, according to Torrance, is still trying to come to terms with this problem. Dualism led to hard causalism and determinism and attributed to God nothing more than the role of an explanation for those phenomena in the universe that could not be explained by the laws of nature.[44]

Finally, Torrance calls the third scientific revolution, and the period immediately following it, that of *Einsteinian cosmology*, which means an integration of entities such as structure and matter or form and being, indicating thus the end of a scientific era obsessed with mere analytical thinking.[45] Furthermore, ontology is now given, in Torrance's opinion, its proper place because of the efforts to "know things in themselves" or reality as it is, not only phenomena detached from their ontological ground, in an indeterministic or nonmechanical framework. Torrance prefers scientific terms such as "field-structures" and "field-laws" referring to the ontological and indeterministic implications of modern physics in which, in his opinion, there is no longer any room for static concepts and a static understanding of nature. Modern physics is thoroughly dynamic. In short, the actual achievement of the Einsteinian cosmology, according to Torrance, is that it has been able to dismiss dualistic modes of thinking which so seriously impaired understanding of nature during the two earlier periods. This has had the effect that the patristic understanding of Incarnation became far more conceivable than during the overtly rationalistic periods of the earlier cosmologies.

In spite of its simplicity, Torrance's periodization is a challenging way of reading Western history and defining its characteristic features. However, some points of criticism must be offered. First of all, Torrance gives the impression that "the Judeo-Christian mind" were an attitude that was born in considerable isolation from the intellectual context of the ancient world. Torrance states that the Early Church Fathers, when fighting against what Torrance calls dualistic habits of thought, were determined to transform the intellectual bases of Greco-Roman ways of thinking.[46] This assertion suggests that he does not view Christianity as having been affected essentially by the all-pervading Platonic philosophy of that time. In Torrance's opinion, quite the contrary was the case. In this respect he seems to differ radically from a more modest approach which admits that Christianity did not emerge and develop its doctrines in an intellectual vacuum but, inevitably, consciously or unconsciously, assumed elements of the ideological and religious climate of that time.[47] It has even been plainly argued that the Church Fathers became "platonized" in their theology in the sense that they assumed the language, ideas, and arguments of the Platonists.[48] In this light, Torrance's line of thought proves to be exaggerated.

This feature leads to the other point to be made with reference to Torrance's periodization, namely his *preferential reading of history*. Torrance's position as a theologian whose authority lies in patristic thinking may explain his strong claim concerning the conscious "revolution" produced by Christian theology in the Greco-Roman intellectual world, as he was seen to suggest. However, one cannot escape the feeling that, for Torrance, systematic coherence matters more than historical precision. He draws lines through history in a manner that inevitably makes careful and exclusive selections. Torrance's preferential view rests ultimately on two interrelated aspects already mentioned, namely the doctrine of the Incarnation and the phenomenon he calls *dualism*. This pair of concepts, Incarnation and dualism, form the content of the criterion by which historical development is ultimately evaluated. The first effect of this theory is that the history of Western science is considered as being in a state of constant tension and struggle where these two aspects try to win primacy over each other. Another result is a critical attitude, though not total negligence or indifference, toward significant strands in the history of ideas and, accordingly, of theology, which, nevertheless, have played a crucial role in the course of the centuries.[49] Torrance does not seem to have paid proper attention to the Latin Middle Ages, during which the relation between theology and science was very intimate.[50] References to the era of the eighteenth-century Enlightenment are not abundant in spite of its significance for relations between science and theology.[51] A modern example of Torrance's preferences is the fact that he names the last of the cosmological periods *Einsteinian* while the present cosmological outlook based on modern physics is commonly called a *quantum cosmology*.[52] This choice of phrases aptly reflects Torrance's position and preferences within the context of science and his particular enchantment with the ideas of Albert Einstein, an issue that will be discussed later with regard to its relation to the doctrine of the Incarnation and to the phenomenon of dualism.

In spite of the obvious problems inherent in Torrance's periodization, mainly linked to his restricted evaluation of history and his subsequent tendency to underscore the relevance of his preferentially made selections, this sketch has some undeniable advantages. In its simplicity it offers a ready introduction to the field of the development of scientific thinking, in general terms, and the relations between science and theology in particular.[53] At the same time it highlights persons whose scientific work has embodied important leaps forward in science and who can justly be called significant figures in the history of Western science, although, of course, other preferences can be justified as well.[54] And further, the periodization in question is firmly based on the development of *physical* science, indicating thereby its novel orientation in the modern intellectual climate. In fact, Torrance holds that there is a time lag between the revolutionary new science, led by physics with its ontological and epistemological implications, and its influence upon our forms of thought and life. Torrance mentions three areas where this time-lag is most apparent: the *modern or popular mind* still operating within the Newtonian framework, the *social sciences* still working with mechanistic concepts, and, third, the *technological society*, which now lives a life of its own and poses a threat both to nature and to the human race.[55]

But what is the relation of this periodization to the three ideas to be analyzed in the course of this work? The answer is that the ideas are a prerequisite necessary for understanding Torrance's argumentation within the scope of the three periods. The *idea of contribution* belongs primarily to the period of Ptolemaic cosmology, explicating Torrance's belief in the positive impact of the incarnational theology of the Early Church upon the emergence of empirical science. Secondarily, it is detectable in the two remaining periods as well, on the assumption that the Incarnation and the *homoousion* are conceived rightly. Only then can theology make contributions. The *idea of compulsion* is the most important, prevailing in all of the cosmological eras as an indication of the realist attitude in which reality is allowed to dictate what is known and conceived of it. The incarnational theology of the Early Church was a product of this kind of realism amidst the hostile environment of dualism. During the Newtonian era this realism became even more actively threatened by dualism due to its close attachment to science. It was not until the rise of Einsteinian cosmology that dualism was, and is, gradually overcome through the findings of modern physics, restoring the compulsive nature of reality in scientific enterprise. Finally, the *idea of reminder* in Torrance's thinking is related primarily to the Einsteinian period, emphasizing the impression that in modern times it is physics that has something essential to teach about ontology and epistemology to other disciplines. In the case of theology this means that all dualistic modes of thinking should be abandoned and a realist attitude toward the object of inquiry should be restored, which, in practice, directs theologians to start from the doctrine of the Incarnation, which Torrance regards as succinctly expressing *the* reality in which God can be met and known.

This proves that Torrance sees the three periods as the historical scene for a network of tendencies explicated in his three ideas. This also reveals Torrance's somewhat *programmatic* approach because he uses the history of Western thinking and his periodization to emphasize the focus of his own intention to assign the doctrine of the Incarnation a primary role, a role that, in his opinion, was unfortunately lost when dualism gained an ever wider foothold in Western intellectual life. This intention cannot be understood without taking Torrance's Barthian context into account, and in this he joins Barth's efforts to meet the challenge posed to Christian theology by eighteenth- and nineteenth-century theologies, which have usually been labeled as "liberal." In them, rational religion weighed more than any traditional adherence to revelation, resulting in an emphasis upon the moral aspect of religiosity, on the one hand, and upon the subjective states of feeling in the believer, on the other.[56] It should be well noted that Torrance's novelty in comparison to Barth is his systematic effort to integrate the natural sciences and their achievements, especially those in physics, as an additional argument to support the necessity and validity of the Barthian program.[57]

It would be a gross error, however, to conclude that Torrance has taken such enormous pains to understand developments in modern physics merely in order to find extratheological reasons for his theological program. There is no point in doubting his sincere willingness to be involved in genuine dialogue with representatives of the natural sciences and to make efforts to contribute to the interdisciplinary discussion.[58] In this approach he has retained his truly Barthian stance, and therefore his views of the relationship between science and theology can be seen as an "experiment" as to how and to what extent such a strong theological position can make a genuine contribution to the dialogue. Its success, of course, can ultimately become apparent only in the process of the continuous dialogue itself, and therefore a single work such as the present one can merely introduce insights as to possibilities and detect problems inherent in Torrance's model. As such, however, it may help us to focus on issues that, despite obvious defects, deserve to be taken seriously and may prove their fertility in further discussions.

Torrance's contribution to the discussion between theology and the natural sciences has to be viewed from another angle, too, namely his view of the unity forming the basis of all rationality, a unity that Torrance describes as "harmony" and even "beauty."[59] In his opinion, what he calls dualism has severely impaired our conception of the world so that our knowledge of God and of the world is deeply divided and a radical separation between form and matter, the empirical and the theoretical, the physical and the spiritual, has taken place.[60] In other words, dualism is a dangerous threat to the very essence of the universe, which is characterized by harmony and beauty. Torrance's belief in the unity of the universe implies that the dialogue between theological and natural science "will play the most effective role, for through it there will surely emerge a new openness to the creative impact of God, the unifying Source of all rational order and beauty in the universe."[61] Thus, Torrance does not hide his articulate theological position, whereby he would like to contribute to an ever-widening conception of God as the ultimate Ground of all rationality.[62]

But does not this approach suggest strongly that Torrance is seeking, as it were, a *theological theory of everything* in which God retains the role of the Ultimate Explanation of all creation? It has been claimed that throughout its history physical science has constantly sought a theory that would unite all aspects of physical reality under the roof of a single grand explanation. From this point of view, every step forward in physics has meant a stronger explanatory power in theories which have been able to combine the special cases of former separate theories. From the ancient Greeks through Copernicus, Galileo, Newton, and Maxwell, to Einstein and Bohr, and to the contemporary geni in physics, the ambition to find a unification theory has been an element in their motivation. At the moment such a theory would, according to physicists, emerge if the theories of relativity and quantum mechanics could be appropriately united.[63] If such a theory were finally found, it is claimed, our conception of God could not remain unaffected.[64] A more skeptical account of the possibility of such a theory is based on the conviction that whenever a grand unifying explanation of the universe is sought, the motivation behind it is unexceptionally religious.[65]

As noted earlier, Torrance's insistence on the harmony and unity of all knowledge in the universe is strongly theological. However, we would miss the point if we straightforwardly conceded that thereby Torrance is constructing his own model for a great unification theory where God is included as the ultimate explanatory Principle. This argument would not be correct because Torrance is articulately emphasizing that both disciplines have their separate fields. A slight ambiguity, nevertheless, is to be detected. On the one hand, Torrance is very plain in his conviction that the direction of thought in scientific enterprise is both "toward God" and "away from God" so that no "cosmological synthesis" of the Creator and the created order is involved.[66] Theology and the natural sciences both have their respective and diverse fields of inquiry, and in this manner they are quite independent of each other. Torrance seems to reject any attempt to offer God as an explanatory premise for the natural sciences but holds that the freedom of creation, endowed on it by its Creator, guarantees a cosmology that is formulated according to the terms posed by the physical reality itself.[67]

On the other hand, however, Torrance subtly suggests that theology has something to remind itself of as well as the natural sciences, namely "the transcendent relation of the Creator to what he has made and continues to uphold through his sustaining power and rationality."[68] God as the Source of all things, then, appears to have some explanatory role in Torrance's ideas of the contribution of theology to the sciences. In the final analysis, this contribution has to do essentially with the concept of *contingency* as the counterpole of *determinism*. There are two possibilities in interpreting the way in which Torrance makes his claim regarding this contributive role of theology. The first one explicitly emphasizes God as a premise in an explanatory argument within the natural sciences. However, as was seen above, this is strictly contradictory to the claim made by Torrance that theology is a science *toward* God and that the natural sciences are moving *away from* God. Therefore it is reasonable to conclude that the second way of interpreting Torrance's apparently ambiguous thought of the contributive role of theology is more consistent with his insistence on the independence of the disciplines. According to this interpretation, Torrance's claim that theology can remind the natural sciences of the contingent nature of reality is another effort to underscore the insufficiency of the natural sciences to offer an exhaustive explanation in a deterministic and mechanistic manner, without the need to take into account what is conceived to lie *beyond*. In other words, the limits of a purely rational inquiry are emphasized. This reading seems to offer a credible account of Torrance's view of the contributive role of theology, setting it on the same level of human rationality as the natural sciences. From this point of view it is also understandable that Torrance demands the attentiveness of theologians with respect to the results of the natural sciences, results in accordance with which "theology also must develop adequate accounts of its own convictions about Incarnation and resurrection, for example, as well as creation with proper respect for their empirical correlates in the spatio-temporal structures of empirical reality."[69]

Torrance's view of the relationship between the natural sciences and theology can be looked at from yet another point of view. Torrance's suggestions concerning the theologically relevant ground for the theology-science dialogue have already been characterized as original because of his insistence on the significance of the doctrine of the Incarnation.[70] Torrance's approach in attempting to unite creation and redemption or the doctrines of creation and the Incarnation can be claimed to present an ambitious effort to give a Christian face to the philosophical concept of God, which has so largely influenced Western understanding of the Divine Being. In the post-Renaissance centuries up to the present day, speculations concerning the existence and action of God, mainly based on natural theology, have increasingly, in the advance of the natural sciences, paved the way for agnosticism and atheism. "The God of the gaps," formerly so badly needed to fill points of ignorance, has been removed from the premises of scientific reasoning to the periphery of imprecise religious feeling and thinking. This had the effect of restricting belief in God only to the limits of the inner states and the moral disposition of human beings.[71] The problem of the relation between a general idea of God as entertained by philosophical thinking and the Christian idea of God as the Supreme Being who is deeply interested not only in the course of the universe but also in the life of individuals and their salvation has undoubtedly troubled Torrance a great deal. His solution is to commence the discussion not from the doctrine of creation but from that of the Incarnation or, in other words, from that understanding of God's being and activity which has, in his opinion, assumed a comprehensible human form in the incarnate Son of God. This solution implies that for him a purely philosophical conception of God is implausible and that it can have relevance for humankind only if a complementary and determinate view of God as revealed in the Incarnation is given primacy. In short, it can be stated that one of the crucial motivations in Torrance's efforts to build bridges between theology and the natural sciences is his willingness to Christianize the quite neutral conception of God entertained in purely philosophical speculations, and to search for fruitful ingredients for theology-science relations on the basis of special revelation and Christology. The discussion in the subsequent pages will penetrate deeper into his argumentation.

As far as I know, the insights presented by Torrance have not been much commented on in scientific circles. In general terms, natural scientists can be divided into three categories according to the position they have taken with regard to the theology-science dialogue. Those in the scientific community who seem to have assumed an indifferent attitude toward such a dialogue do not usually refer to proposals put forward by persons such as Torrance, supposedly because of their reluctance toward considering theological positions in general. In their thinking, theology and science have nothing in common worth mentioning. To the second group belong scientists who express a certain interest in saying their word on theological questions, stimulated by their discoveries but primarily orientated by their general attitude toward human religiosity and belief in a deity. Only scientists of the third category are more or less convinced of the benefits of a genuine dialogue with Christian theology. Their motivation may lie in a personal religious conviction, but that does not seem to be a necessary precondition for entering into discussion.

Explicit responses to Torrance's proposals and insights from the first group are difficult to find. It may just be the case that there simply are none. Comments on Torrance from the second category of scientists are not abundant, either. An example is the biologist Alan Olding, who has criticized Torrance for arguing for the existence of God with a kind of argument from design and found this solution implausible. Olding's learned account of theology-science relations is somehow quite indicative of a typical disposition in biologists in that constant efforts are made to convince the reader of the redundancy of God because of the apparent needlessness of the concept in explanations.[72] Accordingly, the sufficiency of mechanical causation is emphasized.[73] The third group of scientists seem to make no exception in the number of explicit responses. Chris J. Isham reminds us of the healthy correction suggested by Torrance whereby God is not discussed only as "the ultimate ground of Being" but also as the incarnate One, a feature in Torrance's thought described earlier as an effort to give the God of theology-science discussions an articulately Christian face. Isham, however, is apparently confused by Torrance's argumentation, especially with regard to the problem of time and its relation to the Resurrection. After briefly describing Torrance's thinking in this respect he contends rather impassionately that "I think it will be rather a long time before theoretical physics has anything useful to add to that."[74] Another Englishman, John Polkinghorne, refers to Torrance quite often with appreciation while maintaining a polite distance.[75] He makes use of several insights proposed by Torrance but does not accept them necessarily as such but develops them further to make his own original contribution in a manner which reveals wide learning and expertise in both fields.

Although comments on Torrance and his thought from natural scientists do not fill the air, in explicit responses both appreciative agreement and critical disagreement are to be found. Silence, respectively, seems to be the dominant state of affairs. Reactions from philosophers or theologians are far more common and sometimes harsher, too, as will be seen in the course of the work. Torrance's ideas are in any case original to such an extent that they deserve wider attention not only within theology but also within the natural sciences. If the present study can help in directing attention to and clarifying issues raised by Torrance, it has at least partly met the challenge posed by the present theology-science dialogue.

3

The Impact of Theology on the Natural Sciences

The Idea of Contribution

> I want to show that far from theology being based on natural science, the opposite, if anything, is nearer the truth!
>
> Thomas Torrance, *The Ground and Grammar of Theology*

One of the most important themes in Torrance's theological enterprise is his claim that the natural sciences are basically indebted to theology as far as their proper understanding of reality and modes of thinking are concerned. In his opinion, the Christian doctrine of the Incarnation, developed during the first Christian centuries and culminating in the Christological debates of the Nicene era and especially in the application of the concept of *homoousion*, meant that all subsequent scientific efforts were provided with a solid basis that replaced the former application of Hellenistic modes of conceiving the reality. In order to gain a deeper view of this feature in Torrance's thought, here called his *idea of contribution*, it is useful to take a broader look at theology-science relations from this specific perspective of the contributive role of Christian theology.

The Background

During the twentieth century, remarkable research has been done on the reasons that led to the development of Western science.[1] One of its features is the suggestion that religious motivation, whatever its detailed content is thought to be, has been a major force behind those whose scientific work or whose favorable attitude and authority meant a remarkable leap in the rise of modern science. Another feature has been that, when applied, this religiously orientated *idea of contribution* readily reveals denominational preferences, with the implication that not only religious feeling in general but adherence to a particular confession produced fertile conditions for the emergence of science. This idea is not necessarily linked with Christianity, as is well illustrated in André Neher's study. He underscores the positive impact of Jewish religiosity and thinking on the rise of modern science.[2] However, usually it is the Christian religion that is involved in this discussion. Sometimes the idea of an accentuated religious impact on science has been allied with outspoken nationalism, as in the case of the Frenchman Pierre Duhem, who held that the birth of modern science can be dated to the year 1277, when the bishop of Paris condemned more than two hundred previously authoritative articles on theology and natural science based on Aristotelian cosmology.[3] What was at stake in this episode was the question of

God's sovereign omnipotence, which, the ecclesiastical authorities feared, was exposed to constant danger as a result of the Aristotelian view of causes inherent in matter, thereby leaving no significant room for God's purposive activity.[4] Duhem's Roman Catholic convictions led him therefore to accredit his own country and confessional position with the honor of liberating scientific thinking from deterministic features, a view that is based, it seems, more on patriotism than on convincing historical evidence.[5]

The motivation behind the so-called *Merton thesis* cannot be reduced so readily to the confessional appeal of its inaugurator, but it offers another example of how a religious position has been thought to affect the development of science. A national, that is, English, standpoint is emphasized by Robert K. Merton, an American sociologist (b. 1910), who maintains that post-Reformation Puritanism created such an intellectual climate in England that it made a career in experimental science not only possible but also commendable, based on the Calvinist doctrine of election and its associate ethic. While an "intramundane asceticism" was encouraged, practical actions were dedicated to God and his honor, injecting a keen interest in everyday life and making this the realm in which God's election was to be acted out. The Merton thesis asserts that this "monastic" life without monastery led to both utilitarianism and empiricism.[6] Merton's evidence for his thesis lies primarily in his analysis of the religious background of the members of the Royal Society in the first decades of its existence, an approach criticized, for example, for its narrow time span and for the original conception of science involved in his argumentation.[7]

Both Duhem and Merton connect their idea of contribution to a single event or a relatively short period of time, giving the impression that the reasons for the remarkable leap forward in Western science could be postulated by pointing to a carefully selected brief historical occurrence. In this respect Stanley L. Jaki, a Hungarian-born Benedictine priest, represents an exception since he does not see any single historical phenomenon as a definitive beginning for the development of science but rather a whole era or longer period of time. Analyzing the appearance of early modern science in Western Christendom and its nonappearance in non-Christian cultures, he makes the claim that a specific Judeo-Christian doctrine, that of creation, provided the culture of the Christian West with the kind of intellectual climate that made the development of science possible in the first place.[8] By the beginning of the thirteenth century, belief in the biblical doctrine of creation had been thoroughly integrated into Western civilization and had leavened it with valuable insights that could not but have a strongly positive bearing upon the emerging natural sciences.[9] In this way Christianity was the necessary precondition of Western science.[10]

Jaki's argument has met with not only appreciation but also resistance. It has been pointed out that one cannot conclude in a historically plausible way from the fact that modern science originated in Christian Europe that Christian belief was the decisive factor in its development.[11] David C. Lindberg, one of Jaki's critics, particularly from the historical standpoint, clearly maintains that it would be "crude distortion to maintain that Christianity offered major stimulus to scientific activity. But it would also be distortion to create the impression that there was no Christian involvement in natural philosophy or that the church retarded or crushed science."[12] Lindberg does not acknowledge any articulate idea of contribution but tries to find a balance between the two extremes of fierce denial and uncritical assertion of the contributive role of Christianity with regard to the development of science. One might point out that Lindberg has replaced the idea of contribution of Jaki and Hooykaas with his *idea of balance*, the starting point of which is the conviction that a middle way between the two poles should be followed. However, this historically sound and certainly non-apologetical approach cannot, in the ultimate analysis, provide any definite solution to the apparent problem detected by both Hooykaas and Jaki, namely, the appearance of empirical science exclusively in the Christian world. In spite of the danger of one-sided and narrow apologetic tones, this question deserves proper attention. It should not be bypassed by commencing such a study with the fixed methodological premise that the challenge of the extremes cannot be taken seriously but a compromise must be found.

Thomas F. Torrance is one of those scholars who have found this "contribution approach" appealing, which is apparent in the fact that he has made further effort to reaffirm the relevance of this idea of contribution. Jaki's argument is crucial for the further reason that Torrance has reformulated it from his own standpoint, based on the centrality of the Incarnation and the concept of *homoousion*. This fact gives good reason for designating Torrance's view a *modified creation argument*, representing not only the doctrine of creation but also that of the Incarnation as the core of the idea of contribution. In other words, the doctrine of creation alone did not provide a sufficient basis for a favorable intellectual climate where scientific thinking could emerge, although, it has to be admitted, it formed the necessary basis for it. Torrance maintains that it was not until the creation was reflected upon in relation to the doctrine of the Incarnation that fertile soil for scientific development was found.[13] Therefore the period of Nicene theology between the ecumenical councils of Nicea in 325 and Chalcedon in 451 is of utmost importance for Torrance and his argumentation.

In this connection it is useful to note that the patristic understanding of creation and the Incarnation has not always been the articulate heart of Torrance's idea of contribution. In writings published in the 1960s, patristic theology is not totally absent, but particular stress is laid upon Reformation theology.[14] Torrance contrasts it with the theology of the Middle Ages and claims that the Reformation thought made it possible for empirical science to become liberated from the bondage of medieval scholasticism.[15] But which branch of the Reformation is at issue here? The answer is the thought of John Calvin, which does not imply that Luther, for instance, had nothing to contribute in this respect.[16] Calvin's contribution to the rise of science consists firstly of "mutual relation between the knowledge of God and the knowledge of ourselves."[17] Secondly, Calvin succeeded in formulating a new kind of questioning, *interrogation*.[18] Thirdly, the problem of a sign and the thing signified found a solution in Calvin's view of the nature of theological language, which "is indicative, not descriptive, of God and it is to be understood only as we allow it to refer us beyond itself to God in His transcendent reality. It was by developing this view of the relation of language to being that Calvin became the father of modern biblical interpretation."[19] And finally, the Reformation is credited for reviving the biblical view of God as Creator and of the creation as being totally dependent on its divine Maker.[20]

The last point in particular refers to the later development that took place in Torrance's thought. The notion that at the Reformation a certain biblical belief or conception was *revived* anticipates his later view in which a strong emphasis upon the prominent role of John Calvin is replaced by ever-increasing appreciation of patristic understanding and its contributive character. This feature is particularly evident in the writings published in the

1970s and 1980s. While previously the Reformation and especially its Calvinist branch occupied in Torrance's thought a more central role as a major contributor to the rise of modern science, leaving patristic theology with only a minor status, the situation changed drastically. "I would not want to depreciate the impact of the Reformation, or of the Renaissance, upon the rise of modern science, for it was undoubtedly very considerable in this regard, although I believe that the roots of this influence go rather further back in the history of Christian thought."[21] But how far back in history? Definitely, until the patristic period of the Church, for "it would not be very difficult to show through an analysis of the history of thought that the classical approach to the objective intelligibility of the universe which lies behind all our Western science and culture developed together with the Christian doctrine of the Trinity."[22] I would not claim that in Torrance's earlier thought there was no idea of the contribution of patristic theology or that, later, the contributive role of the Reformation totally vanished. The relation between these two contributive factors in Torrance's development is best described as a transformation in *emphasis* which has shifted patristic theology to a primary position while the Reformation contribution has not been neglected but rather removed to a secondary position.[23]

The shift in Torrance's thought from a denominational position to another one with wider credibility and less confessional apologetics is remarkable. It is a thoroughly *ecumenically motivated* occurrence in Torrance's thought, or in other words, a clear representation and result of his longtime contacts and activities in the ecumenical movement. This approach is certainly one which no longer confirms the idea of contribution of the Duhem-Mertonian type but has been able to break down the barriers of denominationalism and confessionalism, leading to a wider understanding of the contributive role of the Christian faith. Therefore his idea of contribution must be seen as a serious effort to interrelate scientific thinking and the totality of Christian theology in a manner which at the same time seeks ingredients of ecumenical rapprochement.[24]

The success and validity of Torrance's modified creation argument rests on two premises, the first of which is naturally that Jaki's argument is sound and relevant. However, it is a question that does not belong to this discussion but must be studied elsewhere. It is sufficient to restate the point made earlier that any criticism of Jaki's view, which does not take seriously the problem of the emergence of science in the Christian world, and its nonemergence in the non-Christian world cannot, accordingly, offer any definitive solution to it.[25] This means that the possibility of the validity of Jaki's argument has not yet been demolished but continues to present an inspiring challenge. On this basis it seems agreeable to pay close attention to Torrance's modified creation argument, which to be successful requires as its second premise that the view of the contributive role of the doctrine of the Incarnation is justified. This original insight on the Incarnation as presented by Torrance determines the course of the subsequent discussion.

The Epistemological Contribution

The Negative Role of homoousion

As was shown above, Torrance's idea of contribution is based on the conviction that Christian theology has contributed substantially to the development of modern empirical science, not only implicitly through its general effect on Western culture but directly and explicitly through some of its crucial doctrines. As was noted, in Torrance's opinion, especially the Christian doctrine of the Incarnation, developed during the first Christian centuries and culminating in the Christological debates of the Nicene era and in its application of the term *homoousion*, meant that all subsequent scientific efforts were given a solid basis that replaced the former application of Hellenistic modes of conceiving reality.[26] Thus Torrance creates a considerably strong tension between what he calls Christian and Hellenistic frames of mind. He describes vividly the explosive expansion of the Church and its proclamation from a peripheral state in Palestine to the center of the intellectual world of that time at the heart of Greco-Roman civilization.[27] The following discussion is a theological one, but precisely through such an approach the contribution of theology to the development of science, as seen by Torrance, takes shape.

The tension between Christianity and Hellenism is one of the most characteristic features of Torrance's understanding of the development of theology and science, a dichotomy in the history of ideas, as understood by Torrance, which repeats itself in various forms in later eras. Ideologically the confrontation of Christianity and Hellenism meant that Greek dualism had to face the Jewish worldview, based on the Bible, which in Torrance's opinion represented a nondualist, unitary conception of the universe. The concept of *homoousion* played a central role in this confrontation, so argues Torrance, by creating a quite new understanding of the relation between God and ourselves. The concept of *homoousion* as a vital epistemologically contributive factor in Torrance's thinking is therefore best analyzed by referring to both its negative and positive aspects as discernible in his thought.

The conflict between the Hebraic and Greek "frames of mind," as expressed by Torrance, was one in which theology and mythology were engaged in a struggle for supremacy. The important point to note in order to understand Torrance's idea of contribution through the concept of *homoousion* is that the Hellenistic world was in the firm grip of mythological thinking, with the basic thought that the center of an epistemological process lies in the human mind, while Hebraic or theological thinking took God as its point of reference.[28] In other words, for Torrance the Greek Fathers represented a genuine theological attitude (direction from God to the humans) in contradistinction to Hellenistic religious movements, in which human beings were given a central place in the relationships between them and deities (direction from the humans to God/gods).

The sharp dichotomy made by Torrance raises at least two questions. First of all, is his conception of the term *mythology* adequate, and secondly, can his claim concerning the nature of early Fathers' thinking be justified? The first question is easier to answer because, as we have seen, Torrance himself gives his own definition of what he means by the terms *myth* and *mythology*. He simply states his conviction that any religious thinking that takes as its starting point the human mind leads inevitably to mythology, the obverse of theology. Further light on Torrance's understanding of these terms is shed by his conversation with certain Bible exegetes, in particular Rudolf Bultmann, whose demythologizing program Torrance criticizes relentlessly.[29] The way in which he analyzes Bultmann's position is highly revealing as to his own understanding of the nature of mythology. Torrance's interpretation of the German Bible scholar Bultmann is that the aim of myth is not to "present an objective picture of the world as it is, but to express man's understanding of himself in the world in which he lives."[30]

Torrance's reluctance to accept Bultmann's conception of myth is easy to understand. Of special importance is the view that in Bultmann's case myth is not allowed to represent objective reality but only men's and women's understanding of themselves. Torrance's evaluation of Bultmann seems to be based only on his demythologizing program, and he does not pay attention to those views of the German Bible scholar which are not, after all, so alien to Torrance's own intentions. Consider, for instance, the direction of communication in the God-human relation as indicated by Bultmann in the following quotation reflecting on our readiness to listen to God: "The first condition for readiness is this: we must silence all other voices; everything we say to ourselves, everything other people say to us. For we want to say what *God* says to us. And if we take this seriously, there is room for but one voice."[31]

In any case, Torrance's way of reading Bultmann reveals that, for him, myth and mythology are something quite implausible. This is not, however, the only way of defining myth and mythology. Contrary to the usual understanding of myth as something unhistorical and therefore also unreal, another view is put forward where emphasis is laid on the story it tells in its sociological context, and the importance of its contents is underscored.[32] A direct stand on whether the events described in a myth really took place in history is avoided because this mode of approach is not considered fruitful.

Although Torrance cannot accept this kind of use of the term from his own point of view, his definition is undoubtedly sharp, and, above all, it illuminates his own arguments remarkably.[33] With regard to the definition of the terms *myth* and *mythology*, Torrance makes his point quite clear: for him *myth* is not just a "technical" term by which we as humans try to understand nature and give meaning to our own position within it. Instead, acceptance of a myth means abandoning the things it contains and closing one's mind to objective reality. Therefore for Torrance, a myth does not lead to genuine knowledge but holds us captive in endless circles of our own self-knowledge where the only reality under investigation is a projection of our own mind. Therefore his position vis-à-vis mythology is absolutely negative. This problem is closely connected with Torrance's view of realism, a theme that will be examined more widely in the next chapter. This also means that he cannot contribute to the theology-science discussion, such as referring to the positive value of the term *myth* as a human effort to give meaning to the universe around him.[34] Because a myth is centered on our human understanding of God/gods, of ourselves, and of the universe, Torrance cannot but exclude it methodologically from his efforts to establish connections with natural sciences.

It is certainly true that those theologians and philosophers who accept myth as a legitimate way of searching for meaning in the world have something important to contribute to the discussion between theology and the sciences. Their premises, therefore, do not lie in such an explicit theological understanding as is the case with Torrance. He wants to stand by his conviction as to the nature of a myth or a mythology in such an absolute manner that it can only confirm the impression that, seen in the light of his idea of contribution, Torrance is not interested in theology-science discussions at any cost. Instead, he wants to demonstrate as emphatically as possible that it is theology which played the primary role in the development of Western scientific thinking. How fruitful this approach is for interdisciplinary relations is a matter of interest, and its success is best evaluated within the theology-science dialogue itself. It is evident, however, that Torrance's own system needs this argument in which the negative side of the epistemological contribution of early Christian theology to the rise of scientific thinking is emphasized: all mythology is absolutely rejected because it cannot provide us with real and genuine knowledge concerning the reality in which we live. It is human-centered and therefore it cannot but remain a projection of our own inner mental processes. All this leads one to reason that Torrance cannot detach himself from the kind of interpretation of the terms *myth* or *mythology* that is represented by Rudolf Bultmann. In other words, he accepts the conceptual framework lying behind the demythologizing process as belonging to the basis of his own system. This cannot but result in a counterattack against all possible interpretations of myth or mythology, even those that attempt to see their positive value.

Torrance's argument rests on the conviction that the rejection of all Hellenistic mythology actually took place in the Early Church Fathers' theological doctrine of the Incarnation and especially in the concept of *homoousion*. Thus the second problem mentioned earlier comes to the fore: can Torrance's claim concerning the Hebraic mind-set of the Early Church Fathers be justified? In other words, how strictly can we draw the dividing line between the Hebraic and Hellenistic mind-sets or between the Fathers, especially the Alexandrian Fathers whom Torrance appreciates most,[35] and other contemporary religious ideas entertained in the Mediterranean area? This question is considerably more difficult than the first one discussed above, because here Torrance has to leave historical accuracy and precision with regard to the history of thought in order to be able to present his ideas concerning theology-science relations as clearly as possible.

Usually the relation between Hellenism and Christianity has been seen as a far more complicated matter than Torrance suggests. Alexandria as one of the intellectual centers of Antiquity has been characterized as a melting-pot of diverse contemporary religious, philosophical, and other ideological trends. A very strong Judaistic movement found its place in the city, and there is therefore no need to neglect the argument of Hebraic influence on the development of Christian theology—one only has to remember the fact that the Greek Old Testament translation, the Septuagint, was produced there. But the important point to note is that what Torrance calls the *Hebraic* frame of mind did not live in a vacuum but was deeply affected by what Torrance calls the *Hellenistic* frame of mind.[36] In other words, such a demarcation line between Hellenism and Hebraism as Torrance explicitly attempts to draw is not very credible but rather gives the impression of oversimplification and neglect of some very essential points in the subtle relationship between Hellenism and Christian theology. That the prevailing Hellenistic culture had a far greater influence upon the formation of Christian theology than is allowed by Torrance is evident even in the crucial term *homoousion* itself, which was originally used in diverse forms by Christian Gnostics in their accounts of the relation between God, the world, and human beings.[37] The issue is a broad one to the extent that it cannot be included within the scope of this discussion. It is sufficient to point out Torrance's tendency of dealing with the matter, the aim of which must be seen in his willingness to argue for the supremacy and originality of Christian theology, or, in brief, of his idea of contribution.[38]

The concept of *homoousion*, expressing the consubstantiality of the Father and the Son, was born in the situation previously described. This means that, according to Torrance, the term is a product of both theological thinking and the Hebraic frame of mind. Furthermore, this means that if God chose to reveal Himself in this reality of space and time under its conditions by becoming incarnate in His Son Jesus Christ, who is *homoousios* with His Father, this is absolutely the only adequate way of acquiring knowledge of God. All other ways to the knowledge of God are thus strictly excluded because then this knowledge would not be in strict accordance with the "nature of things," as Torrance argues, the nature revealed in the Incarnation.[39] Torrance interprets the Greek Fathers' intentions, saying that "if we try to reach knowledge of God from some point outside of God, we cannot operate with any point *in God* by reference to which we can test or control our conceptions of him, but are inevitably flung back upon ourselves."[40]

It is not difficult to notice the point in Torrance's emphasis on the exclusive character of God's self-revelation in his incarnate Son. He is arguing that the Nicene Fathers and Athanasius in particular, with their application of the *homoousion* as an epistemological principle, ruled out all other channels of coming to a knowledge of God. It really should be remembered that, for Torrance, Athanasius is the prime example of a scientific theologian and that scientific methodology in his thought means simply thinking in strict accordance with what the nature of things really is.[41] In theology this means concentrating on the Incarnation. But applying this "scientific method" in exercising "scientific theology," as Torrance understands these terms, inevitably leads to the exclusion of all that is called natural theology.[42]

Torrance's reluctance to accept natural theology as it has been exercised in past centuries can be traced, in the first place, to his Barthian background and additionally to his philosophical standpoint in which he compares the premises of natural science to those of natural theology: in neither of them is God included.[43] Whereas natural science tries to move "in an opposite direction to theology in accordance with the nature of its subject-matter," natural theology "starts from the same premises and the same phenomena as natural science and seeks to move toward God."[44] In Torrance's opinion, the actual problem lies in the fact that those who entertain natural theology have to use what Torrance calls "logical bridges" between this world and its Creator.[45] Torrance thus confirms his basic theological structure according to which everything in theology hinges upon the movement from God to the humans, even knowledge of him. Outside of that vertical axis from above, there is no theologically valid knowledge of God. For Torrance, any movement from human beings to God inevitably leads to severe distortions.[46] This pattern of thought is in strict accordance with Torrance's total theological system which in the orthodox Reformed manner underlines the absolute majesty and sovereignty of God even in epistemological issues. In fact, Torrance explicitly sets himself in line with Calvinistic tradition on this issue.[47]

Such an indebtedness to Calvin on the part of a Reformed theologian should not come as any surprise, and therefore Torrance's interpretation of the Reformer, although its justification is certainly not beyond dispute, does not bear further essential relevance for the present argumentation. It is sufficient to note that Torrance consciously sets himself in line with Calvinistic epistemology and especially the Reformer's conviction that true knowledge of God cannot be defined in terms of natural theology. More important and interesting, however, is Torrance's eagerness to find reasons for his reluctance with regard to natural theology, that is, for the negative aspect of his idea of contribution, as early as the theology of the Early Church. Although nowhere directly stated, Torrance wishes strongly to suggest that Arius and his followers, the opponents of Athanasius and other

Church Fathers who adhered to the concept of *homoousion* as the expression of the consubstantiality of the Father and the Son, applied their own kind of natural theology.[48] That is, as Torrance understands it, the Arians wanted to build "logical bridges" between this world and God, with a starting point in themselves and in what they were able to see and understand, a procedure in which the Son of God inevitably was left outside the realm of the divine indivisible essence. Torrance connects Arius closely with what he calls dualism, a "habit of mind" resulting in radical dichotomies and closed-mindedness everywhere, and thus explains that Arius based his theology on this kind of dualistic thinking.[49] A lengthier discussion of Arianism and dualism will be provided later, and for the moment it is sufficient only to point out that Torrance links together Arius, dualism, mythology, and natural theology.

Is, then, Torrance's claim, that for Athanasius the *homoousion* was an expression that excluded every hint of natural theology, justified? Richard Muller criticizes Torrance, maintaining that the "insight into the work of the Logos before and beyond the union with the flesh led Athanasius to develop a broad and positive view of the role of natural theology."[50] Here Muller refers to Athanasius's apologetic writings *Contra Gentes* and *De Incarnatione* and founds his argument on the claim that God's preexistent Word, *Logos asarkos*, through whom the universe was created and who dwells everywhere in the creation, gives human beings an idea, however vague, of God. This is what Muller understands by the phrase "natural theology." For Muller, therefore, natural theology is something based on the activity of the *Logos* or the Word of God, and so he can argue, from his own premises, that Athanasius knew natural theology and applied it.

For Torrance, Athanasius's *Contra Gentes* offers a very different interpretation. He maintains that this apologetic work has been occasionally misunderstood when it has been interpreted in terms of traditional natural theology: "No attempt was made there to find a way of reaching God by logical reasoning, but rather to point out a way of communing with the regulative and providential activity of God in the rational order of the universe, in which our minds come under the force of truth of God as it bears upon us in its own self-evidence and shines through to us in its own light."[51]

This quotation affirms what was earlier presented as Torrance's view of what natural theology is: a human effort to reach God through logical inductions or deductions with the human being as the center of the epistemological process. For Torrance, Athanasius's theology of the Word is not representative of natural theology because the Word and his activity mean for him God's revelation always and everywhere. The Word is at the heart of what Torrance calls "positive theology." Furthermore, this seems to suggest that Torrance does not draw any distinction between *Logos asarkos* and *Logos in flesh*, as does Muller. The contradiction between him and Torrance derives primarily from a differing understanding of what natural theology actually is, implying that Muller's criticism misses its mark. This does not mean, though, that Torrance's own interpretation of Athanasius was without its problems. A basic difficulty in his reading of the Church Father is the strictly expressed link that he draws between Athanasius, *homoousion*, and the rejection of an independent natural theology. In other words, what was Athanasius's own position with regard to what was later called "natural theology"? And further, is Torrance's use of his writings to back the epistemologically negative side of the idea of contribution justified?

First of all, it has to be noted that Athanasius did not use the concept of *homoousion* as consistently as one might think when reading Torrance's writings.[52] This alone need not, however, make Torrance's argument inadequate, for the idea of the consubstantiality of the Father and the Son is clearly apparent in all of Athanasius's works, despite direct mention of the term. A far more problematic account is Torrance's effort to see Athanasius as inimical toward any natural knowledge of God.[53] The issue becomes apparently more complicated if Athanasius's own understanding is traced. In his early work *De Incarnatione* the Church Father argues that, since God knew the weakness of mankind, he had provided the works of creation whereby men could know God.[54] In spite of this, humans had not taken God's revelation seriously but neglected it. While men did not look upward to God but downward to created beings, God Himself chose to descend from on high and assume human form in order to let these negligent men know Him in his earthly body.[55] This is to say that, according to Athanasius, God has indeed bestowed, in principle, His creation with the actual possibility of knowledge concerning Himself, but that, due to the corruption of mankind, human beings are not able to reach this knowledge without His further assistance. Therefore the Word took humanity upon himself and descended to the same level as we humans.

Torrance seems to have interpreted Athanasius correctly in that he surely holds the Word of God as the basis of our knowledge of God. But equally evident is the fact that Athanasius considers this epistemological process something that took place before the Incarnation while the Word of God was the Agent in the act of the creation, thus enabling humankind to share in his knowledge of God.[56] In this respect Muller's view, described earlier, in which he drew a distinction between *Logos asarkos* and *Logos in the flesh* is certainly adequate. For Muller, the activity of the Word not incarnate forms the basis of his understanding of natural theology, while Torrance, as much as he concentrates on the Word in his theology, can appreciate only the incarnate Son of God who is *homoousios* with the Father as the right way to a genuine knowledge of God. Turning to the Church Father, it should also be remembered that in Athanasius's thinking the *homoousion* is presupposed not only epistemologically but also primarily soteriologically: God became a human being, and Jesus Christ as a human being is *homoousios* with the Father because in that way God's image in humanity was renewed.[57] Therefore, it appears that to relate correct epistemology strictly to the concept of *homoousion* and to give the Incarnation such an exclusive character with regard to the possibility of gaining knowledge of God through the creation, as held by Torrance, is a slight overinterpretation. Torrance's unconditional attitude is apparent in his claim that all our knowledge of God has its source in Jesus Christ the *incarnated Logos*.[58]

But whence the ambiguity in Torrance's thought? On the one hand, he seems to interpret Athanasius correctly in that he discerns the epistemological side in his usage of the term *homoousion*. But on the other hand, his reading is one-sided in that he considers the Church Father as having categorically abandoned any appropriate knowledge of God through the creation strictly, unconditionally, and in principle. It should be noted first that the whole question of natural theology in Athanasius's thought is hardly anything else but an anachronism because natural theology as an intellectual enterprise in which we as humans try to move toward God with the help of our own reasoning did not appear in its articulate form before the Middle Ages.[59] Torrance's Barthian background cannot be left without attention, either, because within it he continues Barth's arguments against natural theology, although in a modified form. This indicates that Torrance not only gives patristic answers to modern questions but offers modern solutions to problems of the patristic era. The ambiguity in Torrance's thought would certainly disappear, or become less at least, if he did not set natural theology and revealed theology in such strict and uncompromising confrontation with each other. To follow Athanasius, the two modes of theology cannot be contradictory in themselves simply because both of them are based on the activity of the Word, and this means, to use Torrance's terminology, that what he sees as natural theology can also be understood as revealed, and that it is not necessary to think of it as anthropocentric: as described, it remains Word centered. Although the Incarnation brought a suitable way for humans to know God, the value of the creation as a source of such knowledge was not thereby diminished.[60] Torrance appreciates this insight of Athanasius's, but he speaks of it not as a general revelation but as *integrated natural theology*.[61] The situation could be quite different, then, if the notion of general revelation were appreciated more in Torrance's system. If the creation were considered as belonging to the realm of revelation, too, then there would be no need for Torrance to construct such a procedure for an integrated natural theology as he feels compelled to do.[62]

This gives us reason to argue that Torrance's point is most convincing when he emphasizes Athanasius's own position, that is, the necessity of the Incarnation for our correct knowledge of God. Far more arguable, however, is his application of Athanasius's position to reject what he calls natural theology, a scheme which is more Barthian than genuinely Athanasian. The fact is that natural theology, understood as deriving from general revelation, continues to present an inspiring challenge in the theology-science dialogue so that its programmatic exclusion in the manner of Torrance seems quite risky.[63]

The Positive Role of homoousion

So far, we have followed only that line in Torrance's thought in which the negative aspect of the epistemological idea of contribution is apparent. It was suggested that Torrance's Reformed Barthian background makes him interpret Athanasius's view of the *homoousion* in an exclusive manner, rejecting all other means of attaining knowledge of God. In other words, Torrance explicitly prefers abandoning *all* knowledge of God whatsoever outside the scope of the *homoousion*, suggesting that beyond that concept every piece of knowledge is not only insufficient but also more or less untrue. This negative epistemological aspect of his idea of contribution is of vital importance for the later discussion of his thought concerning the relationship between theology and science. Just as essential, however, is the *positive* aspect he gives to the epistemological role of his idea of contribution.

There is a phrase which Torrance constantly uses in order to describe the epistemological process in scientific thinking and in which both the person who knows and the thing known are united. His idea of the *disclosure model* contains two remarkable points: first of all, it explains the nature of reality as something that is *hidden* or *transcendent* and that is to be disclosed in order that we might attain knowledge. But then there is the second aspect, the human "knower," who wants to be in deeper epistemological touch with transcendent reality but in which he can succeed only indirectly through these "tools" or models.[64] This kind of understanding is closely linked to Torrance's view of realism, an issue that will be examined with greater precision in the next chapter. At the moment it is sufficient to underline the substance of the phrase: a disclosure model is a medium through which an independent reality can be grasped by the human mind when it is directed into contact with it.

As stated, Torrance holds that in a disclosure model both the objective and the subjective poles coincide. The objective pole is somehow transcendent and therefore needs to be revealed, and the subjective pole has to receive the disclosure in order to be true to the reality under investigation. The interesting point now is that a disclosure model is "forced upon us by the intrinsic intelligibility of some field as we inquire into it."[65] Torrance strongly suggests that this kind of model cannot be just a conventional construction with which we try to penetrate into our fields of study, but rather it is a model which bears in itself, even momentarily, an aspect of certainty precisely because it is imposed upon the human mind. This also suggests that this model is not primarily a description of human experience in the face of reality but rather of reality itself, which needs to be revealed to us in order that we might give it articulate expression. A disclosure model, then, is a product of reality itself. From this it may be concluded that a model can be attributed with the term *disclosure* only if reality is seen to be disclosing itself through it, or, in other words, if reality is considered as being able to reveal itself. The Christian notion of revelation has undoubtedly influenced Torrance's understanding of disclosure models to the extent that the question of the doctrine of the Incarnation and the concept of *homoousion* and their relation to Torrance's view of a disclosure model has to be paid further attention.[66]

It is now argued that according to Torrance a disclosure model is the most adequate way, within realism, to genuine knowledge, representing the means by which the gap between an independent reality and a human knower is bridged.[67] As far as the idea of contribution is concerned, the question should now be asked whether Torrance sees the patristic term *homoousion* and the idea of consubstantiality as such a disclosure model which could also be applied in the natural sciences. If it were not a disclosure model that could be used in all sciences, then his idea of the contributive role of the *homoousion* would remain trivial, with solely a rhetorical or analogical significance at most. If, however, it is understood as a genuine disclosure model that has a substantial contribution to make to scientific thinking in general, then Torrance could consistently claim relevance for his idea. The problem is made quite ambiguous by the fact that Torrance does not directly give the *homoousion* the status of a disclosure model, which, as a matter of fact, could be expected to lie in the basic structure of his thinking when his view of the contributive role of the patristic term is remembered. An answer can be found by a closer glance at his usage of the phrase "disclosure model."

In his older writings, such as *Theological Science* or *God and Rationality*, Torrance does apply the notion of disclosure models but with a rather reserved tone. Speaking about the natural sciences and mathematical representation of human understanding, he states that "we may treat that representation *only* as an explicatory model or a *disclosure model* through which we interpretatively apprehend the reality we are investigating and not as a descriptive formula or as the equivalent of some ontic structure in reality itself."[68] The attention is directed to Torrance's evaluation of the ability of a disclosure model to reveal reality to the human mind: it does not bear a one-to-one correspondence with reality but offers *only* an interpretative tool for grasping something of its inner structure. Torrance, in fact, denies such a strict correspondence between our modes of thought and reality because reality as such always transcends our ability to understand it. The important point for the current argument is that in this connection Torrance is not suggesting any link with the *homoousion*. Slightly different is the tone in later writings, where a disclosure model is seen as "the cognitive instrument we use precisely in so far as it is appropriate" and as directing "us away from its own representation to the objective invariance we discern through it."[69] Torrance's reservation concerning the appropriateness of a disclosure model is maintained, but now it is given the explicit status of *the* cognitive instrument to be used, as far as it is in consonance with the reality investigated. Any direct reference to the *homoousion* as a disclosure model is still lacking, but now it is explicitly admitted that such a model offers, even though as a mere representation, access, however limited, to reality. A disclosure model, then, offers a human knower a way whereby he or she can enter into dialogue with the reality he or she is investigating with the reservation that such a model does not contain truth in itself or bear a strict correspondence with reality. It can be true or consistent with reality and with truth only insofar as it refers to them in an appropriate way. A relation of correspondence is replaced by a referential relation which deprives human constructions of any final validity but leaves the "door open" for reality itself to exercise its coercive influence upon the human mind.

The essential issue with regard to the *homoousion* is now the question whether this patristic term can be seen as a disclosure model in the sense Torrance gives it, that is, with a referential and interpretative but by no means final or self-sufficient role as a product of the self-revelation of reality. Torrance's view of the *homoousion* as the very nucleus of Christian theology both epistemologically and ontologically suggests that he considers this patristic term as having such absoluteness and centrality that it, at first glance, does not bear the characteristics of the disclosure model described above.[70] In other words, the *homoousion* seems to be given a higher status in an epistemological process than that of a mere disclosure model, an impression that becomes evident as soon as one becomes acquainted with Torrance's thought. Accordingly, this means that the actual contribution of the *homoousion* to other disciplines remains on the analogical level, pointing to an epistemological similarity or parallel in the procedures of theology and the natural sciences. But this conception does not seem consistent with Torrance's intentions to underline not only some external parallels or analogies but, in the final analysis, connections between the disciplines. Therefore it is reasonable to conclude that for him the *homoousion* has to bear such characteristics that it can be classified as a disclosure model also applicable in the realm of the natural sciences, as far as its substance is concerned. As was stated above, the difficulty lies in the fact that Torrance nowhere explicitly attributes the *homoousion* with the character of a disclosure model. However, in his writings a trend can be discerned in which it is possible to interpret the *homoousion* as representing a mode of a disclosure model. The most convincing evidence for this identification can be found in a statement where Torrance refers to the nature of the *homoousion*, saying that it is not "somehow sacrosanct and beyond reconsideration, for all theological terms and concepts fall short of the realities they intend."[71] In this case the *homoousion* fulfills the demands set by Torrance's definition of a disclosure model.[72] It is not, in principle, an absolute and final concept, one that excludes critical reevaluation but one that remains open to further precision.

In the last sentence the words "in principle" should be emphasized because it seems evident that despite his willingness to consider the *homoousion* as an open concept leaving room for critical questioning, in the ultimate analysis, Torrance is reluctant to suggest what kind of reevaluation there possibly could be with regard to the *homoousion*. Therefore one can claim that although the *homoousion* meets the requirements of a disclosure model as set by Torrance, they cannot ultimately be identified because of the normative character and status the patristic term has gained over the centuries. In other words, the *homoousion* is in essence more than a disclosure model—it is "a fundamental dogma, which once it comes to view becomes normative for all faithful theological statement, for it enables it to be made in true correspondence with its proper object and in consistent relations with other faithful statements."[73] Thus the *homoousion* is certainly an absolute concept which cannot, in practice, be altered or replaced or even given any deeper reevaluation without severely damaging the current self-understanding of the Christian Church. This leads us to make the claim that in this respect Torrance's thought remains imprecise because, as was shown, on the one hand he considers it as a revisable concept but on the other he allows it as a normative authority.

It was argued earlier that in order to offer a real epistemological contribution, the *homoousion* has to be more than a mere indication of analogy between theology and the natural sciences, applicable in sciences only vaguely on the basis of a remote methodological similarity. Therefore it was further maintained that Torrance implicitly suggests that the *homoousion* is actually a disclosure model in the service of scientific thinking which seeks to gain a firmer grasp of the reality it is investigating. But due to the ambiguity detected in Torrance's thought in this respect, Torrance cannot, after all, consider the *homoousion* as a pure disclosure model, either. Something more is at issue. Rather, for him it is *an authoritative and normative methodological tool* which, as far as its actual substance is concerned, can be applicable in the realm of the natural sciences as well.[74]

The conception of norm and authority leads us to another important feature of Torrance's view of the positive epistemological aspect of his idea of contribution. The question is of the role of a human knower in the epistemological process and of the relationship between faith and reason. Torrance maintains that the process of formulation of the concept of *homoousion*, as well as the whole theological enterprise of the Church Fathers, took place under certain epistemological conditions in which faith played a crucial role, "not as a subjectively grounded but as an objectively grounded persuasion of mind."[75] For Torrance, then, faith provides the guarantee that objectivity is maintained and preserved and that under such circumstances can the human knower act in the only possible way, namely in submission to the authoritative objectivity he is investigating. Thereby true scientific thinking and faith are linked together extremely closely.[76] This means further that faith is primarily something very rational and as such an unavoidable principle in all epistemology.[77] He does not wish to become more involved deeply in the reason-faith debate because he considers it an unfortunate artificial problem in epistemological discussion.[78] In this sense, Torrance's appreciation of the role of faith seems to be in accordance not only with the classical theological trend, where faith is seen as a presupposition for knowledge and understanding, but also with his own Reformed tradition, in which the role of rationality is strongly emphasized.[79]

Torrance thus constructs a simultaneous coexistence of faith and reason in any epistemological process, suggesting that even the concept of *homoousion* is a direct result of this kind of scientific approach, uniting the two elements. The crucial point is that in all scientific knowledge, theology included, the human mind conforms to the nature of things as they are in their inherent intelligibility.[80] This means that Torrance would like to see faith as an essential part of the contribution that theology has made to the development of the sciences.[81] But the problem now is, what exactly is he referring to when he underlines the relevance of faith so emphatically? Is it a general epistemic attitude, coarse and profane, that could be attributed to the work of any scientist who feels "awe" before his object of inquiry? Or is Torrance's faith explicitly religious in nature, or even Christian? Naturally, the work of the Early Church Fathers was Christian in character, but whether the faith that Torrance would like to see as a contribution to scientific thinking is Christian or merely secular is of the utmost importance. To find out the real nature of the faith element in Torrance's idea of contribution, it is necessary to further analyze his thought in this respect and reflect on his claims to the actual "revolutions" in science that he himself appreciates most.

Referring to one of his authorities in the field of philosophy of science, the physical chemist Michael Polanyi, Torrance claims that "belief has to do with the elemental interaction between persons and realities other than themselves, entailing a recognition of their independent reality and truth."[82] In this case belief is not necessarily religious, to say nothing of Christian, but only a very general frame of epistemological attitude in which the reliability of reality is taken for granted.[83] With regard to the idea of contribution there is, however, another aspect to be taken into account, namely that Torrance is not content with stating the relevance of such a general frame of belief. His further argument tries to present evidence that the rise of the doctrine of the Incarnation with its principle of *homoousion* and recent developments in physics, with their relevance for ontology as well as epistemology, are due, not primarily to secular *intellectual* faith in general, but to *religious* and explicitly *Judeo-Christian theism.*

Torrance's account of his fellow countryman James Clerk Maxwell, one of those whose works meant a significant breakthrough for physics, is informative in this respect. Namely, Torrance maintains that the old mechanistic conception of nature was overcome in Maxwell's scientific work because the scientist was deeply affected by his personal Christian belief.[84] Thereby a direct connection between Christian devotion and scientific thinking is strongly emphasized, and Torrance presents the argument that the breakthrough and rise of modern science, as he understands it, is a direct contribution of scientists with a commitment to the Judeo-Christian view of God.[85] Torrance appreciates Maxwell's orthodox and emphatically Christological faith when he explains the significance of the fertile interaction between theological and scientific thinking he sees in the work of his fellow countryman.[86] It has to be noted that Torrance does not see here any *analogical* connection between the theology of the Early Church with its concept of *homoousion* and the significant results of modern physical science. The question is not of remote parallels but, in fact, of an identical epistemological event where Christian conviction has a decisive role to play. Torrance does not even hesitate to remind us of Maxwell's explicit belief in *Christ*, thus confirming his claim that the *homoousion* has had a contributive role even in modern times. It is useful to emphasize that, in Torrance's opinion, through the Incarnation God does not directly give birth to science but, rather, the doctrine of God's Incarnation is fertile enough to produce a mental attitude that has aided in the development of scientific thinking.[87] Historically, however, Torrance's evaluation of the relevance of Maxwell's religious conviction is not without problems.[88] Our concern is not, however, to penetrate deeper into Torrance's line of argument, while it is sufficient to demonstrate the pattern with which he constructs his argument to support his idea of contribution. With regard to Maxwell, this much can be said: in him Torrance easily finds an example to use as a confirmation of the idea of contribution. Maxwell's Christian conviction can be proved without difficulty, in whatever manner its significance is evaluated.

Greater problems, nevertheless, emerge when Torrance tries to do the same with the thought of Albert Einstein, who undeniably wrote a great deal about the religious task of a scientist. Einstein's understanding of what religious belief is was not quite similar to Torrance's interpretation of it.[89] The physicist appreciated a religious attitude toward the universe, an attitude by which he understood a deep and humble wonder before the mysteries of the universe, but he distinguished it sharply from any hint of a belief in a personal god. According to him, such a belief is characterized by fear and expectations of a reward from a supreme Being. Einstein traced a great deal of evil discernible in the world to the way in which established religions have exercised their power. He held that belief in a personal god is an indication of naïveté, while a scientist, according to him, must have a general religious feeling that leads him to respect the harmony of natural laws and the deep intelligibility observable in them.[90]

Torrance is well aware that Einstein explicitly rejected any belief in a personal god, and he understands Einstein's approval of the view of God presented by Baruch Spinoza.[91] However, what makes his point interesting is that he wishes to analyze Einstein's conviction in favor of his efforts to prove the validity of his idea of contribution. Torrance points out that a person who has "a realist understanding of natural law," who refers to the "reliability of nature," who sees a particular theory as "'closer to the secrets of the Old One', in so far as it penetrated into the inner truth and intelligibility of the universe," and who even explicitly uses theistic terms, gives reasons to suggest that Einstein was not, after all, so far away from theistic belief.[92] Torrance's point is that Spinoza's view of the relation of impersonal divinity to the universe was that of strict necessity, making it knowable through "logico-deductive operations alone, but this is very far removed from the way Einstein understood the rational order of the universe."[93] Torrance agrees that Einstein's view of God, as the scientist explicitly and publicly expressed, contains certain inconsistencies with regard to theism because he could not take his distance from the Newtonian deistic understanding of God as "inertia, namely that something acts itself which cannot be acted upon," leading to the conviction that in his absolute transcendence and majesty God cannot be affected by anything that comes from the outside of him.[94] But, on the other hand, Torrance points out that the Jewish heritage or "frame of mind" helped Einstein appreciate Spinoza's "rejection of Cartesian and other forms of dualism, and his unitary conception of the universe with its inherent rational harmony."[95]

There is no need to enter further into the problem of the nature of Albert Einstein's religious beliefs and Torrance's interpretation of them. The preceding discussion only provides confirmation of our argument that Torrance accredits the Judeo-Christian view of the world and the concept of *homoousion* (what we discern with the senses is of the same essence with reality itself in its ontological depth) with a strong contributive role whereby faith and reason are united and the human knower is placed in his proper position in the epistemological process, namely that of obedient submission. To give reasons for this argument Torrance needs Einstein and the above-mentioned interpretation of his views on religion.

The Ontological Contribution

The Ontology of Creation

In the previous pages it was shown that in his idea of contribution Torrance considers the doctrine of the Incarnation and the concept of *homoousion* as epistemologically significant both in their negative and positive aspects. But the contribution of Christian theology to the later development of empirical science, according to him, is not only epistemological in character but is also of crucial importance with regard to the perception of the ontological nature of the universe. A convenient way of examining his idea of the ontological contribution of the *homoousion* is offered in his efforts to link the ontologies of God and the universe together through the doctrine of the Incarnation and the patristic concept of *homoousion*. It should be constantly recalled how Torrance gives the doctrine of the Incarnation primacy over that of creation so that the latter is interpreted by means of the former. When explaining the origin of the creation *ex nihilo* he states that "the birth of the creation hinges upon the *homoousion*."[96]

This approach means that the ontology of the universe does not depend solely on the act of creation *ex nihilo* but, with an essential addition, on the Incarnation. The creation was carried out by the Word of God, who is eternally *homoousios* with the Father and who then became incarnate, assuming human form in space and time. The most striking consequence of this approach with regard to the ontology of the universe is that it is not seen from the standpoint of the Creator-creation axis, the being of the universe deriving only from the fact of its createdness, but from the conviction that this Creator took human form upon himself. In fact, without the Son of God there would be no understanding of God as Creator.[97] This is to suggest that creation alone is not a sufficient basis for an ontology of the universe but needs the contribution of Christology, the doctrine of the Word who became incarnate, in order to be relevant.

It should be remembered, however, that such a Christological ontology is not an original feature of Torrance's theology alone. In patristic times the *Logos* was seen as an ontological basis for the existence of the universe, and Torrance undoubtedly desires to continue this tradition. Athanasius himself, for example, saw no difficulty in holding that the incarnate *Logos* exercises the same power over the creation as in his preexistence.[98] Torrance, nevertheless, goes further when drawing out the implications of this Christological ontology. He maintains that the doctrine of creation, seen through that of the Incarnation, has posited certain basic ideas concerning the ontological nature of the universe without which the development of empirical science would not have been possible.

Torrance deals with the basic ideas contributed by the theology of the Early Church by using a formula consisting of three factors. Two of them have to do with the ontological rationality and freedom of the universe, and in the third one the notion of contingency is introduced in a way that requires explanation.[99] On the one hand, referring to the third factor, Torrance can occasionally speak of the "ordered unity" or "rational unity" of the cosmos, while, on the other hand, in some other connections the notion of "contingency" takes the role of the third factor. (It has to be noted, however, that in Torrance's lists the aspect of rational unity or contingency is always mentioned first, rationality second, and freedom third. This suggests that the latter ones have to be viewed through the perspective of the first aspect, whether it is designated "ordered unity" or "contingency.") But when the concepts of "ordered" or "rational unity" are preferred, Torrance attaches the attribute "contingent" to the first two factors, rationality and freedom, while in the second case, contingency as the third factor, they are used without such an attribute. In other words, Torrance uses two triplets to describe his idea of the ontological contribution of the theology of the Early Church to the rise of modern science. The first triplet, used in some earlier books, consists of the terms "contingent freedom"—"contingent rationality"—"ordered/rational unity." The second triplet to be found in later literature consists of the concepts "freedom"—"rationality"—"contingency."

This feature suggests that Torrance's thinking has developed in this respect, with the result that in the ontological side of his idea of contribution the notion of contingency has gained a more independent status. It is no longer considered a mere attribute defining the content of the further ideas of rationality and freedom of the universe but has in fact grown into an interpretative key concept with a distinct existence of its own. This reveals that contingency is seen as being so important that it deserves this new status as the main ontological contribution of the incarnational theology of the Early Church. Accordingly, the notion of the rational unity or ordered unity of the cosmos, formerly the very first and independent aspect, is placed in the context of the rationality of the universe referring thus to the objective and ontological pole of an epistemological process. Therefore it seems appropriate to analyze the three contributive aspects along the lines of the option giving contingency an independent and emphasized position and to view the notion of the ordered unity of the universe together with the notion of the rationality of the cosmos.

Torrance draws a sharp distinction between the Greek and the Christian ideas of contingency, granting the latter the role of the originator of a unique content for the concept, explaining that the Greek interpretation of contingency was "the polar antithesis to what is rational, the logically and causally necessary."[100] Due to the direct identity of the necessary and rational, contingency was defined as pure chance, irrational, and unintelligible and therefore as something to be despised and totally excluded from what was understood as rational knowledge. According to Torrance, this philosophical attitude contained two presuppositions, the first of which was plainly theological in that it constructed a necessary and timeless relationship between the world and God. Belief in God as Creator was thereby excluded, and eternal, immutable divine reasons were clamped upon the universe to form the basis for its intelligibility. The second presupposition had its origins in Greek dualism, which was drawn between intelligible and sensible, form and matter. In this distinction the form was given the role of the bearer of rationality, and therefore also of necessity, while matter was nothing more than "a sensible element which is only accidentally related to the necessary."[101]

Torrance's main argument in this respect is that it was the incarnational theology of the Early Church that meant a radical shift in the notion of contingency. The ontological status of the universe was altered by the new understanding of contingency set forth by patristic Christian thinking. On this basis Torrance maintains that the possibility of the emergence of empirical science was established only through the Christian interpretation of contingency, leading to the crucial conviction of the idea of contribution according to which the primary reasons for any development in the field of the natural sciences are characteristically theological, not philosophical.[102]

Two questions emerge from this ontological aspect of the idea of contribution with special respect to the notion of contingency. Firstly, how does Torrance relate contingency and necessity in his interpretation of patristic theology and its contributive effect? And secondly, how does he attempt to make this new notion of contingency the merit of incarnational theology? The question of the interrelation of contingency and necessity in patristic theology, as seen by Torrance, is due to the new view of the God-world connection. While Greek thinking was unable to establish the priority of God over necessity, the latter grew into an overall system of divine reasons assuring the intelligibility of the universe. Contingency was then considered something inferior to necessity. Referring to the philosophy of Immanuel Kant as a regression back to this Greek notion, Torrance claims, "Since the contingent exists only in a series of causal connections between contingent events reaching back to what is unconditionally necessary, contingency is only a manifestation, at two or more removes, of necessity. This really represents a lapse back into the old Greek conception of contingency as having an in-built relation to necessity."[103]

Torrance's actual claim is, then, that the Christian theology of creation, seen through that of the Incarnation, reversed the order of necessity and contingency, depriving the former of its primary status and transferring it to the latter so that contingency is the basic ontological feature of the universe and perceptible necessities, for example, in the laws of nature, which derive their meaning and intelligibility only when they are submitted to this contingency.[104]

It is useful to point out that Torrance views the concept of contingency primarily as the counterpart of necessity.[105] In this sense Greek patristic theology undoubtedly contributed a novelty, but it has been convincingly pointed out that its elaborate application occurred only a thousand years later when John Duns Scotus first radicalized the concept of contingency through the notion of *synchronicity*: *p* is contingent if *–p* is possible *at the same time*, implying that contingency as a whole is based on God's free will and his possibility of continuously choosing another alternative.[106] This implies that dating the actual origin of the contingency, with fruitful effects on the development of science, to the patristic era has to operate with a slightly different interpretation of contingency than the one inherent in the notion of *synchronic* contingency.

The other question mentioned above has to do with Torrance's effort to link the doctrine of the Incarnation and the contribution of the basically contingent nature of ontology. It seems that for Torrance the creation is not sufficient to safeguard his argument in favor of his conception of the contingency. He admits that the view of the universe as created out of nothing has its roots in Jewish theology in which God as a sovereign and transcendent Being is "the source of all that is outside himself."[107] This is not, however, a sufficient reason for a contingency that would mean such a strong contribution as he supposes. It is also worth noting that for Torrance *creatio ex nihilo* is not a specifically Christian doctrine but has its origins in the religion of the Old Testament. Only the Incarnation can, in Torrance's opinion, provide such aspects or insights as to make contingency truly effective.[108] The following analysis seeks an answer to the question of how successfully Torrance can give reasons for his claim that the doctrine of the Incarnation was an original contribution on its own terms. The value of his modified creation argument depends largely on the prospective results. In other words, then, what is the original ingredient that only the Christian theology of the Incarnation could provide?

It seems evident that Torrance himself cannot give a very strong and convincing answer. He states that "it was Christian theology which radicalized and deepened the notion of contingence" by the fact that the very God who had created the universe took humanity upon himself, thus submitting himself to the conditions of contingency and thereby assuring that the created reality is real for himself, too.[109] An ambiguity in Torrance's argumentation is to be found in precisely this respect: on the one hand, he maintains that it was the incarnational theology of the Early Church, through which the act of creation was viewed, that could make the contribution of a new ontological base of the universe in contingency. The doctrine of creation was not sufficient, as stated earlier. But, then, on the other hand, Torrance has difficulty in finding convincing arguments for this claim, so that he can only refer to a *radicalizing* or a *deepening* of the notion of contingency. The result is that Torrance cannot present too persuasive evidence for his view of the original contribution of Christian theology to the understanding of contingency as the basis for a new view of the ontology of the universe in distinction from the Greek understanding. If the Incarnation is just a catalyst in the process of formation of such an ontology, the creation as, originally, a Jewish doctrine must retain its primary role. It is conceivable that the *Logos*'s entering the created reality as the Incarnate One implied the affirmation of the reality of the universe even for the Creator himself, but there remains the question of what is the connection between this reaffirmation and the idea of contingency or its radicalizing and deepening.[110]

A far more consistent approach is the purely theological position presented by Torrance. Here contingency is considered as seriously injured by a tendency hostile to being itself, a tendency that had to be removed in order that the creation could be saved.[111] This implies in traditional terms that sin has corrupted the creation to the extent that God himself needed to take action in order to bring salvation to the universe. However, as was seen, Torrance is not content to draw mere soteriological implications from the Incarnation but wants to mold them into an interpretative basis to provide a foundation for his argument in favor of his idea of contribution. In other words, the Incarnation meant not only the reordering of the contingency of the universe onto the right path originally given to it by its Creator for its salvation but also, and even more strongly, the restoration of the contingency of the universe in such a unique way that it made empirical science possible.

To sum up the discussion so far, Torrance's understanding of the contributive role of the doctrine of the Incarnation with regard to the contingent character of the ontology of the universe has two aspects. The first has to do with the contingency-necessity dichotomy. Torrance maintains that the incarnational theology of the Early Church substantially transformed the mutual interrelation of the two concepts by transferring the priority given to necessity by Greek thinking to contingency. Torrance's conviction could be formulated such that the Christian theology of the Incarnation made contingency a substantial feature of the ontology of the universe, and accordingly necessity was dropped to an inferior accidental position, while in Greek thought, according to Torrance's interpretation, just the opposite had been the case. In the new situation necessity is realized through the "laws of nature," which, however, have their ground in the contingent nature of the universe. So these laws cannot have an ultimate validity in themselves but only in the contingent ontology behind them. This reveals that Torrance cannot even implicitly escape using the substantial-accidental dichotomy in his own interpretation of contingency, which could be supposed to be the case because of his reluctance to accept the matter-form or form-being dichotomy that, according to Torrance, "has become almost second nature to us."[112]

The other problem was seen to lie in Torrance's way of interrelating creation and the Incarnation. It was maintained that he finds himself in difficulties when trying to define the actual positive contribution of the doctrine of the Incarnation when compared to that of creation. While the latter holds that creation was effected out of nothing, the former had nothing original to contribute to the ontological contingency of the universe which could be viewed as a necessary basis for the later development of empirical science, as Torrance claims with his idea of contribution. Therefore he is compelled to regard contingency as only radicalized or deepened through the Incarnation, a notion that hardly makes the contribution of the incarnational theology of the Early Church as significant with respect to the rise of the natural sciences as Torrance makes out. All this is to suggest the conclusion that, Torrance's idea of contribution presupposed, the actual contribution of the theology of the Early Church does not lie primarily in its original doctrine of the Incarnation but in the Old Testament view of God as the Creator who out of nothing brought forth the universe. The merit of the Early Church can be seen in its safeguarding and further refining of this doctrine for the purposes of the proclamation of the Church. The significance of the doctrine of the Incarnation should not, of course, be denied or its role underestimated. However, even Torrance cannot proffer adequate evidence for the claim that its contribution was highly unique for the development of later empirical science because the idea of the *Logos* becoming a human being does not lead so readily to the conviction that the contingency of the universe is thereby given something essentially new as compared to the doctrine of creation. In this respect his modified creation argument proves to be unconvincing.

Torrance has, however, better reasons for making his claim concerning the second aspect of the ontological side of his idea of contribution, which concentrates on the rationality or intelligibility of the contingent universe. The Incarnation of the Λόγος has such a close correspondence with the diverse *logos* philosophies of Antiquity, regarded as the basic principle of rationality, that Torrance's conviction with regard to the contributive role of the Incarnation in this respect seems more interesting than the previously described view of contingency.[113]

In accordance with his strong emphasis on rationality, Torrance relates the idea of the intelligibility or rationality of the universe to the doctrines of both creation and the Incarnation.[114] The doctrine of creation meant that the universe was conceived as a grand harmonious system in which one specific order prevails.[115] This rational unity or harmony is the feature of the whole world, not only of earthly phenomena, so that a scientist can trust in the intelligibility of any phenomenon even in the remotest corner of the universe. Torrance thus supposes a kind of rational monism in which any mode of intelligibility is denied which could not be, at least in principle, comprehensible in human terms. His notion of one, all-pervading order leaves no room for any other kind of rational structure than the one a human being can now conceive of. It has to be admitted that Torrance accepts variations (in his phrase concerning the "multi-variable" order) on the phenomenal level of this order, but ontologically he can acknowledge only one order or rationality.[116] From a theological point of view, the case could not, in any case, be very different from his interpretation since the view that God created everything has been here driven to its ultimate conclusion as far as the question of the intelligibility of the universe is concerned. This means that in Torrance's opinion the human being cannot be removed from his or her central universal position as the being who gives articulate expression to the intelligibilities of nature.[117]

If one wishes to seek a cosmological principle in Torrance's thought, it is his notion of the rationality of the universe that could be designated as such.[118] It asserts that the cosmos bears the ontological character of being able to be known. But its createdness further implies that the universe is not self-explanatory as if it contained the reasons for its existence.[119] Its contingency upon God also includes its intelligibility, so that Torrance can speak of the contingent intelligibility or rationality of the universe.[120] Just like the being of the universe, its created or contingent rationality, is totally dependent on God the Creator and his uncreated rationality.[121] The question to be asked at this point is, how do these two modes of rationality, created and uncreated, relate to each other? Since God created the universe and gave it a rationality of its own, to what extent can there be seen an analogical correspondence between the two modes of rationality? Torrance maintains that "the intelligibility which he [God] has conferred upon the universe is not an extension or an emanation of his own but a creaturely intelligibility utterly contingent upon his own, yet somehow coordinated with it as the universal medium through which he may be known."[122] Torrance avoids using the term *analogy* in describing the two rationalities, implying thereby that no definite conclusions can be drawn from the intelligibility of the universe as to the nature of God's own rationality, which remains transcendent.[123] Torrance's Barthian position is thus reaffirmed: there is no natural theology leading to God and no reliable knowledge of him, thus also excluding the possibility of any argument presented in favor of the existence of God. It is not impossible, however, to trace an implicit *argument from intelligibility* in Torrance's thought. As was shown, Torrance explicitly rejects any "logical bridges" from human beings to God or from the intelligibility of the creation to that of God, but in spite of this position the idea of a *contingent* intelligibility, if accepted in the form Torrance suggests, inevitably leads one to admit that this contingency must be referred to its originator, who is God. In other words, if the non-self-explanatory character of the universe is agreed upon, the existence of its transcendent origin has to be supposed as well.

Because Torrance denies an analogical correspondence between the two rationalities and because he, nevertheless, has to link them together in order to safeguard the conception of the creatureliness of the universe he makes use of the very vague term *coordination*. He does not provide any precise definition of what he understands by this concept, but from his thought it can be concluded that the notion of order is the common factor: the rationalities are coordinated so that there is a resemblance between them as to their tendency to avoid chaos and disorder. The two rationalities, then, can be viewed as being connected by a kind of umbilical cord through which rationality and order are channeled from the uncreated to the created side. This connection is essential; without it the created would lack any meaning. While the created rationality cannot correspond to the transcendent uncreated rationality, it can only refer to it as its ultimate basis and ground. In other words, their relation is referential and open, not strictly defined as in one-to-one correspondence.

So far, the intelligibility or rationality of the universe has been reflected upon on the basis of Torrance's understanding of the doctrine of creation. The question concerning the relation of the doctrines of creation and the Incarnation also needs to be investigated in this connection. The idea of contribution asserts that while the doctrine of creation has given the universe a rationality that makes its structures knowable and intelligible, the doctrine of the Incarnation had a further positive impact upon this view. One ambiguity, however, bothers the reader in this respect. Torrance writes: "If in the Incarnation it was the Creator Logos of God, the one eternal uncreated Autologos (Αὐτολόγος) or Self-Word, who became flesh in Jesus Christ, then all the logoi (λόγοι) or rational forms pervading the universe were thereby revealed to be created."[124]

The same difficulty as in the case of the question of contingency is apparent here: the actual contribution of the doctrine of the Incarnation remains obscure. Torrance claims that the Incarnation of the *Logos* meant the revelation of the fact that all that exists is created. He implicitly states that the Incarnation disclosed the ultimate character of the rationality of the universe by revealing the createdness of "all rational forms" and their contingent nature as well. Thus the Incarnation brought not only the revelation of God but of the creation, too, or more precisely, its contingency and dependence on God. It is nevertheless difficult to see where the originality of the doctrine of the Incarnation lies in this respect because the Old Testament view of the one and only God as the transcendent Creator bore in itself the same idea of the createdness of the rationality of the universe.

In one sense, however, Torrance's argument does not sound so implausible. The claim about the doctrine of God's becoming a human being as a contributive factor to the view concerning the ontological rational nature of the universe would, indeed, make more sense if it were considered primarily as an indication of God's deep involvement with his creation and his willingness to sustain its created rationality by submitting himself to its conditions. In fact, this is what Torrance does when he says that "thus not only has God created the world out of nothing . . . but He has confirmed it and established its relation to Himself through the Incarnation of His Son within it."[125] The idea of the Incarnation as the confirmation of the created universe and its rationality seems more plausible than the view, in which the Incarnation meant the revelation of the createdness of the universe's rationalities. Thus the Incarnation could be seen as an act of God's deep solidarity in which the reality of created rationality was reaffirmed as relevant even for God himself, precisely as Torrance suggests. Despite this, one cannot avoid the impression that Torrance wants to link the doctrines of the Incarnation and creation so closely together in order to emphasize the contribution to the view of the ontological rationality of the cosmos, that, because of the lack of evidence, he is using the concept of confirmation. This cannot but imply that the doctrine of the Incarnation, compared to that of the creation, could not contribute anything strikingly original in this case either.[126] Implicitly, then, Torrance has to admit that while the *Logos* was already active in the creation, the rationality of the universe can be claimed to have there its sufficient reason.[127] This is not to say that the doctrine of the Incarnation had no radical uniqueness or contribution to make, because the notion of the incarnate God was certainly an essential intellectual newcomer in the Hellenistic world, precisely as Torrance maintains. The argument proposed here is only to make the claim that he cannot present convincing evidence for his view that, compared to the creation, the Incarnation was a radical novelty from the standpoint of the idea of contribution. Or, in other words, Torrance's modified creation argument has not so far proved its plausibility.

The third factor in the ontological side of Torrance's idea of contribution is his insistence that the doctrine of the Incarnation set forth the notion of the freedom of the universe, an essential intellectual ingredient needed in empirical science. Torrance makes the assertion that the Christian view of the freedom of the creation, dependent upon God's own freedom, released Christians from any fatalistic beliefs in necessity or determinism which, according to him, were so characteristic of the pagan mind.[128] Thus freedom is also closely linked with the notion of contingency as the opposite of necessity. In line with his idea of contingency, Torrance rejects any interpretation in which the exclusion of necessity or determinism is seen as having led to a worldview in which chance or arbitrariness was given a legitimate place. Instead, for him freedom in this sense means orderliness and consistency.[129] From the standpoint of the sciences, as Torrance clarifies, this notion of the contingent freedom of the universe granted them the double conviction that they are free to investigate nature without having to fear any intrusion into the territory of divine powers, and secondly, that the cosmos itself, in its freedom, cannot be interpreted as an overall net of necessary causes and effects.[130] Torrance gives reasons for this contribution of the theology of the Early Church by referring to God's love, which he sees as the solution to the inherent problem in his view of contingency as neither arbitrariness nor necessity: "The full concept of contingency of the creation carries with it the idea that God is related to the universe, neither arbitrarily nor necessarily, but through the freedom of his grace and will, when out of sheer love he created the universe and grounded it in his own transcendent Logos or Rationality."[131]

This passage illuminates the fact that Torrance seeks reasons for the creation in God's grace and will and, ultimately, in his love, which is also the ground for the freedom of the universe. A question concerning the nature of this freedom is raised when Torrance's separate accounts of the issue are compared with each other. On the one hand, he claims that this freedom is granted to the creation as distinct from and limited by the transcendent freedom of God. The relation between the two freedoms, created and uncreated, is thus clear: God's freedom is the only ground for the freedom of the universe.[132] On the other hand, however, Torrance can state that God "does not grudge his creation a genuine freedom any more than he grudges it a reality distinct from himself, but on the contrary grants it to share in an appropriate way in his own freedom."[133] As to this latter view, two questions emerge. First of all, is there not a shift in Torrance's thought when he does not see the freedom of the universe only as grounded upon or limited by God's own freedom but now, in the text lastly quoted, sees it as sharing or participating in God's own transcendent freedom? The second question concerns the mode of this participation, which Torrance calls "appropriate": what is such an appropriate sharing in God's own freedom? The latter question remains unanswered because of the unclarity in Torrance's own use of the term. There is no hint in his discussion as to what might be the difference between appropriate and inappropriate in this connection. But the first question can be analyzed further.

It seems obvious that a certain development in Torrance's thought has taken place with regard to the ultimate nature of the freedom of the universe as an aspect of his idea of contribution. The notion of participation has grown to occupy a basic role in his interpretation: in an earlier writing any such notion is omitted, and the thought of a *limited* freedom, bounded by the transcendent freedom of God, is preferred. Later, however, the interpretation of participation gains a more crucial position, while the idea of a limited freedom is not totally abandoned but intertwined with it.[134] Ambiguity in Torrance's thought seems thus to be inevitable: how can he relate the limited freedom of the universe to God's own freedom with the concept of participation if, at the same time, the notion of differentiation is maintained?

Torrance's answer is Christological and based explicitly on the doctrine of the Incarnation: "The fact that he who freely created the universe has once and for all become incarnate within it, means that as the Creator God will freely to coexist with his creaturely children, and therefore that the continuing existence of the universe is ontologically bound to the crucified and risen Jesus and destined to partake in the consummation of God's eternal purpose in him. It also means that the whole universe is brought to share in the unlimited freedom of the Creator, although always in a differentiated way appropriate to its creaturely reality and contingent nature."[135]

The patristic notion of participation as the mode of the relationship of human beings with God has here found its cosmological application with regard to the idea of the freedom of the creation. It is truly freedom as far as it shares in the freedom of God, although the former does not become the latter. Torrance wants to underscore his conviction that what is creaturely cannot be transformed into what is divine—the ultimate and absolute distinction is preserved, just as it is maintained in our salvation as we participate in the Triune life of God. This means, therefore, that Torrance directly applies the notion of θέωσις and θεοποίησις as concepts used in the context of humankind's salvation, to his idea of the freedom of the universe.[136] But the question has to be asked again, where does the actual contribution of the Incarnation lie with regard to the freedom of the universe? Certainly, by referring to patristic sources, Torrance can give reasons for the new ontological status of the creation brought by the act of God's becoming a human being. In spite of the notion of participation in connection with the freedom of the universe, Torrance cannot adequately clarify his point that the doctrine of the Incarnation contributed to the rise of empirical science by granting it its unconsciously applied background presuppositions.

The problem is identical with those discovered earlier in the discussion of Torrance's ideas of the contingency and rationality of the universe, although, it has to be admitted, his argument seems better reasoned in the last case of freedom. The basic trouble in all the three points can be claimed to be found in the connection made between the incarnate and the preexistent *Logos*. While the doctrine of creation asserts that God's Word was already active in the event of creation, giving the universe its ontological status as a contingent, rational, and free cosmos, it is not easy to find a legitimate place for the incarnate Word as having made an essential contribution in these three respects. Therefore, as was shown, Torrance has to use rather imprecise terms such as "radical" or "deepening" in order to underline the positive and original role he wishes to give to the Incarnation as a crucial Christian doctrine, not only from the point of theology itself but from that of the natural sciences, as well.

The difficulty thus described is not so articulate in another, though in this respect marginal, argument in which Torrance links the contributive role of the Incarnation not primarily with the descent of the *Logos* but to the love of God. Torrance maintains that the Incarnation was an indication of God's love toward his creation and further that by becoming a human being himself God demonstrated its reality and worthiness.[137] This appreciation shown by God to the creation through the Incarnation offers another approach to the idea of contribution as applied by Torrance that is even more fertile than the one based on the activity of the *Logos*, because now it becomes plain where the essential significance of the Incarnation lies, namely in the reality-affirming love of God, who allows himself to become ontologically involved in the creation. While it is difficult to see the all-important relevance of the Incarnation for the rise of empirical science through the three "masterful ideas," as Torrance asserted, in comparison to the doctrine of creation, it is possible to give a more positive evaluation of the same idea of contribution if considered from the standpoint of God's inherent love toward his creation in the Incarnation as a reaffirmation of the reality of the universe. In other words, it is not too troublesome to think of the contribution of incarnational theology if it is seen as an act of God's appreciation of the universe thereby making scientific inquiry possible. The doctrine of the *Logos* is not, of course, totally absent from this line of Torrance's thought, either, but, compared with the more explicit approach previously described, the *Logos* itself now derives its meaning from the love of God and does not stay "alone" as if it were detached from the inner life of God.[138]

As a summary of the preceding discussion, it can be argued that, in spite of his willingness to posit a strong positive role for the Incarnation as an essential contributive factor for the development of the natural sciences, Torrance cannot present convincing evidence for his claim. The ideas of contingency, rationality, and freedom of the universe were shown to be based on the Judeo-Christian doctrine of creation, as Torrance himself admits, but he cannot explain precisely in what way the Incarnation could add anything essential to these insights in order to display the contributive character in the way Torrance maintains. This mainly philosophical approach was demonstrated to have a parallel in Torrance's more theological standpoint where the love of God is considered the ultimate basis for the activity of the *Logos*, an approach with the hint of a promise to solve the problem described. It was argued that the contribution of the Incarnation could be more easily apprehensible if it were consistently linked, not primarily with the doctrine of the *Logos* but with the idea of the love of God, who by becoming a human being expressed his own appreciation of his creation, thus giving it worth and dignity of its own as well as the challenge to be investigated. This line of thought, however, does not occupy such a central position as does the approach based on the "three masterful ideas," a fact that can be interpreted as demonstrating Torrance's willingness to express himself to a scientifically orientated audience and explain his idea of the contributive significance of the incarnational theology of the Early Church.

Space and Time

Closely related to Torrance's view of the ontology of the creation, his idea of contribution includes an important detail which is deserving of separate study. Referring his thought to modern natural science, it can be argued that the centrality of the doctrine of the Incarnation is the primary cause for the fact that Torrance does not concentrate on the problems of quantum physics but rather on the questions of space and time and, accordingly, of Einstein's theories of relativity.[139] This is due to his conviction, as suggested earlier, that God's exclusive self-revelation took place in the concrete space and time of this temporal reality as his Son became incarnate to live under its conditions. Torrance claims that the Word became flesh in those structures of the universe that are also the proper field of investigation for the natural sciences. In other words, both theology and the natural sciences operate within the same medium, that of space and time, making efforts to express their distinct conclusions on the basis of what they have learned in it.[140]

For Torrance, space and time are not only a passive scene or stage where the human mind investigates nature—rather, they have a more active role as "bearers of contingent order and intelligibility in which all created realities share."[141] This is the most important point of contact with the ontology of creation described in the previous section. The contingent intelligibility, the substance, is realized in space and time, the form. Since order and intelligibility are, as it were, implanted into nature so that they are constantly to be found there, Torrance asserts, space and time provide the common ground for interaction and dialogue for both theology and the natural sciences as well as for their independent inquiries.[142]

Torrance holds that because space and time are created realities themselves, God does not exist in a spatial or temporal relation to the universe.[143] In other words, he is neither in space nor in time although he is their Creator and therefore absolutely free with regard to them and not subject to their conditions and necessities. In this created space-time medium, Torrance accredits us as humans with a religious task in every scientific investigation. He holds that it is our duty to read the "books of nature" because, without human involvement, nature would not be understood, as far as the natural sciences are concerned.[144] As for theology, Torrance reminds us that God has decided to reveal himself in this universe where he has established a reciprocal community with human beings who are then able to discern God's Word in an intelligible manner particularly through the Holy Scriptures.[145]

Space and time are not, therefore, a neutral sphere where the natural sciences and theology can encounter one other. Torrance's conviction is that they have a specifically theological slant because of their createdness and the fact that God himself in his own being as the Son, who is *homoousios* with the Father, has entered this space and time.[146] This suggests that for Torrance the religious task of human beings to provide the dumb nature with words that we can understand is not based on the task itself; in other words, it is not a duty to be carried out with an imprecise religious feeling of awe and wonder in the face of the mysteries of nature. In this respect he differs considerably from Albert Einstein, who did emphasize the religious nature of the scientific enterprise but did not link it to an explicit belief in a personal God.[147] For Torrance, this religious task of human beings is basically grounded on the createdness of the universe and its confirmation in the Incarnation. Therefore he suggests that space and time also make natural scientists carry on an inquiry based not only on intuitive religious feeling but on a duty accredited to them by Christian faith.

Torrance is certainly making a very strong claim here, and he can be criticized for entertaining theological monism or even reductionism, turning natural scientists into theologians. This does not serve even Torrance's own explicit intentions, since he generally wishes to emphasize the distinctive nature and independent status of the two disciplines. This theological monism, however, seems an inevitable result of his premises whereby he links the *Logos*, the rationality of the universe, and the creation so closely together, as was seen in the preceding pages, that all rational human enterprise is of necessity related to God. This is certainly a theological position that can be reasoned, as Torrance does, from the patristic theology of the *Logos*. However, the empirical effect of this approach depends wholly on the reception given to it on the part of the natural sciences, and there its positive evaluation remains obviously questionable especially due to Torrance's insistence on contingency as independence which is nevertheless dependent on its transcendent origin, that is, God.[148]

In Torrance's opinion, space and time are theologically characterized not only because of their createdness but also because of the Incarnation, as suggested above, in the act of God's becoming a human being under the creaturely conditions of space and time. The idea of contribution in his system makes the assertion that it was the incarnational theology of the Early Church that radically changed the contemporary view of the nature of space and time. The role of space in this respect turns out to be crucial in Torrance's thought. He makes use of a problem known in the philosophy of physics as that of the receptacle (absolute or container) versus relational view of space. In general terms, according to the receptacle notion, space is a container with strict boundaries or an empty place that needs to be filled.[149] Objects existing in it are seen as independent subjects occupying their own location without direct connection with each other. The relational notion, on the other hand, holds that "all spatial and temporal assertions should be seen not as attributing features to space or time or space-time, but rather as attributing some spatial, temporal, or spatiotemporal relations to material objects."[150]

Torrance agrees with the general definition of the receptacle notion of space described above, but, with regard to the relational view, his interpretation bears a specifically theological character with his insistence on the centrality of the doctrine of the Incarnation. Therefore he modifies the relational notion of space in order to clarify his understanding of the nature of the Incarnation whereby the Son of God, *homoousios* with the Father, took human form and descended from heaven to become flesh under the conditions of space and time.[151]

Torrance rejects the absolute, container notion of space on theological grounds because it does not allow, in his opinion, the possibility of God's becoming incarnate under the conditions of space and time. He suggests, in other words, that the idea of the impossibility of any incarnation of the *homoousion* type is rooted in an inappropriate understanding of space.[152] The decisive argument, however, put forward by Torrance in this respect is that it was the theology of the Early Church and especially the doctrine of the Incarnation that gave birth to the relational notion of space through the work of the Church Fathers, who felt compelled to rethink God's relation to space and time. The idea of contribution thus gives theology a further special role in two respects. First, it asserts that the incarnational theology of the Early Church in a remarkable way originated a novel understanding of space. Torrance traces the history of this view via Leibniz and Huygens back to Plato, whose philosophical impact on the theology of the Early Church in this respect he does not deny. Instead, he insists that the Church Fathers essentially transformed the primitive Platonic conception of the relational notion of space in order to make it more applicable in the interpretation of the Incarnation. Second, and also more interesting, Torrance claims that this relational view of space considerably anticipated the later emergence of the view of space expressed in the theory of relativity of Albert Einstein.[153]

These two aspects of Torrance's idea of contribution with respect to space are best analyzed by asking how he actually defines the relational notion. It has already been shown that, in line with the history of physical thought, Torrance supposes the existence of two major views that are usually seen as exclusive alternatives and that he can agree with the generally accepted definition of the absolute or receptacle notion of space. On the other hand, Torrance sees the relationality of space as something that indicates the relation of space and time to that which it embraces or "includes." This means that space always exists for something so that it never exists alone.[154] On the other hand, however, it was already suggested that Torrance gives a highly theological account of the relational notion of space in order to secure a legitimate place for belief in the Incarnation of the *homoousion* type. But what is this theological interpretation like?

In the traditional interpretation of the relational view of space, as expressed by one of its most eminent proponents, Gottfried Leibniz, it is maintained that space cannot be conceived as any sort of material or stuff but, instead, as a plain system of relations between existents.[155] The relationality of space is thus defined through a horizontal axis between the existents of this space. For Torrance this seems to be of minor importance, for his theological approach makes him emphasize the relationality of space through a vertical axis, namely as a relation between the creation and God. Referring to the Nicene theology, he explicitly claims that "Since space is regarded here from a central point in the creative and redemptive activity of God in Christ, the concept of space as infinite receptacle, or as infinite substance, or as extension conceived either as the essence of matter or as a necessity or our human apprehension, or certainly the concept of space as the first unmoved limit of the container, all fall away. *And in their place emerges a concept of space in terms of the relations between God and the physical universe established in creation and Incarnation.*"[156]

This understanding of space undoubtedly serves Torrance's theological intentions to seek for scientifically reasoned opportunities of grasping the possibility of the Incarnation. The crucial point in his relational notion of space lies in its unboundedness: while the receptacle notion bears in itself the idea of radical discontinuity or disruption between the creation and its Creator, the latter being situated far away from the realm of the former, the relational notion, as seen by Torrance, leaves the door open for God's dynamic activity in the universe by rejecting any limits or boundaries. The vertical relation between the creation and God therefore becomes the determining factor which essentially defines the relational notion of space, not the horizontal axis between existents.[157]

The second question concerning the contributive role of the incarnational theology of the Early Church with regard to the problem of space and time is related to Torrance's opinion that the relational notion of space described above anticipated in a considerable way the general theory of relativity, published by Albert Einstein in 1915. Torrance, in fact, gives the impression that there is a continuum between the contribution of the Early Church and the theory of Einstein.[158] He states that the relational notion of space "was given its supreme expression in the space-time of relativity theory when Einstein, following out a line of thought from four-dimensional geometry, found he had to reject the notion of absolute space and time."[159]

Torrance's argument concerning the contributive character of the incarnational theology of the Early Church is based, therefore, on the assumption that the theory of general relativity has in an eminent way brought to light the idea of a relational space which was first formulated in a revolutionary way by the Nicene Fathers when they felt compelled to give rigorous consideration to the relationship between God and the world in the doctrine of the Incarnation and the Son's *homoousion* with the Father. This argument has, however, two major problems. The first has to do with Torrance's own understanding of what the receptacle notion of space involves. It was shown that for him it is primarily a theologically reasoned concept expressing his view that the relationality of space derives its character from its relation to God and not, as in the traditional form of the view, from relations between existents of space. Only if this premise is agreed upon can Torrance's argument proceed convincingly.

The other problem is no less evident. While Torrance holds that the relational notion of space has its brightest expression in Einstein's theory of general relativity, he assumes that the two are necessarily connected so that adhering to this theory leads one inevitably to presuppose the relational notion and to abandon the absolute notion. The case is not, however, so simple. Among philosophers of science the question of the relationship between the theory of general relativity and the absolute/relational notions of space has not been definitely settled.[160] This proves that there is no inherent necessity in the theories of relativity to adhere to the relational notion only, but they also allow interpretations in which the absolute notion is presupposed.[161] Torrance seems to neglect this fact in his opposition to the absolute notion of space and this makes his argument concerning the contributive character of the incarnational theology of the Early Church as an early pioneer of Einstein's theory of relativity, with regard to the concept of space, rather daring in that he consciously connects the relational notions held by the Church Fathers and Einstein.[162]

From this can be drawn the conclusion that Torrance defends the relational notion of space because it allows more weight for his theological argument in which the theme of openness is repeated in favor of the doctrine of the Incarnation. While absolute space shuts God out of the world so that he cannot enter into any serious intercourse with his creation, relational space with its limitlessness and openness upward permits this and makes it easier to conceive of the Incarnation. The problems of this approach were, however, seen as evident. First, the definition given by Torrance to the relational notion of space is not unproblematically in accordance with the generally accepted explanation and, second, he too hastily gives the theory of relativity the credit of being an undeniable exponent of the relational notion.

Although the question of time was tangentially referred to in the previous discussion, it deserves some separate consideration.[163] It was shown that just as in the case of the ultimate nature of space, that of time is closely connected with the incarnational theology of the Early Church. Torrance holds that God's incarnate *parousia*, its actuality and reality, urged the Church Fathers to formulate and to apply *relational* modes to the understanding of space and time.[164] But how does Torrance define this relationality? With regard to space, the starting point for the current analysis was found in Leibniz's philosophy and in his view that space is nothing else but "an order of coexistence." In similar way he defines relational time: "Time is an order of succession . . . instants, consider'd without the things, are nothing at all; . . . they consist only in successive order of things."[165] At first glance, this is how Torrance too defines time—it does not stand alone as an absolute entity but always has a relation to things existing in time. Time is always time for something.[166]

This "for something" was found to be a characteristic, though a minor one, of Torrance's understanding of space, too, but it was argued that the actual relationality of space in Torrance's thought is based on its openness toward and relation to God. Now the question has to be asked, does this apply to Torrance's view of time as well? He claims that "just as we think of our creaturely being as contingently grounded upon the eternal being of God, so we must think of our creaturely time as contingently grounded upon the eternal time of God."[167] This implies that, for Torrance, there is no time if it is not seen as grounded on God's eternal time, which gives it its constancy and reliability. While, on the other hand, he asserts that time is always for something, he can now, on the other hand, make a theological modification of it by insisting that God has a time for us and for our time.[168] The idea of contingency as independent dependency is thus repeated here with the additional slant of temporality—what is created, as time is, is necessarily contingent in character, having its ground somewhere else but not in itself.[169] The Incarnation, then, means that the eternal God has from his own time entered the created time of the universe, implying further that the relationality of time consists ultimately of the interrelations between these two times and not primarily between the existents in time.

Some consideration should be given to the question of why Torrance needs the assumption that God also has his own time. The answer is easily grasped from the passage cited. If there were no time in God's own life, created time would lack any value whatsoever. This line of thought is, however, difficult to conceive because could not God, as the Creator of time, and the positive value of created time be thought of without the idea of God's own time? In other words, it is hard to see the need for such an emphasis on God's temporality. A more fruitful answer seems to lie in Torrance's constant insistence upon the *orderliness* of the universe. If time as "an order of succession," expressed in Leibnizian terms, is given the role of the bearer of order, it is evident that to avoid any idea of irregularity or disorder in God's own eternal life, a distinct kind of time has to be assumed as prevailing in it, too.[170]

A more important question is, however, can this mode of the relational notion of time be considered as an early anticipation of Einstein's theories of relativity as Torrance suggests?[171] It seems that Torrance is right when he stresses the novel understanding of time in the theory of relativity, perhaps even more right than in his insistence upon the confirmation of relational notion of space in connection with it.[172] In particular, the special theory of relativity has made a lasting contribution to the physical understanding of time, arising out of the apparent contradiction between the constant velocity of light and the common intuitive conception. While in the latter case, as a result of physical experiments, light has been observed to move at a constant velocity irrespective of the velocity of the source of that light, the common understanding assumes that if a light signal is emitted from a moving system, such as from the roof of a train, its total velocity is the sum of its own speed and the speed of its source. The special theory of relativity has proved this to be incorrect and has shown that the speed of light is always constant.[173] This, however, does not necessarily lead as readily to such a relational view of time as Torrance suggests. What can be said with certainty of Einstein's relativistic concept of time is that it makes the notion of simultaneity questionable. Naturally this involves the conception that movement affects our understanding of time or particularly the time-scales that are used to measure it, and in this respect the rejection of an absolute understanding of time can be accredited to the theory of relativity. Thus time cannot be measured absolutely, according to a uniform scale valid everywhere, but only relatively, the relations between the existents taken into account. But the question remains, can Torrance make convincing use of Einstein's theory of relativity as evidence for his argument concerning the contribution of the incarnational theology of the Early Church with particular respect to the relational notion of time?

It has to be admitted, as explained, that the theory of special relativity remarkably transforms the classical and commonly applied understanding of time. Einstein's insights, however, do not contain such remarkable similarities to the patristic view of time, as described by Torrance, that this unreserved connection could be made. The theological definition of time as owing its existence to the eternal time of God, the former deriving its relationality from this relation to the latter, supports the claim in which, contrary to what Torrance maintains, he is himself adhering to a new, modified absolute notion of time, the time of God, as the controlling and all-pervading duration that gives meaning and constancy to created time.[174]

To sum up the previous discussion of Torrance's idea of contribution, the following can be concluded. His modified creation argument, according to which the doctrine of creation alone was not a sufficient reason for the rise of empirical science but it needed the doctrine of the Incarnation for that purpose, was found to be questionable in its essential parts because Torrance fails to introduce convincing evidence for his claim. In the epistemological aspect, considered in both its negative and positive roles, it was found that Torrance interprets Athanasius's views of the new possibility of knowing God, granted to humankind through the Incarnation, along Barthian lines whereby any natural knowledge of God is programmatically rejected. As to the positive aspect, it was found that the *homoousion* in Torrance's thought is a strong interpretative model—not merely a "disclosure model," to use Torrance's phraseology—which is practically settled and irreplaceable, contrary to Torrance's own explicit suggestions. As such it is offered to the service of the natural sciences, combining faith and knowledge. To be effective, this faith has to be, in the final analysis, Christian in character, allowing the mutual meeting of reality and its manifestation in the *homoousion* manner, where what is observed is seen as "of the same being" as reality beyond. Torrance gives reasons for this appraisal of the significance of personal Christian conviction by referring to James Clerk Maxwell and Albert Einstein. In the former case the connection between Christian faith and remarkable scientific enterprise is testified to, but when interpreting Einstein and the role of his religious insights Torrance's intentional approach was found to be questionable.

As to the ontological contribution made by the incarnational theology of the Early Church, the "three masterful ideas" of contingency, rationality, and freedom of the universe were analyzed. The main problem lies in Torrance's inability to give sufficient evidence that it was the doctrine of the Incarnation that could inject these ideas, which, in his opinion, were needed for the rise of modern empirical natural science. No less problematic was found his view of space and time and his claim that the doctrine of the Incarnation in a remarkable way anticipated the relational notion of space and time, which, according to Torrance, found its ultimate exponent in Einstein's theories of relativity. The similarities between Torrance's construction of the concept of space and time and that of the theory of relativity were found to be rather vague, thus implying that such a close link between the patristic understanding and Einstein's views as Torrance would like to draw cannot ultimately be made.

As an overall evaluation of Torrance's idea of contribution, indicating his belief that it was the incarnational theology of the Early Church with the *homoousion* that contributed considerably to the rise of empirical science, it can be stated that to be credible this argument needs a much firmer evidential basis than Torrance is able to offer. This means that his modified creation argument does not work credibly. As a contribution to the discussion between theology and science this approach certainly has its advantages, as it offers interesting insights of common concern. But seen from the perspective presupposed by Torrance himself, namely that this precise Christian doctrine contributed to the development of natural science, his argument is, due to the lack of more convincing evidence, highly interesting and undeniably attractive to the Christian mind but far from persuasive and conclusive. Therefore his idea of contribution should preferably be seen as a suggestion, proposal, and invitation to reflect more fully on the impact of Christian theology on the natural sciences in the search for a common area of discussion rather than a fully settled, closed, and final starting point or basis for such an encounter between the sciences.

4

The Power of Reality

The Idea of Compulsion

> We must submit ourselves modestly, with our questions, to the object in order that it and not we ourselves may be the pivotal point in the inquiry.
>
> Thomas Torrance, *Theological Science*

In the previous chapter it was pointed out that for Torrance the *homoousion* and the incarnational theology culminating in this concept represent a strong and remarkable contribution to the development of empirical science. Torrance was seen to draw a sharp distinction between what he calls *mythological* and *theological* attitudes toward the object under investigation. In his terminology the first concept has its starting point in the human being and in his or her self-understanding, while in the latter the theological reality outside the human being is given an authoritative position. This approach must be seen as a theological application of Torrance's philosophy of science in which *realism* plays the most crucial role, expressing the most fundamental philosophical orientation in his thought.[1] This realism has to be seen as a counterreaction toward those philosophical currents that, in Torrance's opinion, have the tendency to drop reality in its ontological depth out of bounds of scientific inquiry, first of all *observationalism* and *phenomenalism*.[2] A realist attitude, then, does not aim to investigate only observations or phenomena of reality but reality itself, which is believed to be approachable. In his view, there is no genuinely *scientific* thinking without this realist attitude in which it is demanded that human knowledge strictly conform to the nature (κατὰ φύσιν) of the object known.[3]

In order to be properly understood, this conviction of Torrance has to be studied further, not only from a general philosophical standpoint but especially from a theological perspective. Since Torrance's thought raises the question, what happens if science and knowledge are not allowed to live "in accordance with the nature of what is known," or in other words, if the *homoousion* is not allowed to be realized, his idea of the counterpole of realism has to be given a lengthy discussion as well. The threat of dualism as expressed by Torrance will substantially clarify his position as a rigorous exponent of what he calls "theological realism."

Theological Realism

The Context

As was stated, Torrance has made efforts to apply his view of realism in the field of theology and to offer it for the use of other sciences as well. In studies of Torrance his understanding of realism is usually touched upon, occasionally in depth.[4] First, however, it is useful to point out briefly how unsettled and many-sided the concept of realism appears to be in contemporary discussion in the philosophy of science. The aim of this brief account is not to present a comprehensive picture of different modes of realism but only to show the embarrassing situation in which one finds oneself when trying to gain a clear picture of what realism actually is. The first problematic detail is that realism in the philosophy of religion usually has quite a different connotation from the use of the concept in the field of metaphysics and epistemology. In this religious context realism simply states that "there is a transcendent divine reality independent of human thought," a view opposed by nonrealists.[5] The discussion between theism and atheism is an example of this approach. But realism can also appear in other contexts, for instance, in the old controversy between realism and nominalism in which the reality of universals was either affirmed or negated.[6]

Realism is also seen as an attitude in theories of perception. In this respect the least acceptable alternative among philosophers seems to be *naïve realism*, a name Robert Audi gives to a common-sense view that is generally held. According to him, its naïveté derives from its inability to take seriously the possibility of sense distortions as illusions or hallucinations but assumes that a sense experience is in total accordance with reality.[7] A more positive position, at least toward some sorts of naive realism, is entertained by Jonathan Dancy, who makes a distinction between direct and indirect naive realism. In the first case it is assumed that objects can retain all their properties observed by a human inquirer even outside the event of observation. In its indirect form naive realism assumes that there are nonphysical entities in the human mind corresponding to objects in reality, thus making the awareness of them possible factually but indirectly. Dancy contrasts both forms of naive realism with a similar dichotomy occurring in what he calls scientific realism. In its direct form it supposes that objects can retain only some characteristics outside the event of perception, while the existence of other features is dependent on being observed. Exponents of scientific indirect realism, for their part, believe that the physical, indirect objects can bear only "primary" features distinct from the "secondary" ones, which are sensory properties belonging solely to the direct, nonphysical object in the human mind.[8] The difference between the two modes lies in the couplet direct/indirect: the former holds to the direct relation between the object and the observer whereas the latter assumes the existence of some "intermediary" mental constructs which make the perception possible in an indirect way. Dancy's classification is an apt expression of the complexity of realism in the theory of perception.

In scientific theories realism plays another kind of role as a view in which theories are seen as trustworthy indicators of reality, whereas instrumentalism considers them as mere external tools to gain knowledge. Another counterpart of scientific realism is conventionalism in which theories are understood as based on mutual agreement in a scientific community, so that any reference to their truth or untruth is meaningless.[9] An interesting realist effort to construct a compatible mode of seeing the world is pursued by A. F. Chalmers, who suggests the application of what he calls "unrepresentative realism," an alternative that meets the requirements of a realist position in two respects. First, it adheres to the conviction that the external world is what it is, independent of our perception of it, and second, if a theory is applicable to the world, it has this property always, not only in experimental situations. The actual novelty in Chalmer's view is his insistence that the idea of unrepresentational realism does not involve the notion of the correspondence theory of truth and the problems involved.[10] "The unrepresentative realist does not assume that our theories describe entities in the world, such as wave functions or fields, in the way that our common sense ideas understand our language to describe cats and tables."[11] This indicates an inclination to avoid any careless use of the so-called common-sense philosophy and the correspondence theory of truth as the basic structure of a realist attitude. Chalmers's problem seems to be, however, his inability to give definite *positive* characterizations of what his unrepresentative realism is. In this case, definition through negations does not lead to convincing results in spite of the fact that Chalmers makes an effort to define his view positively when he states that theories "are a special kind of social product, although the extent to which they are able to come to grips with the physical world, which is not a social product, is not socially determined."[12]

The subject of these divisions is collected in a concise and useful form by John Hick, who firstly makes a distinction between realism and nonrealism, idealism belonging to the latter group. Realism is then divided further into naive realism, already described, and critical realism, in which the role of the human subject in perception is acknowledged but not so that belief in an external reality per se is diminished.[13] Hick also reminds us that hardly anyone can be a pan-realist, adopting a realistic attitude in every detail and, for instance, believing in "the reality of fairies and snarks as well as of tables and electrons."[14]

It should be repeated that the discussion above is meant to serve only as an indication of how diverse and open to a variety of interpretations the concept of realism is in philosophical debate. One of the problems in considering Torrance's view of realism is that he has applied the concept in all of the four categories, namely realism within the philosophy of religion, theories of perception, debate on universals and discussion on the relation of scientific theories to reality.[15] The flexibility of the concept is discernible not only in Torrance's own usage but also in studies of him and his view of realism. Brian J. A. Gray has underscored the impact of the Scottish Common Sense philosophy as one crucial background framework for Torrance's thought, and, indeed, Torrance himself explicitly wishes to be identified with it.[16] This connection has also been emphasized by James B. Miller, and he relates Torrance's realism to other philosophical strands which he sees as "post-modern" or "post-critical" in opposition to Kantian epistemology.[17] Douglas A. Trook takes Torrance's realism for granted without further criticism. With respect to the nature of realism in Torrance's thought, he claims that one of the reasons for the general reluctance to develop Torrance's thought further is his adherence to "realist metaphysics," by which Trook undoubtedly means Torrance's belief in an ontological reality beyond the immanent one.[18]

A deeper analysis is provided by Brian H. Meardon, who explicitly holds that Torrance represents *critical realism*, critical in the sense that a methodology applied within a scientific discipline cannot be seen as an adequate means of gaining knowledge in another field of inquiry. Reality can be reached but not in a naively uniform way—the existence of different methods and systems of concepts, as well as their ability to lead to proper knowledge, have to be acknowledged. Thus we are not imprisoned in a single mode of inquiry only.[19] This has led Torrance to speak of the complementarity of different methods and systems of scientific concepts, resulting in the conviction that their distinction becomes sharper, on the one hand, and their complementarity deeper, on the other.[20]

A very thorough analysis of Torrance's realism is offered by Roland Spjuth, who also regards it as critical but suggests, too, that there are certain features of metaphysical realism to be found as well. Spjuth's main criticism of Torrance's realism is his insistence that Torrance has not succeeded in resolving the Cartesian subject-object problem but is forced to remain fixed in the dichotomy in question.[21] Spjuth points out that Torrance's emphasis upon intuition "is the rather typical answer of the realistic side of modern debate between realism and skepticism; the reliable bridge, in theology as in all scientific activities, is the intuitive experience in which reality imposes itself upon us."[22]

In all these examples the characteristic feature is that Torrance's realism has been studied as a purely philosophical concept without any deeper penetration into its intellectual background. Torrance's own view of his realism has been taken for granted, and it has been analyzed along the lines of general classifications, resulting in considering it *critical* and sometimes also *metaphysical*. The usefulness of this method is not denied, but it is now argued that this philosophical approach does not provide sufficient ingredients for properly understanding Torrance's realist position and his insistence upon its rigorous application in any field of science. The question to be asked, then, is how does Torrance's view of realism, *homoousion*, and their applicability to any scientific discipline relate to each other?

Before moving further, it is necessary to note that Torrance himself quite often gives the impression that he is himself speaking about realism in purely philosophical terms. This is evident in his arguments against phenomenalism and observationalism, already mentioned above. Torrance links the concepts to post-Newtonian science and the post-Kantian era, claiming that during that era "a huge gap opened up between the phenomenal and the noumenal, or things as they appear to our human observation and things as they are alleged to be in themselves."[23] In this connection Torrance refers to the notion that every thought in the human mind must arrive there through the senses—otherwise it would remain empty.[24] This leads one to the conviction, according to Torrance, that scientific knowledge has to rely on observation or logical deductions from it, so that in the final analysis knowledge does not have its sources in the reality around us but in our own capability of discerning and structuring it. Torrance claims that, in that case, "scientific theories have to be regarded not as having a descriptive reference to structures in being but as convenient or operationally effective arrangements of our observational data."[25]

The passages quoted above indicate the actual reason for Torrance's reluctance to accept what he calls phenomenalism or observationalism as adequate modes of approach in scientific enterprise. Their defect lies in their reliance upon the *human capability* of discerning and observing, on the one hand, and in their powerlessness to penetrate into reality itself, on the other, leaving them solely on a phenomenological level with the conviction that mere phenomena are the object of scientific activity. Thus Torrance identifies the *object* of scientific investigation with *reality* independent of our internal states or our view of that reality, thereby disclosing his dependence upon the idea of objectivity that is found in Karl Barth's theology.[26] According to it, our subjective states can never be supposed to provide the basis for knowledge but must be replaced by reality itself.[27] From this perspective Torrance's view could be termed *objective realism* as distinct from *subjective realism*, a dichotomy that comes near to what was seen as direct or indirect realism. Torrance's approach has not been able to get rid of the problem of Barth's objectivism because it cannot give any definite answer as to the criteria to be applied to be certain that the object of knowledge is reality itself and not any subjective psychological state.[28] This remark alone is sufficient to suggest that designating Torrance's realism as objectivist itself contains a severe problem. Therefore I would maintain that this objectivity cannot be considered the final clarificatory factor of Torrance's realism because it leaves open the question of what kind of impact this objectivity has upon the human mind. Further enlightenment in this respect is needed if the roots of Torrance's realism are to be understood.

The Compulsion

The problem concerning the ultimate nature of Torrance's realism can find an answer in his claim that we have the obligation to obey the commands of the object so that "we think in terms of it, and not in terms of what we think we already know about it."[29] The objectivity of realism thus involves the thought that the object of a scientific enterprise must be seen as dictating the rules by which it has to be interpreted. Torrance's realism does not, therefore, contain only the belief in the existence of reality independent of an observer but also the conviction that, in the ideal case, this independent reality can and must be allowed to determine what we can know about it. In this respect Torrance sees no difference between various disciplines, for he holds, referring to theology and the natural sciences, that in both disciplines the question is of *human* knowledge. This means that we are placed in a central position in an epistemological process but, nevertheless, in both disciplines we as human knowers are servants, not masters, having to obey the commands of reality.[30]

Torrance's realism thus gains rather a strong place in his conviction that reality is able to submit us to its compulsion. To be objective does not mean that the subject, the human knower, has the priority, but the contrary is the case. To be a subject means to be submitted under the compelling power of the object.[31] To be a realist means to consider apparent phenomena or observations as *homoousioi*, of same being with reality itself. Torrance states that as the scientist "knocks at the door of truth through his scientific questioning and finds it opening to him, the truth holds lay of him in its own objective force and sets up its laws in his mind."[32] So-called ultimate beliefs, that is, convictions that do not need any evidence, arise in our mind out of the sheer power of reality.[33] The rationality of any scientific enterprise is totally dependent upon the obedient submission of the human mind to this compulsion and, accordingly, the scientist has to conform his thoughts to this reality.[34] Our human rationality and our willingness to submit ourselves to the "majesty" of reality are deeply interlocked.[35]

In his view of the so-called hermeneutical circle, too, Torrance applies his idea of compulsion because the questions asked by a scientist have to be in accordance with the object under investigation.[36] The compulsive force of reality is not exercising its power only in the factual act of knowing but enters this process in an early stage in the formulation of questions to be put to reality.[37] Torrance's objectivism is at its most apparent expression in his insistence that a human being as the one who asks questions has only a formal role in that he or she only expresses the question in articulate human words. The actual pressure and motive for asking that precise question derives from reality itself. This further means that a human being is allowed to ask not whatever he or she wants but only questions that are considered appropriate to the nature of reality under investigation.[38] Reality, then, is regarded as the first and the last criterion for the adequacy of human asking and curiosity.

All this gives sufficient evidence for the claim that the idea of *compulsion* is the most characteristic feature of Torrance's realism, a feature that makes it hard to classify him strictly as a representative of one of the commonly known branches of realism.[39] This compulsion of reality, however, would not be possible if a perception or observation were not identical with reality itself, if they were not *homoousioi* with each other, and for this reason the crucial realist role given by Torrance to this patristic term cannot be bypassed. The close relation between realism and the *homoousion* in Torrance's thought is based on two aspects. First of all, Torrance holds that the idea of *homoousion* itself originated under "the compelling power" of reality and must therefore be seen as a product of realistic thinking. It was not an arbitrary invention but was pressed into realization in the minds of the Nicene Fathers when they made efforts to express their understanding of God under the compelling force of his self-revelation in Jesus Christ.[40] The second point to note in the relation between realism and the *homoousion* in Torrance's thought is that, in spite of its explicitly theological origins, this patristic term offers a most crucial conceptual tool for any scientific enterprise with its conviction that what is observed is *of the same being* with the reality of the object, or consequently, is reality itself.[41] In other words, the *homoousion* guarantees the functioning of the compelling force of reality. Only from this point of view can Torrance's thought-provoking claim that physics, too, has its own *homoousion* be properly understood.[42]

Torrance seems to be rather uninterested in reflecting further on the nature of thought. It was noted earlier, when the idea of contribution was discussed, that Torrance has a firm belief in the rationality of the universe, a rationality that owes its existence to an intelligent Being who has made the creation intelligible. This suggests that Torrance implicitly sees the creation as the materialization of God's thought and intelligibility although he never articulately enters such speculations. Because we humans, too, are rational beings capable of reflecting on the creation and because this activity belongs to our very essence as creatures of the rational God, being itself is conceiving and understanding. Nevertheless, the nature of thought and conceiving of its relation to the world is not further elaborated in Torrance's writings, but they are seen as self-evident.[43]

In the previous lines it was demonstrated that in Torrance's opinion the idea of compulsion unites his realist approach and appreciation of the *homoousion* in such a way that the compelling power which is seen as an essential feature of reality comes into effect only if the *homoousion* is applied. It was also noted that Torrance suggests the validity of this model for any scientific enterprise, explicitly physics. But the inevitable question to be asked after the preceding proof of the centrality of the idea of compulsion is: How can this view be applied in other disciplines? What possibility is there for Torrance's realism to succeed in the natural sciences? Because Torrance articulately points to physics as another example of a science where the substance of the *homoousion* is applicable, it may offer a point of departure here, too.

As noted, with regard to modern physics Torrance feels particularly comfortable in the company of Einstein and his theories of relativity, for it has allowed him to construct a view of the Incarnation that does not need to follow the lines of the absolute notion of space. Albert Einstein was certainly a realist in the sense that he supposed the existence of a world independent of an observer but he also insisted that the very concept of reality as defined by classical mechanics should be reevaluated according to the results of the theory of relativity.[44] Torrance's appreciation of Einstein has led him to stay on the same front with him in the debate concerning the possibility of a realist interpretation of quantum mechanics, a fact that has evidently prevented him from engaging in further dialogue with modern quantum physicists, especially with those who do not adhere to Einstein's interpretation.[45]

On the basis of what has been stated, it is now argued that the primary reason for Torrance's reluctance to the widely accepted Copenhagen interpretation of quantum physics lies precisely in his view of realism. But what is the feature in Torrance's thought that makes them incompatible? In the Copenhagen interpretation of the behavior of atomic particles the old problem concerning the real existence of objects observed has come to the fore in a manner that remains, in principle, quite agnostic with regard to the existence of an objective reality. An instructive expression of that view is presented by one of the main architects of the Copenhagen school, Werner Heisenberg, who claims, "However, if one wishes to speak about the atomic particles themselves one must either use the mathematical scheme as the only supplement to natural language or one must combine it with a language that makes use of a modified logic or of no well-defined logic at all. In the experiments about atomic events we have to do with things and facts, with phenomena that are just as real as any phenomena in daily life. But the atoms or the elementary particles themselves are not as real; they form a world of potentialities or possibilities rather than one of things or facts."[46]

Heisenberg's point is clear: in experiments we deal with real facts, but this does not allow one to infer that the same could be said about the real existence of these events in themselves, outside the experimental situation. In other words, in quantum physics and especially in its Copenhagen school, the human capability of reaching reality in itself, at least on the microphysical level, is denied.[47] In the strictest sense of the word, only those features that are observed under experimental conditions can be considered as real. If the features really have an independent existence, a scientist has nothing to say about it.

In order to understand Torrance's point, we should pay closer attention to the idea of realism entertained in the current discussion in quantum physics. The notion of the *correspondence theory* of truth seems to play an important role in the conviction that the reality presupposed by a realist position should be defined before a test result can be compared with it for the judgment of the adequacy of an experiment.[48] This means that reality should be known first and the test results thereafter so that comparison can be performed. Only positive results stemming from this comparison guarantee the truth of a theory. In other words, the Copenhagen interpretation of quantum theory seems to cast doubt on the relevance of traditional realism because it views the problem through the idea of minor or greater correspondence between theories and the reality they refer to.

Torrance's point of view differs quite remarkably from this understanding of realism in that it does not presuppose any prior definition of reality to which theories might then refer.[49] Why is this so? We can take a glance at Torrance's view concerning the theory of correspondence and especially at his concern over the matter that correspondence and coherence theories of truth have been radically separated from each other.[50] This view is easily discernible in his way of dealing with theological statements that can be divided into two groups, existence statements and coherence statements. Worth noting is that he does not call them existence or coherence theories of truth, but he aims at reflecting the problem only on the level of statements. Torrance maintains that the former kind of statements "refer to an objective reality above and beyond them, and which are true in terms of that reference," while the statements in the latter group "are related to one another in a coherent framework of knowledge, and which have their truth in their inter-relations or with the reference to a system of ideas, within which they have an appropriate place."[51]

The case of existence statements is of particular importance. The most striking point to note is that in their application Torrance does not presuppose any definition of reality independent of an observer. He relies on reality itself and its capability of exercising its compelling power over the human knower. Thus the relation between a statement or a theory and reality is not one of correspondence but one of *reference*, a term that gives the impression of incompleteness and a certain vagueness, thus denying the possibility of a one-to-one correspondence.[52] Such a referential relation rejects any effort on our part to force reality into the straitjacket of human formulations. Rather than giving definite answers, this relation calls for further attention and deeper humility before the object under investigation. The reality observed is therefore always beyond—it is transcendent. This view reinforces the point that we as humans have no ability to define reality because it always remains transcendent and totally independent of us. This leaves us as knowing observers with only one choice, to submit to the compelling power of this transcendent reality where we can rely on the *homoousion* character of our observations: reality forces us to obtain genuine knowledge of reality through the phenomena we observe. This means that reality, as an open, previously undefined entity, gains absolute primacy while our knowledge must be something secondary. Reality cannot be defined beforehand but is actually disclosed in scientific investigations, and, accordingly, the power of a scientific theory does not lie in human definitions or theories themselves but in reality as its ultimate judge.[53]

Thus, the question of what possibilities there are for Torrance's realism to function in quantum physics can find an answer precisely in the problem of what the theory of correspondence in this connection is thought to consist of. It was seen that Torrance and the Copenhagen interpretation differ at this point. While the latter demands that reality should be defined before any correspondence, viewed as the cornerstone of realism, can exist, the former rejects any effort to define reality beforehand, denying the sheer possibility of such a procedure. It seems, however, that the Copenhagen interpretation, despite its rejection of realism in general, can, at least to some degree, accept the idea of compulsion even in its unrealism. When describing a situation in which the scientist has to evaluate the correctness or incorrectness of a physical theory, Heisenberg states that "it may sometimes take many years before one knows the solution of a problem, before one can distinguish between truth and error; but finally the questions will be decided, and the decisions are made not by any group of scientists but by nature itself."[54] It seems obvious, then, that on a general level nature is given the final word on determining the substance of human knowledge. In this way reality is accredited with certain compelling force.

But in the final analysis Torrance's realism and the Copenhagen interpretation, at least as it is expressed by Heisenberg, remain irreconcilable due to the central role of the idea of *homoousion*. Torrance reminds us that the Copenhagen school is not interested in reality itself in a direct way but only indirectly, concentrating on statistical probabilities and thus making "no claim to represent the field of their material and dynamic structure."[55] Torrance's position implies that, in this case, reality itself is not under investigation but only human constructs on it in the form of statistical probabilities.[56] The *homoousion*, by contrast, demands a direct though subjected contact with reality itself in its ontological depth so that ontology is given absolute priority over epistemology.[57] This makes Torrance's program possible for an application in quantum physics only if the realist position is accepted in principle, an approach that is not, after all, totally unknown within its field.[58]

In quantum physics as well as in theology and other disciplines, the *homoousion*, granted that it is applied in the realist way suggested by Torrance, raises the question of to what extent we are actually able to know reality in its fullness. The idea of compulsion seems to imply strongly that reality is impressed upon our mind with such force that we can only accommodate, accept, and obey in our submission to the demands of reality. How does Torrance, then, relate our capability of knowing, our comments on reality, and the compelling power of reality? Given that we really submit ourselves to this compelling power of reality, can we reach knowledge that can be considered final and uncorrectable?

It has already been shown that rather than speaking of a correspondence relation between a human expression of reality and reality itself, Torrance is in favor of a *reference* relation. Articulate forms of human knowledge can only be seen as expressions referring to reality as their object. On the basis of this, the answer to the questions mentioned is offered in Torrance's statement concerning the distinctive characters of sign and symbol: "A truly realist position will be one in which the sign differentiates itself as sign from the reality which it actually bears, and therein reveals a measure of disparateness or discrepancy which is an essential ingredient in its successful functioning as a sign."[59] How Torrance explains the difference between the two concepts is well seen in his claim that whenever the subtle relation of a sign and the reality signified is impaired, the sign bears a tendency to lose its controlling contact with the objective reality it is intended to refer to, thus changing from a true sign into a symbol. Torrance suggests that a symbol does not have very much to do with objective reality, for it tends to turn into a myth that has only an imprecise or indirect relation to reality.[60]

But if the relation of a sign to the objective reality is preferable, what makes it so extraordinary in comparison with a symbol? A sign has to refer to reality only in a differentiated manner, and for this reason it should have a small amount of inadequacy in order to be precise. In Torrance's opinion, a symbol has only an indirect relation to reality while a sign signifies it directly. What, then, constitutes this indirect symbol? The answer must be sought in the close identification of symbolic presentation with a myth. As has been demonstrated, for Torrance a myth is something very anthropocentric, a means to express our own ideas and fantasies, and therefore it can offer no direct access to reality itself.[61] This is also the point where a sign is quite different, in his view. Although he does not offer an explicit account of his understanding of a sign, he gives enough implicit evidence for stating the following.

While a symbol stands close to a myth or at least makes mythological thinking possible, and while a myth means that a human being tries to manipulate reality with his or her own ideas or fantasies, a sign, on its part, remains more faithful to objective reality. Precisely this makes possible the compelling power of reality upon our conceptions and knowledge so that reality is given the opportunity to formulate our understanding of it. In other words, a symbol has an anthropocentric character, whereas a sign is claimed to be reality centered. This provides another indication of Torrance's conviction that realism, in connection with the use of signs in human language, means a dedicated and direct submission to what he calls the compelling power of reality. It has to be noted, however, that Torrance does not want to be seen as holding an "ultra-realist" position where a sign and the thing signified become absolutely identical. Thus a statement about a truth coincides totally with the truth itself, a feature that is not a mark of a realist position in Torrance's opinion.[62]

It has been now clarified that while a symbol as a human, mythological construct, totally misses reality, according to Torrance, a sign is able to express it under its compelling force, not because of its completeness or perfection but just because of its lack of them. Torrance thus maintains that realism does not lead human knowledge to perfection or even to the assumption that articulate knowledge could in an exhaustive manner describe or explain reality. In fact, Torrance makes a distinction between comprehension and apprehension, suggesting that reality in its ontological depth can be reached only in an approximate manner, in apprehension, implying at the same time that we can never hold the whole truth in our hands as if we could comprehend it.[63] The plenum of truth belongs to reality itself.

At first glance, it would seem obvious that with his idea of compulsion Torrance could have solved the problem concerning the relation between subject and object in an epistemological process in a way that safeguards the absolute primacy of the object and provides a legitimate place for the subject. To be a subject means to be subject to the compelling power of the object of reality.[64] In Torrance's opinion, this view has been given its ultimate expression in the concept of *homoousion*, in the conviction that what is observed is of the same essence as reality itself, or simply, observation is reality, identical with it. This leads us to suppose that the human mind has no alternative but to obey the commands of reality quite submissively so that it can be seen as a passive recipient of knowledge imposed upon it by reality. But is this really so? In this connection, it is useful to remember that Torrance has been criticized for giving too loose an answer to the problem of subject and object with his view of *intuition*, which is undoubtedly a crucial term when the two aspects are related as in Torrance's thought.[65] It is now argued that, although the dichotomy between subject and object in Torrance's realism is maintained, he is able to avoid, at least to some extent, the idea that we have only a very passive role to play under the compelling power of reality.

A clue to this approach is given in Torrance's account of Albert Einstein and his significance for epistemological thinking in general. One of the characteristic features in the work of the great physicist, according to Torrance, is its emphasis on intuition as the means by which reality is apprehended rationally though not logically. This is the way *freely invented concepts* are born in the human mind under the compulsive force of reality.[66] Setting oneself under the compulsion of nature is the actual starting point for creating new scientific concepts, such as "field" or "quantum particle" or even "*homoousion*." In his interpretation of Einstein, Torrance refers to his attitude toward "non-logically derived concepts," although Torrance's interpretation bears a measure of inadequacy in this respect. Namely, Einstein himself does not seem to have had such a strict view of the possibility of deduction as Torrance claims, although, it has to be admitted, he was unwilling to give *logic* too great a role.[67] For Torrance, logic as an independent way of gaining knowledge is untenable because it tends to proceed too far into conceptualization of reality so that the connection between logic and reality breaks up.[68] And further, logic for Torrance is always an action of the human being toward reality, however conceptualized its efforts may be, and not an action of *reality* toward the human being. In other words, independent logic does not allow nature to *compel* its "message" upon the human mind, because the latter is so busy constructing its own devices for understanding. Torrance's rejection of logic as an independent discipline bears a remarkable resemblance to his understanding of natural theology as a movement from us to God in opposition to revealed theology as a movement from God to us.

Torrance seems to favor Einstein's view of nonlogical, "freely invented concepts" for another reason, too.[69] The fact that Torrance readily applies his idea of compulsion, underscoring Einstein's critical attitude toward independent logic, is sufficient to prove that for him Einstein is an eminent representative of realism, for the physicist believed in the existence of an independent and objective reality reachable in its ontological depth by human knowledge. At this point, however, Abraham Pais's claim is worth noting, namely that "objective realism" as the most characteristic feature of Einstein's epistemology dates from sometime in his last decades. In his earlier years he was in favor of adhering to several basic epistemological approaches. If Pais is right, this observation means that Torrance does not take into account in his own analysis the younger Einstein, who was the actual inventor of the theory of relativity. Pais points out that the physicist urged scientists not to adhere "too strongly to any one epistemological system." Therefore he described a true scientist as a realist, an idealist, a positivist, and even as a Platonist or Pythagorean. "Elements of all these 'isms' are clearly discernible in Einstein's thinking. In the last thirty years of his life, he ceased to be an 'unscrupulous opportunist,' however, when, much to his detriment, he became a philosopher by freezing himself into realism or, as he preferred to call it, objective reality."[70]

But how Torrance views Einstein's intentions in his epistemology of "freely invented concepts" is a more weighty question to be raised. He does not explain sufficiently in what respect these concepts are called "free." The problem is that, since these concepts arise in our own mind under the "compelling power" of reality, how can they be claimed to be free? The heart of Torrance's argumentation lies in his conviction that these concepts are freely invented, because their process of birth is free of all kinds of necessary connections, or because they are *contingent*, so that they are independent while depending on the ontology of reality. Therefore, extrinsically, toward objective reality, epistemology in Torrance's opinion is bound to obedience, but intrinsically, as an act of the human mind, it is free from what Torrance holds as necessary connections, apparent especially in independent logic.

Although Torrance appreciates Einstein's view of the role of intuition as the means whereby one can reach knowledge of the objective reality, he nevertheless finds the physicist's inability to define that intuition more exactly quite unfortunate.[71] Therefore he considers the views of Michael Polanyi, the British scientist, open to further elaboration. Torrance describes how Polanyi "insisted on thinking out further the heuristic process in which we reach knowledge of the real world which we cannot infer from what we already know, and has succeeded in carrying further the coordination between our ideas and experience, which Einstein did not consider possible."[72] Torrance explains Polanyi's approach in a manner that bears significant resemblance to Einstein's criticism of independent deductive logic from phenomena as a relevant way to reach reality. For Polanyi, according to Torrance, knowledge of reality cannot be reached by inferring it from the knowledge we already possess. In this heuristic process, as Torrance calls it, all genuinely new knowledge comes from "outside," not from those standpoints that are already present in the human knower, although their significance as ideas and experience are essential in the process itself. In addition, this connection of the human mind with the world "outside" can be understood in knowing itself.

Torrance is fond of Polanyi's concept of the "tacit component," by which Polanyi understands many elements in human knowing, all related to unconscious, subsidiary knowing, or, better expressed, awareness, thus introducing a strong personal slant to every act of knowing. Therefore objects cannot be observed independently of the inquiring subject.[73] Torrance interprets Polanyi as claiming that subjectivity plays a central role in our knowing, above all in what is called its tacit dimension.[74] The knowing subject is therefore, according to Torrance, implicitly aware of much more than he or she can articulately and explicitly express, and it is this implicit side of human awareness that has its impact on every act of knowing and our whole epistemological orientation.[75] Torrance's interpretation thereby confirms his insistence on the inadequacy of human expressions concerning the nature of reality as well as the approximate character of scientific knowledge.

The question concerning the relation between subject and object has to be viewed on this basis, and the claim is made that this application of Polanyi's concept of personal knowledge must be considered Torrance's effort to solve the subject-object problem, raised by his insistence on the absolute supremacy of objective reality and implying a rather impersonal role of a receiver of knowledge for the human inquirer. In other words, Torrance explicitly states that the human self is always present in every single act of knowing, thus being in *relation* with the object of knowledge. It is, nevertheless, significant to note that, in Torrance's opinion, this personal coefficient can affect only knowledge itself, not reality in its supreme independence. We are called to obedience to reality and to its compelling power, but in the midst of this submission we must constantly check our standpoint and knowledge to bring it more and more into line with reality. On the basis of the previous discussion concerning the relation between subject and object, it can now be argued that, in Torrance's realism, the role of a human being in the process of knowing is receptive although not passive.[76] Accordingly, the *homoousion* must not be seen as an indication of the objective side of the process only but as that of our subjective participation as well.[77] It is the human mind that observes a phenomenon under the compelling power of reality, with which this phenomenon is *homoousios*, of the same being.

This outlook indicates that, on the one hand, we are primarily receptive, for we cannot make claims on how reality should behave. The compelling power of reality acts through intuition upon the human mind, demanding obedience from our side. On the other hand, however, Torrance holds that we have a crucial role to play in the process of knowing, not only because we are the objects of reality, reality that exercises its persuasive compulsion, but also because we have to respond to the demands of reality. "Personal acts of will and decision are of course involved, but in response to the demands of truth which we cannot in good conscience or reason avoid or refuse."[78] While it has been argued that with this dual approach Torrance has been able, at least to some extent, to avoid the problem of naive objectivity, assigning the human being the role of a merely passive recipient of knowledge, his account of the personal nature of knowledge, its "tacit dimensions," and the role of unawareness do not, in the final analysis, lead him farther than halfway in this direction. He allows the human knower an active status in scientific inquiry, not only that of a passive recipient, but the area of this activity remains quite restricted. Questioning, making decisions and evaluations, and drawing conclusions do belong to the scientist's domain but only along the lines demanded by reality.[79]

Torrance seemingly does not feel obliged to apologize for his strict adherence to objectivity even when it imposes problems upon the role of the knowing subject. His standpoint can therefore be characterized as a foundationalist one in the sense that he presumes the existence of objects, serving as the basis for reasoned knowledge, which objects never need to be evidentially proved themselves. His argument is therefore open to the criticism generally addressed to *classical foundationalism*, namely that one can never know for certain that the knowledge assumed is in line with reality.[80] This makes Torrance quite indifferent toward the question, how can we know that the knowledge we have really accords with reality? His reliance upon the self-evident compelling power of independent reality cannot avoid certain features of a circular argument: we know that our knowledge accords with reality if we let its compelling power affect our understanding, and we know that the compelling force acts upon us when we adapt our knowledge to reality.[81]

This major feature of Torrance's realism, if considered a defect and thus leading to an apparent dead end, cannot but remain a target of continuous criticism. As shown at the beginning of this chapter, this is what certain scholars have done in their studies of Torrance. But two problems are disclosed in these interpretations. First, they have not taken seriously enough the compulsion aspect of Torrance's realism, and second and more important, they tend to reflect only his purely philosophical thought.[82] The first problem has been met in this section in which the discussion has followed general epistemological lines. It has been argued that the idea of compulsion, together with the concept of *homoousion*, provide the most characteristic and also the most original features of Torrance's realism. It was shown that these two elements result in the irreconcilable contradiction between his approach and the widely held view within quantum mechanics, the so-called Copenhagen interpretation. Torrance's appreciation of a referential relation between articulate knowledge and reality was emphasized, rather than any notion of correspondence. He also has a preference for apprehensive, approximate knowledge instead of comprehensive and exhaustive knowledge. Finally, a critical analysis of Torrance's ideas concerning the actual role of the human subject in the process of knowing reality with its compelling force was offered, and it was argued that, although the subjective element is given a proper position and not seen as a mere passive recipient of knowledge, Torrance is not, after all, able to avoid the implications of his strong objectivist position. This approach results in a situation where the problematic relation of subject and object remains both strict and imprecise at the same time: the difference between them is emphasized relentlessly and rigorously but the difficulty lies ultimately in the inaccuracy of the demarcation line between them.

In this respect Torrance's realism certainly contains problems, so well underscored by some scholars. However, we must ask whether or not it can be evaluated more positively, for example on the lines described above, where the human subject is factually given a positive role, however subordinated it may be. The role of intuition in an epistemic process deserves far wider attention, and a straightforward rejection of it does not seem to offer any fruitful alternative.[83] Within the general framework of realism, Torrance's approach could provide an alternative worthy of further consideration. Admittedly, one of its defects lies in his stubborn adherence and reference to Einstein and, to a minor degree, to Polanyi. But on the whole this should not be any final obstacle for taking its challenge more seriously, let alone for the reason that, despite its apparent problems, the "common-sense" approach has not lost its popular appeal.

If the first problem in Torrance scholarship is the lack of sufficient attention to the idea of compulsion and its relation to intuition, the second and more relevant defect is the effort to analyze Torrance's realism in a purely philosophical manner. This approach cannot, of course, be totally forgotten, for Torrance himself deals with his own views rather abstractly, undeniably, giving the impression that his speculations are ultimately philosophical in character. We now argue, however, that the roots of Torrance's account of realism, and its apparent problems, do not lie in his interpretation of the general philosophical context of realism as such but essentially in Christian theology. This is an insight that cannot be bypassed if an adequate picture of his realism is to be obtained. This claim is based on the presumption that Torrance's use of what he calls theological realism is not ultimately a philosophically reasoned realism applied in the domain of theology but essentially the obverse: it is theologically reasoned realism offered for the use of the other sciences and backed only subsidiarily by purely philosophical argumentation.

Election

Torrance's use of the idea of compulsion strongly reminds one of the Calvinist emphasis upon God's unconditional and absolute majesty and sovereignty which is manifest in the creation.[84] In the light of this divine reality, our role as submissive and obedient creatures is evident, involving the problems previously described with regard to the relation between subject and object. Torrance's account of objective knowledge, strong realism and his unshakable belief in the ability of reality to impose exact knowledge of it upon us are features that are not difficult to classify as implications of Reformed theology. But something more profound is at stake here, not just the apparent connection with his view of realism and the general atmosphere of Calvinist theology with its constant reference to the majesty of God. We now argue that Torrance's realism and its close link with the concept of *homoousion* and the idea of compulsion are ultimately grounded in one specific Reformed doctrine, that of election.[85]

Torrance only very rarely refers to the doctrine of election in this connection.[86] In these cases the nature of theological knowledge of God and its possibilities are discussed, but in spite of the theological nature of Torrance's accounts they can essentially illuminate the wider context in which he wishes to apply his idea of compulsion. Torrance refers to several interpretations by some New Testament authors of the significance of Jesus' words or actions as follows: "In such statements the New Testament witnesses are both reporting and thinking inside the inner movement of the Truth, yielding their thought, and therefore their speech, to its inner *compulsion* or *election*, for it is only through penetrating into the coherent structure or pattern of the Truth in that way that they are able faithfully to translate what they hear and understand, and genuinely to report."[87]

Two details are of particular interest in the passage. Firstly, Torrance claims that the divine reality or truth, the object of the New Testament writers' attention, was not only something on a phenomenal level but it demanded a deep penetration into the *inner* structure of the object. This implies that only in this case, through such a penetration, would the compulsion be effective and could genuine knowledge of the significance of the events recorded in Scripture be reached. Reality was, then, assigned the determining role, according to Torrance. So far the nature of the idea of compulsion, as analyzed earlier, is evident. The second point to be made is his way of linking the terms "compulsion" and "election," giving an impression of their identity or interchangeability. Precisely here arises the question concerning the justification for this identification. What does this linking together of these terms mean? And more important, what does this procedure imply for the discussion of realism in general?

Further information is provided in Torrance's analysis of Calvin's insistence that God permits us to know his Being by condescending to the same creaturely level as ours. He emphasizes two characteristically Reformed doctrines, namely those of *accommodation* and *election*. Of the latter Torrance says that it "rejects every projection of man and his creaturely forms into the eternal and divine, and teaches the Incarnation of the divine purpose (πρόθεσις), the projection, as it were, of the divine into the human, in Jesus Christ, and the establishing in Him of true relations between God and man and man and God."[88] One point is already here quite evident: election is closely connected to the Incarnation, establishing the point of meeting between human beings and God. The compulsion works through the election by excluding any effort to project human conceptions or structures of thought on the divine, leaving it untouched by human manipulation and confirming its superior position as the compelling object of knowledge. Torrance's rejection of natural theology, as described earlier, can therefore be claimed to have its roots in the idea of compulsion, which does not allow any other possibility of gaining knowledge of God than God's self-revelation in the Incarnation. God's election excludes any human intervention and sets us in a position of submission to a given reality. The reality of the universe, as it exists for us, is, then, chosen to bear rationality that is achieved by human investigations. This universe is, therefore, the only actualized one out of all the other possibilities that God could have chosen, the only one that, chosen in the act of creation, remains absolutely exclusive. Torrance can thus reason his view of contingency from the same premises: reality is chosen by God to have a rationality of its own, implying its independence. But because it is chosen, it cannot be self-explanatory or contain reasons for its own existence. In this sense it is dependent on God even in its independence. God's elective will, then, obtains the crucial role in determining what is the appropriate channel for coming to a knowledge of Him. The close link between the idea of compulsion, the doctrine of election, and the rejection of natural theology will receive further light a little later.

It has now been demonstrated that the idea of compulsion in Torrance's thinking has an essential link to the doctrine of election. But how does Torrance relate the doctrine of election to that of Incarnation, and through it to the ontological nature of the universe as the bearer of rationality which compels the human mind to accept it as it is given? For this reason it is useful to know more precisely Torrance's interpretation of the doctrine of election within the Reformed tradition. In this respect Calvin's writings are of primary importance. It is one of Calvin's intentions within the doctrine of election that through it the glory of God is given its appropriate place.[89] God's majesty is manifested, among other things, in predestination, which for the elect means everlasting life and for the reprobate eternal damnation.[90] This means that election is basically *individual*, having individual persons as its target. Thus, God's choosing or rejecting affects every person on the basis of a divine, eternal decree.[91]

Calvin's teaching of predestination remains quite problematic with regard to Christ, whose position in salvation is crucial but not, primarily, because of the atonement realized through him. Rather, for Calvin, Christ is a kind of "test case" or "mirror" before which we can be made aware of our state in God's decree while we are in a relationship with him.[92] In this context Christ's centrality is based on his ability to give assurance to the believer of his being elected or to give a subjective experience of the confirmation of his favorable position in God's eternal decree.[93] Christ's atoning work does not, then, add anything essential to the divine decree that exercises its predestinative power from all eternity.[94] Later, due to intellectual activity during the period of Reformed orthodoxy, the doctrine of predestination reinforced its status as one of the most distinctive features of Calvin's theological heritage, at the same time growing more rigorous and fixed.[95] The significance of this interpretation of the doctrine of predestination is appreciated by those scholars who have delicately claimed that, as entertained especially among the Puritans, it has been particularly effective both in the fields of economy and science, with special regard to their rise and advance.[96]

To understand Torrance's way of linking the idea of compulsion and the doctrine of election, we must emphasize his critical attitude toward the traditional interpretation of predestination as expressed in the confessions of the period of Orthodoxy. He maintains that after Calvin's time Reformed theologians found it appealing to construct an abstract doctrine of predestination in which election or reprobation are viewed independently from the work of Christ. Torrance sees a transformation in the doctrine of God as having taken place in Orthodoxy, so that this new doctrine of God "tends to be formulated independently and then there has to be added on to it the knowledge of God revealed in the Person and Work of Christ."[97] Thus Torrance has detected the implications of Calvin's thinking and sees it as problematic in this respect.[98] Nevertheless, he wishes to argue that Calvin, on the other hand, evaluated the significance of Christ in the doctrine of predestination more positively than might be imagined. In his opinion, Calvin "thinks of Christ as the Agent and the Matter or Substance of election." Torrance's interpretation of Calvin strikes familiar tones when he continues stressing that natural theology reveals its most harmful aspects whenever God is sought for "behind the back of Christ."[99]

Election without taking Christ into account would mean, in Torrance's opinion, relying on natural theology. Torrance wishes to argue that Christ occupies a crucial place in Calvin's own doctrine of election so that the Reformer's theology does not exclude Christ and his work, a view that certainly is not so simple; it was earlier shown that for Calvin God's election or reprobation took place in eternity before the Incarnation, a standpoint that involves serious problems as to the salvific significance of Christ's work. In other words, Torrance wants to make Calvin's view of predestination sound more Christocentric than it does at first sight, at least when considered in the light of the *Institutes*.

The reason for Torrance's reluctance to accept the interpretation of predestination entertained after the Reformed Orthodox era, an interpretation that left no specific room for Christ in the act of predestination, can be seen to lie in its *necessitarian* character.[100] Torrance thus maintains that the idea of individual predestination, based on God's decree issued before the beginning of time, implies that we can find ourselves only in a fatalistic, strictly predetermined relationship with God. Therefore he defines election in another manner: "Election refers to the eternal decision which is nothing less than the Love that God himself is, in action; it is the unconditional self-giving of God in the undeflecting constancy of his Grace which, as we have seen, flows freely and equally to all irrespective of any claim or worth or reaction on their part."[101]

Here the heart of Torrance's idea of compulsion becomes discernible. The notion of an eternal decision made by God is the basis of such a compulsion that achieves its aim in spite of possible resistance or negligence. Nothing can prevent it from exercising its power; no one can resist its force. This compulsion or election is identified with the Love of God, not as an attribute of God but essentially as God himself. The conclusion of this argument is very evident: "The Incarnation, therefore, may be regarded as the eternal decision or election of God in his Love not to be confined, as it were, within himself alone, but to pour himself out in unrestricted Love upon the world which he himself has made and to actualise that Love in Jesus Christ in such a way within the condition of our spatio-temporal existence that he constitutes the one Mediator between God and man through whom we may all freely participate in the unconditional Love and Grace of God."[102] Election, acting not individually but Christocentrically, means, then, that compulsion is Christological in character as well. The Incarnation is the one irresistible act of God's grace that touches everyone and everything in the universe. In this line of thought Torrance is explicitly following Karl Barth's modification of the Reformed doctrine of predestination. Barth holds that God's election has its only object in Jesus Christ, the one elected and also rejected, so that election does not concern individual persons directly on the basis of God's eternal decree but only indirectly through the person and act of Christ. This election of Christ took place in eternity, and as such it is a fact that manifested itself in the Incarnation, the crucifixion, and the resurrection, events that have taken place irresistibly, objectively, and independently of human attitudes.[103]

In the very event of the Incarnation, when the Word of God took our humanity upon himself, he became our Head and we became his body.[104] Barth explains the relation of a human being as an individual and the elected Christ in an ontological manner. The Incarnation transformed us into a new position vis-à-vis God, a position in which Christ's ontological unity with the Father is the basis of our union with God. This relation means for Barth the all-pervading victory of divine grace which excludes all human attempts to reach God and salvation. Barth comments straightforwardly that there is "a sure and certain salvation for man, and a sure and certain risk for God."[105] Barth's view of the powerful implications of Christological election for soteriology is open to interpretations resulting in a form of universalism, the theory that all men and every individual will ultimately be saved. On the one hand, Barth determinedly rejects the notion of *apokatastasis* as an unbiblical doctrine.[106] On the other hand, however, he emphasizes the impotence of unbelief to nullify the decree of God, manifested in the election of Jesus Christ, a fact that makes Barth's final intentions quite obsolete and ambiguous.[107]

Barth's insistence upon the absolute supremacy of God's grace, as manifested in the election of Christ, has one feature of particular importance when Torrance's view of the election and its relation to his idea of compulsion are considered. God's election cannot be frustrated by human unbelief or resistance due alone to the fact that "we resist the very being and existence of God Himself, if we raise even a question concerning the purpose and validity of this election, if we do not recognize that any such question is already answered by the fact that it is God who here decides and elects."[108] The election is the *datum* of the divine decree, the cornerstone of God's gracious actions toward us as humans, and it would not be far from blasphemy to ask reasons for this divine decision. "We must recognise the sovereignty of God and the inscrutability of His election, respecting the Therefore of it which no Wherefore can circumvent. Yet in so doing—if we really have God before us as the Subject of this work—we honour the source and citadel of all equity and judgement: not merely the wisdom which *must* silence the objections of our thinking and feeling, as though all discussion were terminated at this point by force—by a 'higher' force, but still by force; but the wisdom before which we *can* only be silent."[109] God's coercive force in election leaves us with no other option than to submit ourselves humbly to the divine dictates, the central theme in Torrance's idea of compulsion.

So far an effort has been made to clarify the genuinely Reformed background of Torrance's idea of compulsion both in its Calvinist and its Barthian emphases. But how does this Christocentric election relate to the coercive force of reality in general, as seen by Torrance? How can he support his claim that the *homoousion* as the exponent of the idea of compulsion is directly derivable from this theological context and applicable to wider scientific interests? These questions are important due to the fact that it is precisely here that Torrance moves from pure theology to the wider field of philosophy of science, making efforts to apply his views there as well.

Compulsion in Torrance's thought has been shown to involve a power that cannot be resisted and to rest ultimately on the act of election. The shift from theology to the realm of the natural sciences and philosophy does not produce insuperable problems for Torrance, since, in his opinion, election forms the basis for the coercive force of reality in that field as well. Torrance is convinced that we eventually realize our position in the face of a power that cannot be resisted but which exercises its force in such a manner that, in order to be reasonable, we cannot but consent to the demands of reality.[110] The noteworthy point in this respect is, focusing solely on the realm of the natural sciences, that the commanding factor is not God directly but the rationality or intelligibility of the universe. Their contingency makes them, after all, dependent on their Source and Creator.[111] Thus, Torrance's theological position is easily discernible. God has chosen to endow his creation with rationality and therefore it is coercive to the extent that there is no other way to know reality. In other words, God has *elected* his creation to bear rationality that compels the human mind to assent and excludes any other way of attaining knowledge. We should remember that linking rationality and election in this way is not a novelty in Reformed theology, for, according to Reformed rationalistic trends, "predestination applies to all the rational creatures of God, i.e., to angels and to men."[112] Worthy of note is that according to this interpretation rational beings fall into two categories: angels and humans, creatures who can be accredited with personhood.

Precisely in this respect Torrance's theology makes a remarkable addition to this traditional Reformed view by applying rationality not only to angels and humans, making them thereby possible objects of divine election. As has been shown, "the intelligibility of the universe" and its "inherent rationality" are among Torrance's central concepts. Torrance widens the sphere of creatures that are able to be elected by God so that not only men and angels but also creation as a whole is seen as a rational entity. In the context of the doctrine of election this means that God has elected his creation to bear rationality and knowledge and, additionally, that this election does not take place outside Christ but in essential connection with him, for he is the *Logos* through whom the universe was created and who was elected to be incarnate under its temporal conditions.[113] Here the Barthian view of the elected Christ has its application in the field of philosophy of science: knowledge of the universe is possible only because of God's election of Christ and of the creation through him.[114] Therefore there is good reason to conclude that in Torrance's opinion the idea of compulsion and the realism based on it are direct results of this election: this world is the only one realized out of innumerable possibilities, and so now we can only accept this and yield to the coercive power of reality, a power that is not self-explanatory but that is, in the final analysis, identical with the power of God, the divine and sovereign Elector. On this basis any natural theology pursued apart from the Incarnation and the *homoousion* is conclusively excluded: the creation does not reveal only God in the general sense of a Divinity or Creator but explicitly God who is manifest in the Word or *Logos* made flesh.

The reasons behind Torrance's decision to apply the substance of the concept of *homoousion* to the natural sciences, in their efforts to interpret phenomena realistically, can be shown to be thoroughly and radically theological. It is not only because he thinks that it might be relevant in general usage but also because it is the consistent result of his arguments which link the idea of compulsion, realism, and the Reformed doctrine of election. His position implies that the *homoousion* is the only accurate way of obtaining trustworthy knowledge, because God has chosen created reality to bear a rationality of its own, not just on the phenomenal level of observations but at its ontological depth. Another implication of this argument is that a realist attitude is considered as the only possible one whenever genuine knowledge is sought because that is the way God wanted things to be. The concept of *homoousion* thereby unites Torrance's themes of compulsion, realism, and election in such a manner as to elevate it to the position of a key with which scientific enterprises, in his opinion, can be pursued. As a somewhat daring but obvious interpretation of Torrance's position, he suggests a unifying theory of all human knowledge using the concept of *homoousion*, the contents of which is regarded as applicable in every field of investigation. There is, then, no other way of obtaining trustworthy knowledge.[115]

When the idea of compulsion was introduced and analyzed in the preceding pages, the problems with Torrance's position were underscored. The actual difficulty in Torrance's idea of compulsion was seen to lie in the inconclusive role of the human knower in the epistemological process when he confronts reality as a coercive force. Torrance's emphasis on intuition and subsequently on the receptive but not passive role of the human subject under the compulsion of the object does not solve the problem satisfactorily. Now that the connection between the idea of compulsion and the doctrine of election has been studied, the reason for this problem is evident: it is a direct consequence of the problem inherent in the Reformed doctrine of predestination, the problem of the role of human beings and their possibility of making decisions with regard to their own salvation.

Torrance admits that we have the possibility, in fact an obligation, to make a decision with regard to what we have discovered as the truth.[116] This would seem to imply that we should decide matters ultimately concerning us and that thereby we would play a more active role with respect to what we have found as truth. But Torrance does not think so. In line with his Reformed tradition, he states that election is the basis also for our personal decisions so that they cannot be viewed as having their starting point in ourselves. Instead, election guarantees that Jesus Christ is not only God's act toward us but also our act toward God.[117] Election, or God's eternal decision, precedes our decision, which means only participation in the decision already taken by God in the incarnate Christ. We can only accept this arrangement, and our decision for, or participation in, God's prevenient decision is seen merely as a sign of our submission and obedience in the face of compulsive reality.[118] Nothing essential can be added to this divine arrangement by any action of ours, not even by our decision.

The idea of compulsion with its inherent problem concerning the role of the human subject is, then, shown to represent a direct application of the doctrine of election in its Barthian form, seeing Christ as the elect human Person and involving all human beings in the possibility of participating in his election. Epistemologically this does not provide any solution to the subject-object problem or to the question on what criteria the correspondence or the truth of the referential relation between reality and our conceptions of it can be evaluated. Even after this analysis of the theological background of Torrance's idea of compulsion, the question must be seen as inconclusive, with constant tension and ambiguity, and his view of realism as unresolved.

The doctrine of election and the concept of *homoousion* as the theological basis of Torrance's idea of compulsion are the foundation for still another argument concerning the general nature of his realism. While Torrance holds that, due to God's merciful election through the Word, creation has been endowed with a rationality of its own, and while this rationality is able to exercise its coercive power upon the human knower, realistic knowledge can be achieved only if reality is allowed to *reveal* itself.[119] Torrance can speak of this feature of reality as if it were a personal being, referring also to its ability to withhold total openness without which human investigation would be meaningless.[120] It was pointed out earlier how difficult it is to define Torrance's realism or categorize it as one of the usual types. Now, after the preceding analysis based on the insight that the characteristic features of Torrance's realism have to be located in his theological rather than his philosophical stance, it is justifiable to call Torrance's position *revelatory realism*, a phrase that combines the idea of compulsion, the concept of *homoousion* based on the revelatory act of the Incarnation, and the doctrine of election. At the same time it stresses the independent character of reality in a way that is not discernible in other modes of definition, and particularly in that feature of reality which underscores the total dependence of human comprehension upon what reality discloses of itself.[121]

At the end of the previous section, it was argued that Torrance's realism can be called theological, not primarily because a philosophically defined realism is applied in theology but rather the reverse. The preceding discussion is sufficient to prove this assumption correct. Theological realism in Torrance's thought is, first of all, theologically reasoned, revelatory realism that is then offered for the use of other sciences and the philosophy of science. In other words, Torrance's realism is structured thoroughly according to a theologically reasoned pattern, where philosophical concepts are used in a subsidiary manner. This basic theological attitude becomes even more apparent when the most serious threat to realism, namely dualism, as seen by Torrance, is discussed.

The Threat of Dualism

In the previous pages it has been shown that the most distinctive feature of Torrance's realism is his *idea of compulsion*, which was shown to be based ultimately on his Barthian theological position, particularly on the doctrine of election, seen Christocentrically as the election of the Son of God who is *homoousios* with the Father. This position was seen to imply that reality exercises such a power upon the human mind that without our submission to it there would not be genuine scientific knowledge. Because this reality of ours is the only one, chosen as such by God, we cannot resist its inherent compulsion as to how things really are. The *homoousion* is the meeting point where our submission to this compulsion takes place. Now we should recall how closely Torrance links the *homoousion* with realism, both as a direct product of the realist approach of the Nicene Fathers, so Torrance claims, and as a conceptual tool expressing and safeguarding any realistic enterprise in the field of science as well as of theology. We also found out that Torrance attempts to create a remarkable tension between what he understands as Judeo-Christian or Hebraic and Hellenistic frames of mind.[122] Concretely, this tension is expressed in his conviction that the former had to struggle against the latter in order to secure an authentic interpretation of the Christian message, resulting in the contradiction between a realist attitude based on the concept of *homoousion* and a dualist attitude that has been a serious threat to Christian theology and to its implications for subsequent scientific thinking. To express briefly the relations between realism, the idea of compulsion, the concept of *homoousion* and dualism, one may say that in Torrance's opinion dualism prevents any application of the idea of compulsion and the concept of *homoousion*, resulting in the rejection of a realist approach. Therefore, according to Torrance, in the era of the Early Church *dualism* offered fertile soil for the growth of several religious trends that were later condemned as heresies.

According to Torrance, Arianism, viewed as a distinctively dangerous heresy by Athanasius,[123] constructed its Christology on dualistic notions, which had to be overcome so that the correct way of obtaining knowledge of God could be safeguarded.[124] In spite of the fact that dualism occupies such a central position in Torrance's theology, not only in its patristic emphases but also throughout his thinking, he has not paid adequate attention to properly defining it. This should be one of the most crucial tasks when one criticizes something with such ferocity as Torrance does dualism. Therefore considerable energy should be expended to describe precisely his view of this concept. However, first, we shall look at crucial features of some definitions of dualism. After having presented Torrance's view, we must take a critical look at the way in which he analyzes the rise of modern science with regard to the concepts of *homoousion* and dualism, concepts representing two mutually exclusive approaches in theology and the natural sciences.

The Complexity of the Concept of Dualism

The concept of dualism is not an easy one to be defined due to the vast variety of connotations and nuances hidden in it. R. M. McInerny says that "although it is possible to reduce the uses of the term dualism to a finite number of meanings, the term remains vague and of widely varying applications."[125] Here the emerging difficulties in any effort to define the concept are well described; it is easier to find different usages of the term while still finding the concept itself hard to explain.

In every case, however, dualism has always been related to the way people have tried to interpret and understand reality. In P. F. M. Fontaine's words, "It is clear that we need this opposition, and others too, in order to understand the world since we want many kinds of groupings and sets like series and categories which we tend to oppose, often very strongly, to each other."[126] Various attempts have been made to define the difficult term with respect to the origin of the two principles, as in the suggestion by the Italian Ugo Bianchi. He points out that there are two kinds of dualisms, a *radical* one in which the two principles involved are coeternal, and a *mitigated* one which holds that one of the principles has, as it were, emanated from the other, thus referring to the origin of the two principles as the determining feature of the concept.[127] But classifications of different types of dualism purely with regard to the origins of the two principles involved are not very useful in the long run. A more fertile effort for our purposes is Eberhard Simons's definition. He distinguishes two extreme modes of understanding the reality around us, between which he tries to find an appropriate balance. He rejects, on the one hand, pure dualism defined as a view of a "total reality" consisting of two separate and distant subrealities, and on the other hand, he also rejects monism, defined by him as a view adhering strictly to one reality.[128] Simons asserts that efforts to understand historical forms of dualism in its various aspects must be seen as essential ingredients in the human quest for understanding reality itself. By this statement Simons reveals one important feature of the concept, namely that it can be used as a *technical term* to describe a certain way of understanding and conceiving reality but not reality itself.

According to Simons, there is, so to speak, a balance, equality, and mutuality between the two levels of the two principles involved in the concept of dualism: primordiality and opposition. In other words, the principles derive from different origins but in spite of this they are equal, though on opposite sides. The principles involved in a dualistic system bear an ontological mark that cannot but affect their nature—they have an ontological basis anchoring them to the very center of existence and reality. The manner of opposition leaves its mark on dualism as well.[129] Simons does not discuss this point further or give any additional hint of what these manners of opposition might be.

In this respect Günter Lanczkowski penetrates more deeply into the problem. He makes a distinction between three kinds of dualism according to the nature of the opposition that is involved.[130] He illustrates his explanation with examples primarily from the history of religion. In spite of this, his way of making distinctions is worth considering, for they undoubtedly also have relevance from the standpoint of theology and dogmatics. First of all he speaks of *antithetical dualism* which, according to him, is a common feature in all gnostic systems and especially in Manichaeism.[131] The two principles involved are hostile toward each other and there is an eternal struggle between them but not necessarily any final solution.

Another and quite different form of dualism is *complementary dualism*, a good example of which is the Chinese Tao religion with its distinction between two principles, the strong, active, and male *yang*, and the weak, passive, dark, and female *yin*. Such a complementarity is seen as essential to the harmony of the universe and to existence in general.[132] Lanczkowski mentions still a third form of dualism, an *analogical* one which, in addition to the two forms already discussed, is a typical feature of what Lanczkowksi calls "universistic world pictures."[133] This form of dualism is related to the previous one, while both of them are concerned with harmony and harmonious existence. Analogical dualism finds the harmony by regarding a human being as a microcosm living in close relationship with the universe or macrocosm. This suggests that the relationship between the principles involved are closer than in complementary dualism in which in their diverseness they form a unity, whereas harmony in analogical dualism is based on similarities and correspondences in the two principles.

Lanczkowski's useful distinction between different forms of dualism provides a view of the nature of relations between opposing principles. In its antithetical form dualism contains two principles, working like enemies against each other. The opposition of the principles is the most extreme conceivable.[134] Quite contrary is complementary dualism, where the relationship is, in its basic nature, friendly and peaceful. Radically different principles are able to live together so intimately that their separation might mean collapse of a harmonious system. In analogical dualism the question is no longer, strictly speaking, of opposites but of two distinct principles, the relation of which is analogical when the two principles contain remarkable resemblances and similarities.

Opposition is the central theme in P. F. M. Fontaine's definition, too. He claims that in dualism it is not "possible to reduce the terms of the opposition more or less to each other; there are no longer intermediate terms; there is no longer any relationship or connection at all."[135] In Lanczkowski's classification this seems to be close to the antithetical dualism approached from another angle; any intermediate terms to mitigate the tension between the principles are denied. Opposition is also involved in McInerny's definition, according to which dualism refers to any theory "that invokes two opposed and heterogeneous principles of explanation."[136] But now another feature of the polymorphism of the concept in question comes into view. When compared to Simons, the difference between the statuses given to dualism becomes apparent. As was noted above, Simons considers dualism as *view*, a comprehensive and possibly also holistic attitude toward reality that can be seen as a basic factor in one's world picture or worldview in general. But McInerny does not call dualism a view—for him it is a *theory*, a choice of phrase that indicates a rational character in his thought, a character that becomes more evident when he discusses dualism with its epistemological implications.

So far two important variables of dualism have been discerned. The first deals with the degree of opposition between the two primordial principles of separate origin involved in a dualistic system. Lanczkowski's distinction between three categories with respect to the degree of opposition was found to be a useful one. The division into antithetical, complementary, and analogical modes of dualism helps us to understand the different usages of the concept. But dualisms do not vary merely in the degree of opposition of the principles. They also vary in their degree of internalization, in other words, in how far a dualistic mode of thinking has been adopted as an essential element of one's personal attitude toward existence. It was noted that Simons can speak of dualism as an external technical term with which we can try better to conceive the universe and systematize our understanding of it. However, at the same time he regards it as a view, as an integral part, not just as a technical term, of our self-understanding and consciousness of ourselves and of the universe around us. By contrast, McInerny sees dualism merely as a theory or as a concept describing theories.[137]

The problem of dualism and its definitions can be discussed further by referring to G. Mensching, who does not perceive dualism through the opposing principles of a dualistic system. He prefers speaking of *polarities* which are the parts of reality.[138] These polarities, the poles, are not derivable from or to each other. Mensching's way of talking about polarities or poles instead of opposites has one advantage. It does not focus attention only on the two principles as opposites, but also on what lies *between* the poles. In other words, dualism understood in terms of polarity contains a dynamic feature due to the tension between the poles.[139] In addition, the concept of polarity does not carry the same value judgments as does speaking of opposites.

Mensching, like Lanczkowski, is a historian of religion, but his distinction between different forms of dualism is worth noting.[140] First, he speaks of cosmic dualism, which can be further divided into two submodes: the dualism of equal poles is best manifested in the Chinese Tao religion, which we discussed earlier, with its complementary principles of *yang* and *yin*; and the dualism of ethically or soteriologically unequal poles, to which the universal view of the eternal struggle between God and evil belongs.[141] Mensching's second mode of dualism is eschatological, in which world history is divided into several *aeons* or phases. An example is Judeo-Hellenistic apocalyptic, which considered life in the present *aeon* as wicked and troubled by demonic powers but at the same time expected a new *aeon* of healing and salvation.[142] On the level of human existence Mensching mentions the anthropological dualism of mind and body, the most famous representative of which is Plato, who taught that the body is like a prison of the soul or the mind. The last form of dualism mentioned by Mensching is that of numinous powers, in which there is a conflict within a godhead or between different gods in a pantheon. German and Greek religious mythologies provide an example of this, though a more radical form of this dualism can be seen in Indian religion, where one god may have two mutually opposed or contradictory images.

Simons, too, refers to Indian religions and their incapability, due to the lack of reflection, to become conscious of their dualistic nature.[143] His analysis discloses an important feature of the usage of the concept of dualism: it can be seen as a human intellectual reaction to the phenomena with which we have to deal in everyday life when we try to understand them. Dualism is not, therefore, necessarily a conscious choice but rather a hidden, unconscious framework of the thinking mind, the purpose of which is to provide us with easily conceivable structures of the universe and its phenomena.[144]

The above discussion has touched upon different modes of dualism as defined by several scholars. We can conclude that, just as there are different degrees of opposition of the principles and degrees of internalization in dualisms, there is also always a certain field of application, so that we can speak of cosmic or eschatological or anthropological dualisms. But yet another feature can be discerned in these different definitions of dualism. It has to do with the value judgments concerning dualism itself—in other words, dualism is seen as either a good or a bad phenomenon. At this point we must relate the problem to biblical interpretation and theological thinking in general, because without them there would not be very much reason in a theological treatise to speak about kinds of value judgments concerning dualism.

Gloege says that the biblical message of God's will directed toward humans excludes dualism in its strictest form, described above as a view subscribing to two conflicting powers of completely different values.[145] Belief in one God, who is the Creator of the universe, gives no support to mythological or cosmological dualisms. Simons admits that Yahweh of the Old Testament is radically different from his creation and from his chosen people, who in this sense stand "against" him. But there is, nevertheless, no sign of dualism. In the New Testament dualistic frameworks are intimated, for example, in narratives about Jesus' *epiphania*, which imply the end of the old era and the beginning of the new one (the motif of eschatological dualism). However, Simons is not ready to accept this as a straightforwardly dualistic view, because it involves belief in the lordship of one God in all areas of existence. Therefore, in this connection, he prefers using the term *antithesis*.[146]

Gloege goes further by referring to Paul's view of making distinctions such as law-promise, works-faith, and flesh-spirit.[147] In Gloege's opinion, these are not any indications of metaphysical dualisms but simply of God's saving acts which in their justice and holiness break through into this world of sin and death. Nevertheless, dualistic thought-forms become more apparent in Johannine theology, in which strong distinctions are drawn between light and darkness, law and grace, life and death, truth and falsehood, etc.

Simons holds, like Gloege, that Israel's monotheistic faith excludes what he calls "any absolute form of dualism."[148] When discussing Pauline or Johannine theology, Simons does not use the word *dualism* but prefers to speak of "dialectics" or "antithetics." Therefore, the biblical distinctions are not, for Simons, evidence of hidden dualism in the Bible but only necessary conceptual tools in the forms of antithetics and dialectics, with which the biblical message is made more conceivable. This clearly confirms what has been noted earlier, that Simons regards dualism more as a comprehensive view than as a mere technical term. Consequently, what for Gloege are indications of dualism (distinctions between law and promise, light and darkness, etc.) are for Simons only helpful conceptual tools in the form of dialectical distinctions.

The reflection on Simons's and Gloege's thought is sufficient to show that when dualism is considered within the Christian framework it inevitably attains a certain level on the scale of values and truths. And in these cases, though more so in Simons, the judgment is *against* dualism, so that it is seen as an alien element in Christian faith and thinking, an element that should in one way or another be excluded.[149] Gloege is more cautious: he admits that there really are dualistic tendencies in the Bible and especially in Johannine theology but that they are not important as such—they only serve the proclamation of the Christian message without introducing dualistic contents themselves.

Still another point of view is presented by McInerny, who says that "there is little to be said for or against dualism."[150] This irenic and neutral statement is in line with what was noted in McInerny's definition at the beginning, an explanatory aid when trying to understand reality, a theory with its advantages and disadvantages. Thus it can be concluded that from the Christian perspective dualism is generally seen as something alien but useful in certain forms for clarifying some features of Christianity. Therefore, value judgments concerning dualism are mostly negative, sometimes neutral but very seldom or never consciously positive. Negativity correlates with the worldview aspect, and neutrality with the theory aspect.

The difficulty of defining the concept of dualism has been evident throughout the previous pages. In any case, as a summary of our discussion of the views of scholars of different theological orientations, we may construct a rough definition for the concept. By dualism is understood any way of thinking that presupposes the existence of two principles. Most important is the nature of the relationship between them. Three points are essential. Dualism, with the two principles involved, has 1) a certain degree of opposition between them; 2) a certain degree of internalization (dualism as a theory v. dualism as a comprehensive worldview); and, finally, 3) certain implications for value judgments concerning itself (dualism is seen as either good or bad).

Torrance's Definition of Dualism

When Torrance's view of dualism is considered, one problem emerges above others, namely that he fails to define the concept of dualism with sufficient accuracy. Rather, he seems to presume that the reader is well aware of what he is talking about.[151] However, this should be of major importance if only for the reason that general definitions of the concept are so ambiguous.[152] Even if one is well aware of the complicated problems in its definition, what is needed in interpreting Torrance's thinking is an adequate and sufficiently precise definition grounded in his own understanding.

Torrance, actually, has made an effort to define the concept. In it he states the following.

> *dualism*—the division of reality into two incompatible spheres of being. This may be cosmological, in the dualism between a sensible and an intelligible realm, neither of which can be reduced to the other. It may also be epistemological, in which the empirical and theoretical aspects of reality are separated from one another, thereby giving rise to the extremes of empiricism and rationalism. It may also be anthropological, in a dualism between the mind and the body, in which a physical and mental substance are conceived as either interacting with one another or as running a parallel course without affecting one another. In the Judaeo-Christian tradition man is regarded as an integrated whole, who is soul of his body and body of his soul.[153]

Certain points can be picked up from Torrance's description at this stage. First of all, in dualism the question is one of the division of reality, implying that for Torrance there is strictly one reality, not two distinct ones. Even this sort of phrase is enough to indicate Torrance's firm belief in the unity of reality: he avoids talking about two realities but only of its division. In other words, reality remains one even if it is thought of as consisting of two different realms. Second, this division results in two spheres of being that are mutually incompatible. What Torrance means by incompatibility in this connection remains unclear, and further questions arise, for example, are the spheres equally significant and mutually opposed, and if so, what is this opposition like? Or, is there a remarkable inequality between them? Can we speak of a deep and hostile contradiction between the spheres, so that they try to annihilate each other? Or is their coexistence more friendly, more of a complementary nature? After all, what becomes clear from Torrance's brief definition is that this incompatibility is a result of the division of the one existing reality so that the separation is not an inherent feature of reality itself but a consequence of a mental attitude in which the reality is seen through a framework of two spheres. Reality is one, but the human mind tends to divide it into two. Attention will also be paid to the fact that the spheres are those of *being*—they are of an ontological nature—so that being itself on its deepest level, as understood by humans, is affected. In dualism there is, in other words, a single reality that is seen, due to the division concerned, as consisting of two separate modes, principles, spheres, or realms.

Third, Torrance makes a distinction between three, primarily philosophically reasoned, forms of dualism—cosmological, epistemological, and anthropological. The first two are of the utmost importance for Torrance due to their implications for scientific thinking although, it has to be admitted, there are some indications that suggest that the third one is an archetypal form of the other two, that is, that cosmological and epistemological dualisms actually derive their origin from the dualism of body and mind.[154] In cosmological dualism we are dealing with sensible and intelligible realms, "neither of which can be reduced to each other." The philosophical character of this approach to dualism becomes evident in the single fact that Torrance traces this distinction back to Plato, claiming at the same time that this dualism was "hardened" by Aristotle, involving disjunctions between "action and reflection, event and idea, becoming and being, the material and the spiritual, the visible and the invisible, the temporal and the eternal."[155] On the other hand, epistemological dualism, according to Torrance, separates the empirical and theoretical aspects of reality.

The distinction between epistemological and cosmological dualisms is a central theme in Torrance's treatise, especially when applied to the development of science. Accordingly, it should be remembered that the definition analyzed above appears in a book with a greater interest in theological implications for science and vice versa than in theology itself.[156] But Torrance is above all a theologian, and therefore his view of dualism will not be clear if it is considered apart from the theological context in which it first emerged. The definition above, given by Torrance himself, is insufficient, and it does not adequately illuminate the role of dualism in his thought. It can be seen only as a secondary definition with a special interest in modifying theological intentions to a more readable form for scientifically orientated readers. Therefore its mode of approach is more philosophical than theological. As such, it does not tell us much about dualism and why Torrance is so concerned about its effects. For this reason it is necessary to return to the event which gave rise to Torrance's understanding of dualism and its effects in the whole intellectual totality of Western thought.

As has been pointed out, Torrance's idea of contribution asserts that the struggle of the theology of the Nicene Fathers was crucially important not only for the Christian Church but also for the development of Western scientific thinking in general. The struggle that came to exert such influence took place between Athanasius and Arius, between Judeo-Christian and Hellenistic modes of thinking.[157] As has been mentioned, this involved Christology and the position of Jesus Christ as the focal point. Torrance maintains that Arius's theology was essentially dualistic because of the sharp distinction between the *cosmos noetos* and the *cosmos aisthetos*, or between God and his creation. This Arian dualism was the actual background for the development of the orthodox doctrine of the Son of God as carried out by Athanasius and the other Nicene Fathers who fought against dualism in the distinctive form in which it was encountered in the thought of Arius and his followers.[158] The hypothesis now is that Torrance's understanding of dualism is also based on Athanasius's theology, which developed in confrontation with another line of interpretation, named after the Alexandrian presbyter Arius.[159] Torrance indicates his view of what was the philosophical basis of Arianism. Plato's distinction between the two *cosmoses* is said to have been essential to Arian theology. And further, while the Arians radically separated God and the creation from each other, they at the same time came to its epistemological conclusion: a sharp distinction was drawn between the eternal unknowable Being of God the Father and the more knowable *Logos* the Son. This had the effect that in their doctrine of creation the Arians regarded the Son as a first-created principle by which the rest of the universe was created. From this line of thought Torrance draws the conclusion that the Arians "interpreted Christ in terms of the cosmological principle which Athanasius found he had to reject entirely."[160]

From these statements one might gain the impression that both Arianism and Athanasius worked consciously and openly in their attempt to apply philosophical speculations to their distinct theological understandings. What in Arianism for Torrance represents objectionable dualism is the radical division between God and the world which further meant shutting God out of the world of empirical reality in time and space.[161] This division undoubtedly has its philosophical background as Torrance represents it, but how consciously this was done is debatable. As far as Arius is concerned, Rowan Williams has pointed out that Arius "is not a philosopher, and it would be a mistake to accuse him of distorting theology to serve the ends of philosophical tidiness."[162] To follow Williams's reading, then, Arius was sincerely and primarily interested in theological matters, and thus he was most eager to safeguard God's transcendence against all kinds of conceptual mixtures.[163] The rigorous application of this presupposition led to the sole logical conclusion that the Son is a creature of God. This conviction rested on his principle that every metaphor or anthropomorphism should be excluded from human speech about God as unworthy of Him. Therefore, he could not think of the Son as unbegotten or eternal since these attributes had to be features of God, strictly speaking.[164]

Nevertheless, for Torrance Arianism means dualistic separation of God from the world on the basis of Platonic philosophy.[165] Torrance does not criticize the thought of Athanasius but faithfully follows the polemical and anti-Arian thinking of the Church Father.[166] Torrance's selective way of presenting details of the history of ideas is clearly discernible in the fact that he seems to identify Platonism and dualism, suggesting at the same time that Athanasius and the other Nicene Fathers were free from such Platonic influences. This interpretation is, however, quite limited and not at all as simple as Torrance would have us believe. It has been convincingly shown that Platonic thought had a remarkable influence upon Athanasius and the other Greek Fathers to such an extent that the sharp distinction Torrance draws between Platonic dualism and the thought of the Nicene Fathers can hardly be justified.[167]

Nevertheless, in Torrance's argumentation this leads to an important point in the current effort to define his view of dualism. Arianism taught that Christ the Son of God is, on the one hand, merely a created being but, on the other hand, a being higher than "ordinary" creatures like humans. In other words, dualism in Arian thought resulted in the presupposition that the sharp separation between God and the world must be softened by a mediating being, the Son of God, who represents a third kind of being alongside God and the humans, but who, in the final resort, belongs to the created realm. There is therefore no *relation of being* between the Father and the Son, as Torrance expresses it.[168] To put this in another way, Arianism, in Torrance's opinion, because of its basic dualistic attitude, sees God as so transcendent in relation to the world that it cannot admit a real Incarnation; Athanasius submits to the fact of the Incarnation and thus overcomes the dualistic "habits" of Greek thinking. Torrance actually holds that dualistic thinking is the *cause* of Christological problems, which would not have emerged if dualism had been recognized early enough.[169]

Here is the heart of what Torrance holds as Christological dualism: an imprecise mixture of divine and human features coexisting in an unbalanced way in the person of Christ. Accordingly, what for Torrance overcomes these Christological tendencies is understanding the Son of God as being *homoousios* with the Father, at the same time fully human and fully divine, God of God, who took our total humanity in all its aspects upon himself in the Incarnation. Therefore Christological nondualism may be defined as a balanced view of Christ in which, in his person, nothing of his humanity and nothing of his divinity is reduced.

From this it is possible to proceed to another problem of Christological dualism as presented by Torrance. During the Christian era the union of the human and the divine in the person of Christ has raised a great deal of intellectual questioning. Due to dualism, so Torrance argues, it is most difficult to view God as having become a human being without serious damage to his humanity. But what has this difficulty or even apparent impossibility to do with dualism? Torrance answers by saying that dualism makes the divine and the human, the Creator and the creature, the infinite and finite *logical opposites* and therefore mutually exclusive.[170] In other words, dualism has the effect of creating sharp distinctions between poles, the relations of which must then be understood through logic or logical necessities and with the help of what Torrance calls a "logical bridge."[171] From all this it may be concluded that Torrance views dualism and logic as closely related to each other, so that, in order to cope with the problems raised by dualism, human thinking needs logic, a discipline somewhere "in between," to keep together the poles separated by dualism.

The close relation between dualism and logic in Torrance's theology is more easily conceived when it is recalled how he presents the Incarnation as a confirmation of the createdness of the universe and especially of its contingency. In Torrance's opinion, the contingency of the universe derives from the fact that God, who was absolutely free not to create the world, nevertheless created it to have its own freedom and reality. The Incarnation was an act of God by which the contingency of the universe upon its Creator was confirmed, for how could God enter as a human being into the world that was not created by him, dependent on him, and real to him?[172] But, furthermore, belief in the contingency of the created universe bears in itself the inevitable exclusion of all determinism, reliance on fate, or exact logico-causal relations and necessary connections.[173] Thus, according to Torrance, the contingency of the world in the sense defined by the Nicene theologians disappears when dualistic modes of thinking are applied. And accordingly, when dualism is recognized and abandoned, belief in contingency is reinforced. In Torrance's view, contingency is thus the proper feature of the created universe in its relation to its Creator, and whenever dualism separates God from the creation the former tends to remain in his own obsolete transcendence and the latter is trapped in deterministic causalities, logical relations, and inevitable necessities. Therefore, according to Torrance, dualism means the destruction of contingency.

In such a treatment of dualism, it is worth noting that, although Torrance's argumentation is somewhat philosophical, it never loses its basis in the Arian controversy and the victory of the Nicene theology: contingency is based on createdness as revealed by the Incarnation of the Son of God, fully human and fully divine. The simultaneous coexistence of the divine and human as such in the person of Christ is therefore for Torrance not a sign of dualism, but rather the only way to safeguard a real, dynamic, and open (that is, free of deterministic causalities) relationship between God and the world. This conclusion also discloses the reason why Torrance is committed to treating dualism as a threat to Christianity: Arianism showed that dualism has the effect of shutting God out of the world so that there is no connection between human beings and God on an ontological level.[174]

Torrance's way of stressing Christ's divinity and humanity leads us to expect that the Chalcedonian formula might have had a more central place in his argumentation against dualism. Nevertheless, it is the theology of the Councils of Nicea and Constantinople that gain his main attention due to their emphasis on the concept of *homoousion*. No doubt, in this concept Torrance already sees implicitly included the doctrine of the person of Christ that was later expressed in an articulate form in the Chalcedonian definition.[175] His adherence to the concept of *homoousion* in this respect might be founded on several reasons. First, it represents a soteriological aspect which Torrance emphasizes constantly: if there is no identity of being in the Son and the Father, salvation would remain outside the being of God and Christ's atoning work could not have any truly soteriological significance.[176] Second, the *homoousion* does not begin by viewing the mystery of God within the being of God but from the factual historical event of the Incarnation, in which the view toward God is attained under the conditions of space and time. The *homoousion*, then, is not based on speculation but rather on an *a posteriori* movement of thought.[177] Third, the concept of *homoousion* is certainly an expression which conveniently meets the demands of Torrance's arguments in favor of realism; what is observed is of the same being with reality itself so that an observation does not relate to a superficial phenomenon only but to reality in its ontological depth.[178] Apparent phenomena and reality, then, do not live their own separate lives but are actually one and the same. These points summarize the motivation behind Torrance's preference for the *homoousion* over the Chalcedonian formulation of the two natures in the person of Christ as the concept to defend the proper interpretation, as he understands it, of reality both in science and theology.[179]

But we have not yet found any reply to the question concerning the most characteristic feature of Torrance's dualism, a problem that still has to be answered. In the

Christological context, belief in the coexistence of two natures in the person of Christ is not an indication of dualism for Torrance, for if that were the case he would be compelled to abandon the Nicene and Chalcedonian formulations of Christ's being *homoousios* and of his two natures. This feature is highly revealing, because it leads to the conclusion that, whatever kind of dualism is concerned, the mere appearance of two poles does not make the distinction in question dualistic! Just on this point several interpreters of Torrance fail to recognize the basic nature of his treatment of dualism and cannot see deeper into the problem or to the heart of what Torrance wishes to express. Namely, the crucial issue appears to be *the nature of the relation* between the poles involved.

With respect to the Christological question, Torrance refers to early heretical deviations of Ebionite and Docetic types.[180] In the former, Jesus Christ was regarded as a mere creation, while in the latter Christ's true humanity was denied spiritualistically and his suffering was considered unreal because the entire weight of Jesus' identity was placed upon his divinity.[181] According to Torrance, this reduction of one nature of Christ to another, the human to the divine or vice versa, was an inevitable consequence of the dualism lying behind both Christologies, a dualism in which God and the Word were radically separated from each other.[182] Torrance's point is that once dualism is accepted, it *distorts the balance between the poles*, so that one is appreciated more than the other or one pole absorbs the other into itself. With respect to Christ, this means that he is seen either as more divine than human or, conversely, as more human than divine, as was the case in Arianism. In Torrance's opinion, then, dualism presents a threat to Christian faith, and to genuine scientific realism as well, precisely because it does not accept the harmonious and balanced coexistence of the poles involved in the person of Christ or, consequently, in any subject in our empirical reality, the scientific enterprise included.[183]

Before beginning to draw together any conclusions as to what has previously been said about Torrance's own view of dualism, a short look should be given at an ambiguous feature in his use of the terms and concepts. While Torrance usually applies the concept of dualism to what he sees as the way of thinking that poses a threat to Christianity, he nevertheless also uses other terms and phrases that might cause confusion. One example is the word *dichotomous*, used in an identical sense with dualism. So he can speak of dichotomous modes of thinking "which disrupt and distort the very things we seek to understand and explain in any field of inquiry."[184] Dichotomy, division of something into two parts, is thus unacceptable, too, one might think. What is problematic in this case is that Torrance himself has to operate with certain unavoidable dichotomies and with what some might even call dualisms according to Torrance's own definition: God and the world, Creator and creature, divine and human, etc. We should remember, however, that Torrance, as a theologian, is not only compelled to use these distinctions but also to avoid using them in such a manner that he would contradict himself. In other words, dualism is for Torrance something more than just a distinction or dichotomy between two principles: it is rather of a *paradigmatic* character, affecting the very basis of human thinking and the ways of perceiving reality.[185] Therefore dualism cannot be seen as a mere name for certain phenomena involving two principles, but as a deep-rooted and even unconsciously working framework of the human mind. In other words, for Torrance dualism is not a mere technical term.[186] To make his point clearer, he can talk about dualism as something "radical" or "disastrous," thus underscoring its negative significance as a paradigmatic force, rather than just a nominal concept referring to any phenomenon or entity, understood in a twofold manner.[187]

In the preceding pages the strong theological "color" of Torrance's conception of dualism and its culmination in the Christological debates of the Early Church were emphasized. The hypothesis has been that a philosophical approach will not give a sufficiently precise definition of it and that therefore understanding the conflict between Arianism and the Nicene Fathers is the most fruitful way to conceptualize Torrance's position. Therefore, the results of the preceding analysis of Torrance's view of dualism can be drawn together, with regard to 1) the relation between the poles in dualism, 2) the degree of internalization of dualism, and 3) the kinds of value judgments that are implicitly or explicitly applied in Torrance's criticism of dualism.

First, as was noted, Torrance describes the dualistic thinking of Arianism as having the effect of making a radically sharp distinction between God and the world, so that the Athanasian understanding of the *homoousion* was totally excluded: nothing human was to be applied to God, and this took place precisely through the *homoousion*, in the Arians' opinion. This reveals that for Torrance dualism consists of two poles in such sharp contradiction with each other that, as in the case of Arianism, they presuppose a deep and irreconcilable split between them, involving either mutual negation or the logical impossibility of both having equal and simultaneous existence at a certain point in space and time.

Second, according to Torrance, dualism is not just a technical term to describe the existence of the two. The Arian controversy showed how internalized it can be at the deepest levels of human understanding. Once assumed, it has inevitable consequences for one's thinking and modes of conceiving reality. A crucial feature of this considerable degree of internalization is that dualism works mostly unconsciously and in unrecognized form in the depths of the human mind.

Third, Torrance's attitude toward dualism becomes apparent as soon as one reads his treatise on the Arian controversy: dualism is destructive and dangerous for Christianity and therefore it should be recognized and avoided. Its danger lies in its tendency to lock reality into two blocks with no dynamic interaction between them. All fields of human understanding, scientific and theological thinking included, are affected by it. Torrance's interest lies especially in the implications of dualism for theology, since whatever its nature may be—ontological, epistemological, or cosmological—it culminates in Christological dualism, which tends to exclude God from the world, thus making the relationship static, deterministic, and, at the best, only tangential. Therefore, it may be argued that dualism and, most of all, overcoming it, are, alongside the Incarnation and its *homoousion* principle, at the heart of Torrance's theology.

On the basis of what has been argued, it can now be briefly stated what dualism is for Torrance. For him, dualism is a paradigmatic, most often unconsciously applied but deeply internalized way of perceiving reality which is seen as consisting of two principles. In the final analysis the poles are God and the world, or the Creator and the creation, between which there is no real interaction of dynamic relation as indications of the ontological and epistemological openness of God and the universe toward each other, thus offering a negation of the *homoousion*. A distinction or existence of two distinct poles as such is, therefore, not to be seen as dualism, but the distorted or missing *relation* between the poles results in unbalance or total disruption, determining the dualistic character of a distinction.

Dualism and the Origins of Newton's Science

As has been stated, Torrance gives an original interpretation of the history of Western science, basing it on three distinct periods, each with a differing cosmology. This division can be claimed to be based solely on the influence of dualism, seen by Torrance in these different phases of history. The transitional time from one period to another is considered most significant. In short, these changes between different periods are the emergence of Ptolemaic, Newtonian, and Einsteinian cosmologies.[188] Ptolemaic cosmology made a distinction between the intelligible sphere of celestial realities and the sensible sphere of mundane phenomena, thus having the effect of building a remarkable dualism into Greek science, philosophy and culture.[189] Newtonian dualism is characterized as tearing apart absolute space and time and the contingent events within them.[190] The third change is the paradigmatic transition from post-Newtonian to Einsteinian cosmology, in the midst of which we are now living. This new understanding of physics challenges our whole mode of thinking and our forms of conceiving reality, as one of Torrance's statements reads.[191]

According to this periodization, the Arian controversy of the Early Church can be seen as a by-product of the emergence of Ptolemaic cosmology, so that, in the ultimate analysis, it was not only Arianism that had to be rejected but also the whole cosmological frame of mind that was born amidst that new cosmology.[192] Torrance argues that at the moment, when Newtonian cosmology is being replaced by the Einsteinian one, Western civilization faces similar problems, problems emerging from dualism, as were the Fathers of the Early Church when the concept of *homoousion* was developed. Torrance claims that, just as Christian theology with its concept of *homoousion* was able to overcome the dualism of Arianism, in a like manner Einsteinian cosmology is overcoming the older, Newtonian one.[193] Therefore, it is Newton and his influence upon the natural sciences against which Torrance places the emergence of modern physics and its influence upon new modes of thought.[194]

The problem can be specified with the question, how do Newton, his physical and theological thinking, on the one hand, and Arianism, its Christology and understanding of reality, on the other, relate to each other? In order to understand Torrance's firm rejection of Newton and his legacy, seen as contaminated by dualism, there has to be a search for those features in Newton's thought which for Torrance justify his standpoint.[195]

Why does Torrance lay such heavy stress on Newton's significance and take his thoughts as the most remarkable channel through which dualism came to prevail until now, both in science and in theology?[196] When explaining the role of medieval theology as a considerable defender of nondualistic Christian concepts (objective rationality of the universe, contingent intelligibility, contingent freedom), which he sees as derived from Nicene theology, Torrance proceeds to ask what their fate was when the sixteenth and seventeenth-century scientific revolution took place. Torrance replies by referring to Isaac Newton and his immense impact on the basic modes of understanding as applied in the natural sciences until the beginning of the twentieth century.[197] In other words, in Torrance's opinion, Newton's thought represents a culmination point as far as Nicene theology is concerned. Torrance says that Newton "supplied" basic paradigms for science, the significance of which has been immeasurable. When Torrance examines contemporary features in theology and science, his argumentation can be understood only in the framework of Newton, his conception of science and his scientific legacy in general.

But yet another detail can be pointed out as describing Torrance's eagerness to emphasize the close link between modern science, theology, and Newton. Nowhere is this explicitly stated, but it can be argued, on the basis of the way in which he presents his descriptions of Newtonian thought by pointing to its overall significance for theology,[198] that for Torrance Newton represents a scientist in whose person the rigorous approach of a physicist and theological thinking are both present.[199] Therefore Newton offers Torrance an excellent explanatory framework with which to interpret certain tendencies in modern theology.[200] Torrance is, then, interested in Newton for two reasons: first, Newton initiated a new era in the history of science, and second, Newton's own scientific thinking includes strong theological features and implications. The common factor in these two reasons is that neither of them is free from what Torrance calls dualism.

The core of Torrance's criticism of Newton lies in the manner in which the physicist explained rationality and order in the universe. Whenever Torrance describes Newton's dualism, he draws one's attention to the division between, on the one hand, absolute or mathematical time and space and, on the other, relative or apparent time and space.[201] Apparent space and time are experienced by an observer, and for this very reason it is also a relative and variable reality. Therefore another reality is needed—that of absolute, mathematical, immutable space and time in order to give a point of reference to the former, phenomenal reality.[202] This means that apparent reality would remain incomprehensible and irrational, totally dependent on human experiences without taking absolute, mathematically constructable space and time into account. This reveals Torrance's appreciation of Newton's basic intention to accredit reality with a rational and intelligible basis, but also his firm criticism of its application, which forced the physicist to think in dualistic terms.[203]

It is now argued that Torrance focuses his attention intensely on Newton's distinction between absolute and relative space and time—a distinction that Torrance calls dualistic—because it has considerable theological consequences and implications. Expressing his awareness of the close relation between scientific thinking and religious belief in Newton, Torrance suggests that the actual point where theology enters Newton's thought lies exactly in the problem of absolute space and time.[204]

Torrance can appreciate Newton's "firm belief" in the inherent rationality of the universe and also his reluctance to make any effort to penetrate into deeper causalities beyond apparent phenomena or to entertain *hypotheses*.[205] From this belief Torrance draws a direct line to Newton's understanding of God. Referring to the General Scholium at the end of the third book of the *Principia*, Torrance says that the independent ontological status of absolute time and space, and consequently "the objectivity and comprehensibility of the universe, are finally grounded in the eternal and infinite God who 'by existing always and everywhere, constitutes duration and space.'"[206] For Torrance this means that Newton identifies absolute time with God's eternal duration and absolute space with God's infinite presence, resulting in a "receptacle" in which everything exists and moves.

Torrance's interpretation shows Newton as having a receptacle notion of time and space, so that they are understood to form a container, a closed system within the limits of which events take place. But what makes Torrance's reading more interesting is that, in his opinion, for Newton this container consists of God's eternal duration and infinite presence. In the light of Newton's own words this standpoint is not without its problems. He certainly asserts that God is "eternal and infinite, omnipotent and omniscient; that is, his duration reaches from eternity to eternity; his presence from infinity to infinity." But Newton determinedly rejects any idea that the adjectives *eternal* or *infinite* could be used as subject predicates referring to God in phrases like "God is eternity" or "God is infinity"—he is just eternal and infinite.[207] This line of thought clearly does not allow such a ready interpretation of Newton's notion of God's duration and infinity as a receptacle in which things happen as Torrance is inclined to think. It is certainly true that, according to Newton, the universe and human beings included have their existence in space and time, and in this respect they may be held to be *in* space and time. But this approach does not necessarily lead to a strict receptacle notion of space and time in the sense that the existence of the universe should be interpreted as taking place in a container-like vessel named God.[208]

What, then, is the motivation behind Torrance's interpretation? It could be maintained that Torrance sees Newton as holding a view of God as an explanatory factor when the scientist makes efforts to construct a relevant view of the world embracing all phenomena in it so that a consistent overall perception of the universe results. Torrance explicates his views more fully when he describes Newton's way of applying his mechanistic view of the world. The scientist agreed that the universe as a whole cannot be conceived as a mechanical system complete and consistent in itself. Instead, its consistent mechanisms can be grasped only if they are related to "the *counsel* of a *voluntary* and *intelligent Agent* beyond it, the living God who rules over all."[209] Torrance thus explains that Newton had some kind of idea of contingency with regard to the universe. The absolute point of reference, for Newton, is therefore no one and nothing less than the omnipotent God of eternal duration and infinite presence. The world is not comprehensible without taking this God, the intelligent Agent and Creator, into account.

This kind of contingency, as nonnecessity or non-self-explanation, seems to be an adequate way of finding a place for both science and theology on the basis of Torrance's theology. However, he interprets Newton's view of the two kinds of space and time, absolute and relative, not only with the latter having its fixed point of reference in the former as its basis of contingency and source of intelligibility but also with something more theological and ontological. Torrance says that Newton identified absolute, mathematical space and time with the *mind of God* which both contains and lays order upon the created world.[210] This claim is surely more weighty than what was earlier stated with respect to Newton's view of absolute space and time as identical with God's duration and presence. For in the latter case they can be seen only as predicative, whereas absolute space and time as the mind of God are clearly more related to the very substance and inner ontology of God. But precisely here is the heart of Newton's dualism as presented by Torrance. Newton's efforts to construct a rational understanding of the wholeness of phenomena in the universe led him, according to Torrance, to suppose the existence of two kinds of space and time, absolute and relative. In Torrance's opinion, the actual harm occurred when Newton identified absolute time and space with the mind of God, resulting in strict exclusion of God from the relative events in this phenomenal space and time. From the standpoint of Torrance's system, Newton's theological physics, in other words, is another example of separating God and the world from each other, so that any meaningful, active interaction between them becomes impossible. Both are closed to each other, even though God remains the Creator and the basis for the intelligible order of the universe, but only in an inertial, passive way.[211]

Torrance's way of detecting dualism in Newton's thought, based on the distinction between absolute and relative space and time, is certainly an interesting effort to trace a detail in the history of thought to back the overall argument against dualism and its modern origins in Newton's scientific and religious thinking. As has been noted, his discussion rests on the conviction that Newton really did identify absolute space and time with the mind of God. But precisely here three problems emerge that cast a shadow upon Torrance's interpretation and upon the whole of his argumentation in this respect. The first problem is a minor one but not quite irrelevant. Torrance maintains that Newton identified both absolute *time* and *space* with the mind of God. But this is not the case. Only space is involved, as can be clearly seen from the passage in Newton's *Opticks*, the work that gave rise, in the first place, to the whole discussion of the identification concerned. Newton writes: "And these things being rightly dispatch'd, does it not appear from Phaenomena that there is a Being incorporeal, living, intelligent, omnipresent, who in infinite Space, as it were in his Sensory, sees the things themselves intimately, and thoroughly perceives them, and comprehends them wholly by their immediate presence to himself."[212]

Only absolute space, then, forms the *sensorium* of God, a view that is not sufficient for Torrance's purposes, in which the Newtonian dichotomy between absolute and relative space and time as a totality serves his intentions to demonstrate the dualistic origins of modern science. For him, time also has to be involved in his interpretation of Newton's identification of absolute space with the mind of God.[213]

The second problem is no less interesting. Torrance's decision to translate Newton's *sensorium Dei*—absolute and mathematical space and time—by the English phrase "the mind of God" is not unambiguous, for the significance of Newton's *sensorium* in this context is really a matter of debate. What did Newton really mean by this word, which he used, as Torrance correctly points out, in an edition of the *Opticks*? Related to the controversy with Leibniz, it has been maintained that Newton did not intend this phrase to be taken literally but rather as an analogy, referring to God's ability to be intimately aware of the world and events in it.[214] In addition, it has been claimed that this *sensorium*, according to Newton, is not the mind of God itself but the place "where the brain meets the mind, where the physiological processes give rise to a representation of the world in consciousness."[215] Thus a direct identity between the *sensorium* and the mind of God, presupposed by Torrance in his translation, appears to be questionable.

The third problem apparent in Torrance's reading of Newton is, by and large, the weight he has given to Newton's notion of absolute space as the *sensorium* of God. It has been argued that Newton himself was not, after all, interested in the issue to the extent one might think on the basis of Torrance's exposition. Only one explicit mention of the *sensorium* of God in a wider context does not give sufficient reason for such a conclusion, to say nothing of the fact that Newton uses the phrase "as it were" in a diminutive sense before the word *Sensory*, as can be seen in the passage quoted above. Newton himself gives the impression that he is not, after all, too convinced of the connection he makes between the *sensorium* of God and absolute space, a subtlety ignored in Torrance's writings.[216] Torrance makes the direct identification on behalf of Newton, but at the same time he reads more into the thought of the physicist than is justified.

What has been demonstrated with regard to Torrance's reading of Newton strongly suggests that he needs exactly this kind of interpretation for his own purposes. It should be constantly recalled that what he most rejects in the Newtonian system is its alleged dualism, not primarily the distinction between absolute and relative space and time as such, but rather, according to the definition of Torrance's dualism presented earlier, the unbalanced and distorted relation between the two poles. In Torrance's thinking this distortion seems to derive its origin from the identification of the mind of God with absolute space and time, thus making a physical entity more divine than mundane. But why, in the first place, does Torrance criticize Newton's scientific thinking and find its dualism, still influential in modern thought, according to him, so untrustworthy?

In order to find an answer, it is now argued that the roots of Torrance's criticism as well as his problematic interpretation of Newton lie in the view of the Incarnation that the physicist had adopted. The fact is that Torrance cannot discuss very thoroughly Newton's understanding of Christ primarily because the sources Torrance mostly relies on contain very little or nothing relating to Christology. For that purpose, one should pay attention to those unpublished manuscripts that for newer Newton scholars provide the most valuable material for a study of Newton's theology, something that Torrance has not undertaken.[217] Therefore his interpretation starts from restricted source material. There, however, he has sufficient evidence to make the most important claim that if the universe is contained, so to speak, in God and if God is not affected by anything in the universe, this destroys every form of God's interaction with the world. Torrance concludes that this is the reason why Newton eventually rejected the Incarnation, opposing Athanasius and supporting Arius.[218]

Thus the Newtonian dualism between absolute and relative space and time has the effect of closing God and the world off from each other so totally that any "active interaction" between them becomes not just weaker but plainly impossible. Torrance understands that the conception of active interaction between God and the world, in opposition to Newton's thought, has been realized in the patristic, and especially during the Nicene, era in its incarnational theology.[219] This framework of understanding, already discussed in connection with the idea of contribution, becomes evident through Torrance's mode of involving the Incarnation, and it also discloses the reason why Torrance regards Newton's distinction between absolute and relative space and time, on the one hand, and his alleged identification of the mind of God with absolute space, on the other, as dualistic and therefore objectionable: Torrance underlines that, holding such ideas, Newton had to deny the possibility of the Incarnation. He claims that "if God Himself is the infinite Container of all things He can no more become incarnate than a box can become one of the several objects that it contains."[220] Torrance presents Newton as a man of strict logic who cannot establish distance from the receptacle notion of space to which his notion of absolute space, according to him, inevitably leads. A container cannot be identified with a part of its contents–God, constituting absolute space by his mind, cannot become incarnate. Torrance admits that Newton tried to link God and the world through the idea of common space and time but that this "led finally into their separation."[221] Torrance also points out that if the identification of God's mind and absolute space really is made, the distinction between Creator and the creation, space included, fades away, and creation is elevated to the level of the Creator, obliterating the need for God to come down among his creation.[222]

This argument could be convincing if, as has been noted, the identification of the mind of God and absolute space could be justified on the basis of Newton's own thought. However, this not being unambiguously the case, Torrance's effort to reconstruct the physicist's line of thought proves to be superficial and too extreme. To underscore Newton's distrust of the Incarnation, it would have been sufficient for Torrance to abstain from the argument so described and to refer solely to Newton's sympathies with Arius rather than with Athanasius, a fact that is more adequately documented in Newton's writings.[223] Indeed, Torrance does readily make additional use of this further point by stating that "Newton found himself in sharp conflict with Nicene theology and its famous *homoousion*, and even set himself to defend Arius against Athanasius."[224] In other words, Torrance draws a connection between Arius and Newton on the basis that both objected to the *homoousion*, a feature that from an orthodox point of view is sufficient to throw a shadow of suspicion upon Newton's thought. This is evidently what Torrance wishes to do in order to back his claim that dualism is a threat to realism, to the idea of compulsion as well as to that of the Incarnation of the *homoousion* type. Despite the undeniable achievements of Newton's scientific thought, he suggests, the inherent dualism could not but inflict considerable harm on later understanding of the relations between science and theology. This argument rests essentially upon the connection Torrance draws between Arius and Newton.

But the question to be asked at this stage concerns the nature and depth of that connection. It is a fact that Newton indeed sympathized with Arius rather than with Athanasius.[225] The originality in Torrance's interpretation is disclosed when the entirety of his argumentation between theology and the natural sciences is recalled: he makes the claim that Newton's view of the Incarnation, the impossibility of the *homoousion*, is closely tied to the unfortunate influence of dualism upon scientific thinking and the relations between theology and science. Related to this, the connection between Arius and Newton does not, according to Torrance, restrict itself solely to the rejection of *homoousion*, but both men are also seen to share a common view of *spatial concepts*, holding that space has to be understood primarily as a receptacle or container.[226] Torrance does not explicitly forge any closer links between the two men, as if seeing their relation as merely analogous. He suggests that there is at least a parallel in the thought of Arius and Newton, in the sense that both men rejected Athanasius's *homoousion* and supported the receptacle notion of space; for both of them the *homoousion* represented an unworthy way of thinking of God, as if God were divided into several individual persons or as if something created were included in his substance, and seeing space as a receptacle made the Incarnation logically impossible for both of them.[227] In short, Torrance holds that a dualistic understanding of space made both Arius and Newton disregard the Incarnation and the full Deity of Christ, presupposed by the *homoousion*, suggesting that a certain preconceived presumption concerning a physical issue, that is, space, had certain inevitable consequences for the religious understanding of the two men.[228]

The conception of space as a common feature for Arius and Newton is the last explicit link Torrance draws between them. But it can be argued that something more is still at stake, namely the intellectual origins of these spatial and dualistic ideas in their thinking. In an implicit way Torrance can be claimed to build a bridge between the two men, so that his reluctance to accept their anti-*homoousion* views becomes more clearly explained and understood. As far as Arius is concerned, Torrance does not strongly present either Hellenism or Judaism as his intellectual background. Sometimes Arianism is described as an heir of both Ebionite and Docetic Christologies together with Sabellianism, the former representing Judaistic tradition and the latter ones Greek thinking, so that it is quite difficult to determine which one had more influence on Arius and his theology, in Torrance's view.[229] But at the same time, Torrance links Arius and his theology so closely with Greek philosophy that in the final analysis he certainly sees in Arianism more Greek than Judaic influence.[230]

In Newton's case, too, Torrance does not establish direct links with Judaism or Greek philosophy, but, nevertheless, some connections can be seen in the way Torrance presents Newton's thinking. He states clearly that Newton's cosmology with its division between absolute and relative space and time had its origins in the Augustinian dualism.[231] Together with Aristotelian physics it has affected the Western world of thought so that the gulf between God and the universe has remained deep and unbridged.[232] Torrance suggests that Newton absorbed into his notion of absolute and relative space and time the Greek distinction between κόσμος νοητός, and κόσμος αἰσθητός, modified by St. Augustine.[233] Although Torrance links Newton and Greek philosophy through a mediating figure, not directly, the fact remains that he makes the division between the intelligible and the sensible the heart of both Arius's and Newton's systems. From this it can be concluded, with good reason, that according to Torrance both men share not only common features in their opinions on the Incarnation and the *homoousion* and spatial concepts but also a common intellectual background, at least as far as the deep-seated dualism of Greek thinking is concerned, a background that Torrance uses as a factor explaining the forms of dualism he claims to have detected in Arius's and Newton's thought.[234]

Torrance's criticism of Newton's thinking does not restrict itself merely to pointing out the alleged identities between him and Arius. He wants to argue as well that dualism has the effect of leading to *deism*.[235] What Torrance understands by this concept is not quite clear for two reasons, namely because he nowhere says what he actually means by the term, and because general definitions are not at all unambiguous.[236] Nevertheless, a good picture of what Torrance means by deism can be gained by an examination of his analysis of Newton's deism. When he points out Newton's dislike for Athanasius and sympathy for Arius, he claims that Newton placed too much emphasis on God's total transcendence, his inability to be moved or touched by created things, so that dynamic interaction between God and the creation simply vanished and gave rise to deism.[237] Thus, for Torrance deism means God's radical separation or detachment from the universe because of his immutability and impassibility. This view, at least in Newton's case in Torrance's interpretation, is equivalent to the feature previously noted when the division between absolute and relative space and time was discussed: the absolute cannot be touched by the relative. God remains in his obsolete supremacy and therefore his relation to the world is strictly transcendental. He is not interested in mundane individual histories or events or people as personal beings. He is the Supreme and the Absolute One, but hardly capable of sympathetic and genuinely personal relations with his creation.

Torrance quite obviously regards Newton's theology as deistic, leading directly to the dismissal of the doctrine of the Incarnation as an inconceivable system of thought which does not fit the scientific outlook developed and represented by Newton.[238] To view Newton's thinking as deistic is, however, not at all unambiguous. The principal criterion for Torrance to call a religious phenomenon deistic seems to be that the reality of the Incarnation and the *homoousion* is denied. The issue is not only God's revelation in general but specifically God's becoming a human being in the sense formulated by Nicene theology. As a consequence, Torrance assumes as well that any deviation from Athanasius's stand on the Christological question and dualism, or acceptance of a deistic view, are essentially related to each other, the latter affecting the former. He makes, then, the implicit claim that any denial of the possibility of the Incarnation is a mark of deism, a consequence of a deistic understanding of the relation between God and the universe.

Another problem emerges when it is asked if Newton really was a convinced deist in the sense Torrance would like to demonstrate. As stated, Newton can hardly be considered such, if one understands by that term, as is usually done, those who believe in God on the basis of *natural* reason, not on the basis of any divine revelation, for he had a firm belief in the God revealed in the Bible.[239] From this point of view Torrance's judgment of Newton's thought oversimplifies and overinterprets it, because it relies on the idea of the *homoousion* as its main criterion for evaluating the nature of Newton's religious thinking. But inevitably Torrance feels compelled to make these assertions concerning Newton's thought in order to fit it into the context of his argumentation whereby dualism and deism are seen to affect the denial of the Incarnation. This becomes apparent as soon as his evaluation of the relations between Newton's deism and his scientific thinking is considered. Torrance maintains that God's freedom is detached from the freedom of the universe, leaving the world imprisoned within its own necessities. God is seen to have the role of an impassible system which provides the ground for the universe as a closed playground of cause and effect. Torrance draws the conclusion that "deism and determinism go together."[240]

Torrance sets Newton's understanding of God as an inertial absolute, an interpretation that was detected as problematic, as opposing the cosmology entertained by the Early Church Fathers: denying the possibility of the *homoousion* reveals for Torrance that Newton's universe, after all, is not free in the patristic sense, that is, that it has an intelligibility and orderliness of its own with only the limitation of God's freedom.[241] Rejecting the Incarnation in the Athanasian sense means inevitably, according to Torrance, abandoning the contingent freedom of the universe, resulting in its opposite, necessary freedom where God and the universe are *necessarily* related to each other. Whatever takes place in the universe does so out of the necessity God has implanted within it, thus making it a mechanistic and closed system of cause and effect.

The second important point in Torrance's reading is his view that this mechanistic universe, full of necessities, causes and effects, is grounded, not in itself, for it is not self-explanatory even for Newton, as Torrance admits, but in the inertial and impassible system named "God." Torrance draws an interesting picture of Newton's God, in which God is far from personal but rather only a "system" elevated to the status of a divine being. At this point Torrance has in mind what was discussed earlier about his view of absolute space and time as the *sensorium* of God. As was shown, his interpretation of Newton's original thought is problematic, but taken as a premise, it supports the view Torrance wishes to make about Newton's understanding, namely that this idea of absolute space and time as the mind of God correlates with Newton's impersonal view of God.

Torrance does not hesitate to indicate a certain further parallelism between Newton's thought and the form of dualism that the Nicene Fathers had to struggle against. The idea of a closed and deterministic universe was behind the thoughts of those theologians, Arians included, who met their challenge in the view of a contingent, open, and basically open universe entertained by early Church theology. It should be also remembered that this common feature is not only a parallelism but also has an identical basic nature derived from Greek philosophy, as described by Torrance.[242] Belief in impersonal fate derives from belief in an impersonal, deistic deity, and this is what Torrance strongly attributes to Newton. But here again Torrance's purposive way of argumentation and its overinterpretative style is apparent because even to suggest that Newton's God was an impersonal divine power is not justified, as can be easily seen in his reluctance to accept fatalism and pantheism.[243] From the fact that Newton strongly emphasized God's lordship and dominion over all creation and humans' reverent submission to his divine rule, one can conclude that a God who is adored and revered, who has servants and a dominion and who cannot be identified with fate, is surely not impersonal.[244]

Newton as the eminent exponent of modern physical science and scientific thinking in general is the target of Torrance's criticism for one basic reason: in his opinion Newton's dualistic thought leads inevitably to the exclusion of the possibility of the Incarnation and the idea of the consubstantiality of the Son with the Father. The consequence is that God and the world are ultimately separated, so that any dynamic interaction of the *homoousion* type cannot find its place in this system.[245] From the scientific point of view this leads to the rejection of a realist attitude, in which reality is allowed to reveal itself rather than be manipulated by human preconceptions, for example, by the ideas of strict causality and necessity. The universe remains a closed mechanistic system without any connection with the ontological depths of its reality, so that the idea of the *homoousion* of a phenomenon with reality in its ontological profundity is lacking. Therefore, in the final analysis, the compelling power of reality is not given a chance to have any effect upon our conceptions of the universe. This argument of Torrance's was shown to be based primarily on one premise concerning the ultimate role of absolute space and time in Newton's thought. The alleged identification of absolute space with the mind of God was found to be crucial for Torrance's argument. But, at the same, time, its defects were demonstrated to lie precisely in an inadequate translation of Newton's *sensorium* of God, resulting in overinterpretation. Additionally, the complex issue of Newton's alleged deism was seen to render Torrance's argument rather unconvincing.

This discussion, however, has revealed the many-sided intermingling of scientific and theological thought in Newton, in whom it is remarkably difficult to define the demarcation line between the scientist and the theologian. Torrance is well aware of this complex background to Newton's thinking, and, in spite of the previous criticism, there is a more interesting side to Torrance's reading of Newton, a side that bears direct relevance to the relations between science and theology. This aspect can be regarded as Torrance's thesis concerning the causes of the exacerbated relations between science and theology after the period of Newton himself. Torrance points out Newton's view that causal and mechanical explanations are valid when the structures of the universe are investigated but that they lose their applicability when the universe as a whole, or its existence or its initial conditions, is reflected upon. Torrance maintains that it is at this very point that the role of God becomes important. He expresses, rather strikingly and somehow in a contradictory manner if compared to what was earlier said about Newton's deism and determinism, that the relationship between the universe and God implies that, in the ultimate analysis, the universe is an open system.[246]

The noteworthy detail in this reading is Torrance's approval that Newton upheld the idea of the openness of the universe, which otherwise is described by him as deterministic and thus a closed and self-explanatory system. But now Torrance sees in his thinking the theme of contingency, the openness of the universe toward its sole basis and source of reason, namely God. In other words, after all, Newton is not such a thorough representative and proponent of determinism and mechanism as Torrance himself in other connections asserts. In the last resort there is an openness to be found in this "first role of God" and, undeniably, this makes Torrance's position quite ambiguous.

Torrance also tells us that Newton needed God not only as the guarantor of the contingency of the universe and as its rational source but also as an explanation of such irregular phenomena as could not be explained through the laws of motion alone ("the second role of God"). In this way Newton could use God as an explanatory factor for otherwise obscure phenomena or as a guaranteeing factor for harmony and regularity in the universe.[247] Torrance implies here that Newton introduced the idea of the *God of the gaps* into the framework of science, thus assigning to God only an explanatory role.[248]

Torrance continues by stating that this two-fold role of God later had fateful consequences as Newtonian science developed. An increasing amount of phenomena could be explained without any reference to God because the understanding of the processes in nature became deeper.[249] This meant that finally all phenomena were believed to be explainable without taking God into account so that God's second role was abandoned as unnecessary. Torrance thinks that Newton's followers were correct in rejecting it, but what is lamentable in Torrance's view was that together with God's second role his first role, namely God as the referential being for the contingent and rational universe, was also discarded. Thus, the idea of the contingency of the universe was forgotten, and deterministic and mechanistic trends dominated.

This is certainly an interesting and challenging reading of the development of relations between science and theology, and it has also a far more consistent aspect of credibility because this argument is not thoroughly permeated by Torrance's demands for a central interpretative role for the *homoousion* and for incarnational theology in general. However, this is not the main line along which he conducts his thinking but rather an inspiring sideline. The problem Torrance detects in Newton's ideas is, how do these two roles of God relate to each other in the physicist's scientific thought? Is the universe really contingent, dependent on God, and not a closed and deterministic system? Or is it ruled by mechanical and deterministic laws, leaving to God only the task of being an explanatory factor for irregularities and inexplicable phenomena? This problem undoubtedly troubles Torrance a fair deal, precisely because his earlier argument is based on the presumption that Newton, and not only his later "disciples," had deterministic views and a very restricted understanding of the contingency of the universe. Torrance tries to find a satisfying explanation to this ambiguity by stating that Newton was not discerning enough to avoid linking wrongly together God's roles as the Creator and as the Sustainer, or as the Source of lawlike regularities and as the One who has made room for irregularities and nonpredictable phenomena.[250]

We may now ask why this ambiguity causes such a problem for Torrance that he searches for an appropriate explanation for it by suggesting a confusion or a mistaken way of running together two different themes. Could not this ambiguity be explained away by simply referring to the idea that for Newton God is, after all and in every case, the sole explanatory principle for the created universe, whether as its source of rationality, the point of reference for its contingency, or as a universal explanation for those difficulties in the course of space-time conditioned life in the universe that cannot be understood otherwise?

Torrance, however, cannot pass over this question so lightly, because he has found in Newton's thought an important feature with which he can gladly agree: the idea of contingency. Indeed, for Newton contingency undoubtedly meant the relation of relative space and time to absolute, mathematical space and time, the latter being the reference point for the former, the ground and source of rationality for it. In other words, from the very beginning Newton's contingency is God-oriented. The problem that bothers Torrance is certainly the fact that for Newton this God is immutable and impassible, incapable of being affected by the relative space and time he contains. So Newton's contingency, after all, is slightly different from that of the Nicene Fathers as interpreted by Torrance.[251]

Torrance's actual point lies in his claim that whenever God is seen simultaneously as both the reference point of the contingent universe and an explanatory principle for irregularities in its mechanistic and necessary structure, there is always the danger that these features are totally identified with each other. This leads one to the conclusion that Torrance is not entertaining an idea of God that could be used as an explanatory factor when reality is described. Contingency as dependent independence was mentioned earlier as giving the natural sciences their own independent field of inquiry, where the existence of God as the reference point of contingency need not to be taken into account in their own premises.[252] Torrance's analysis of Newton reveals that for him the universe in scientific treatises does not need to be explained theologically, because it tends to lead to the concept of the "God of the gaps." The ambiguity in Torrance's thought can be expressed in the question, is not God as the reference point of the contingent universe, in every case, an explanatory factor?

This problem in Torrance's thought seems, after all, to make him a thinker who tries to guarantee a place for God on the edges of the universe.[253] But the case is not so simple, due to the fact that contingency as dependent independence, as much as it refers to God upon whom it depends even in its independence, is for Torrance not primarily an explanatory concept in our mind but an inherent feature in the universe itself. The attractive analysis of Newton's two roles of God, interesting and challenging as it is in itself as a possible description of the relations between science and theology, leads finally to the doctrine of the Incarnation, Torrance's all-pervading interpretative key, and precisely here it loses its freshness. But in spite of this, the line should be followed to the end. Torrance's argument rests on the conviction that the Early Church understood the contingency of reality as having been revealed in the act of the Incarnation. The Word, through whom the Father created the universe, became flesh under the contingent conditions of space and time. Torrance suggests that if now contingency is abandoned in scientific enterprises, this results in a deterministic and mechanistic view of the universe. This process has the effect of excluding a genuine interaction between God and the world from the facts of reality, and by this interaction is meant precisely the event of the Incarnation, where the contingency is revealed. Thus Torrance constructs, in Newton's case, a "circle of denial," where rejecting contingency means rejecting the possibility of the Incarnation, which in turn means rejecting the contingency to be revealed. Abandoning the idea of contingency means that any ontological interaction of the *homoousion* type between God and the world, in which a phenomenon is viewed as identical with reality, is ruled out. In Christology, if Jesus of Nazareth is not the Son of God, and if the Son is not of the same being with the Father, this God-world relation is not trustworthy any more, because Christ is reduced to a mere historical phenomenon. In epistemology, if a phenomenon does not have any direct connection to reality or if it is seen only as a superficial and accidental expression of a subjective state and not as a trustworthy indicator of objective reality, our efforts to attain knowledge become meaningless. This argument implies further that there is no possibility for the idea of compulsion to determine what is the ontological bearing of reality in its inherent force upon human knowing. The preceding discussion confirms the hypothesis that the *homoousion* is the all-pervading interpretative instrument by which Torrance reads Newton and judges his relevance for theology as well as for the sciences and their mutual relations.

Despite its apparent simplicity and straightforwardness, this approach can hardly do full justice either to Newton's thought or to Torrance's own intentions to reinforce the notion of contingency as a basic concept for both theology and the natural sciences. Torrance's idea of dualism as the opposite of the profound unitary realism in the scientist's thought, affecting the later development of modern science along mechanistic and deterministic lines, proves to be based on insightful ideas rather than on convincing evidence. On these terms, however, Torrance's argumentation might turn out to offer a fertile resource, asking important questions and suggesting some possible referents for solutions in the relationship between science and theology as well as in their history of reciprocal influence. The grand theme of realism discussed in this chapter cannot easily lose its appealing character, and it is precisely in this search for a basis for human knowledge that Torrance's efforts present a continuing challenge, the relevance of which, despite their defects, should not be underestimated. The significance of his insights becomes more apparent when the role assigned by him to modern physics and its consequences for theology are considered.

5

The Impact of the Natural Sciences on Theology

The Idea of Reminder

> There cannot be any question about the fact that the whole of future life and thought will be dominated more and more by pure and applied science. It is within this perspective that theology must learn to think with its own kind of precision and its own disciplined controls.
>
> Thomas Torrance, *Theology in Reconstruction*

In the previous chapter Torrance's view of realism was examined and the idea that compulsion as a mode of election is its constitutive feature was analyzed. He was seen to emphasize the conviction that genuine science can be applied only as the compulsive force of reality is acknowledged as the supreme judge of all human knowing, a force to which every scientist should submit. Torrance's view of the development of Western science and theology, however, maintains the deleterious effect of what he calls dualism, an antirealist trend which prevents the idea of compulsion from functioning in scientific thinking and which thereby leaves us to our own devices and abilities to construct a picture of reality so that the universe itself can no longer exercise its own force.[1] From the standpoint of science, realism was no longer effective, we as humans beings left to construct a view of the world based upon our own speculations and thoughts and subjective states and preferences.[2] From the standpoint of theology, this had the effect of neglecting the reality and objectivity of the Incarnation and the true consubstantiality of the Father and the Son.[3] The subjectivity of religion became the main research area of theology, and the *homoousion* relation between a phenomenon and reality both in science and theology was denied.[4] The emergence of the new physics, with the theories of relativity and quantum mechanics, transformed the situation radically. Torrance's third main argument with reference to the relationship between science and theology is that now, in its turn, it is theology that should learn from the advance of the natural sciences. Particularly helpful in this respect is modern physics, which has had to assume a novel approach to reality, an approach that theology, too, should take with the utmost seriousness if it entertains any intention of being a scientific discipline.[5] This is called Torrance's *idea of reminder*. It should be carefully noted that Torrance does not appreciate modern physics because of its ability to teach theology something essentially new but only because it can remind theology of those basic elements of science which have been neglected for a long time because of all-pervading dualism. In other words, modern science can only show theology the way back to its proper foundation in the Incarnation and the *homoousion*.[6]

In the following pages the actual contents of Torrance's idea of reminder is analyzed with special attention to the question of what features exactly, in Torrance's opinion, make modern physics an example for other disciplines, theology included. After that analysis, a critical glance will be given to the implications of modern physics that Torrance would like to draw for theology.

The Contents

Ontology

Torrance claims that at the moment Western civilization is living in the middle of a deep paradigmatic change in its whole scientific and cultural outlook when dualistic modes of thought are being rejected. "Open-structured" forms of perceiving reality are emerging and a profound transformation of basic ideas is taking place.[7] In the third chapter it was shown how Torrance identifies precisely the doctrine of the Incarnation as the main contributive factor to the introduction of realistic devices and concepts for the use of scientific thinking in general. The centrality of the concept of *homoousion* was underscored. Torrance suggests that now, due to the rise of modern physics, ontological questions can again be appreciated at their full weight so that science and theology share a common field of interest in their search for the ultimate nature of the universe.[8] The problems of space, the question of relationality, and the case of contingency are crucial in this respect, and in the following pages they are discussed with reference to two scientists, Maxwell and Einstein, whom Torrance views as heralds of a new scientific era.

According to Torrance, the first major post-Newtonian breakthrough of a new conception of the relevance of ontology took place in the theory of the electromagnetic field, devised by James Clerk Maxwell and published in 1865.[9] In Torrance's view, Maxwell's work is a prelude to the third stage in his periodization, namely the *Einsteinian revolution*, as he himself calls it, and in this claim his position can be seen to be justified: Maxwell's work meant a new shift in physics even though its full implications were revealed after his own time.[10] What was new in Maxwell's science, according to Torrance, was his application of the concept of field, originally adopted by Faraday and strictly contradicting the static concepts of Newtonian mechanics with its understanding of space as divided into two spheres, absolute and relative ones. In Torrance's interpretation this theory of field meant that the interrelations between bodies are so continuous and so intense that they form a special "field" of forces affecting the bodies all the time. Therefore bodies that are under human investigation cannot be seen as such, independent of space and time, which must be understood as full of events through these fields of forces. Torrance draws this together, explaining that "time as well as space must belong to the vast field of force which is the underlying reality of all phenomena in the universe."[11]

Torrance's view of the concept of field contains three points that illuminate his arguments for a healthy restoration of ontology in the core of science. The first ontologically relevant aspect in Torrance's interpretation is his conception that Maxwell's theory is closely related to the concept of space, making his understanding of spatial concepts one major element in his contribution. Torrance wishes to underline the distinction between Newtonian and Maxwellian views by pointing out that, in fact, all space *is* a field of force, and this gives him reason to argue explicitly that Clerk Maxwell discarded any absolute notions of space and time both in the Newtonian and Kantian sense.[12] With regard to Newton's understanding of space as a static entity, Maxwell's theory certainly is a novel enterprise, and it may be admitted that Maxwell did not hold on to the absoluteness of space and time in the Newtonian sense, involving immutability and obsolete transcendence of space, not to mention identifying them with the "*sensorium* of God." But in one detail of Maxwell's understanding of the ontology of space, Torrance's point is more ambiguous.

The problem comes to the fore in the fact that Torrance does not pay proper attention to the archaic concept of ether still to be found in Maxwell's thinking. Maxwell had to interpret the problem of the effect of bodies on each other from a distance with the conception of an ethereal medium filling space.[13] With the idea of ether, a kind of receptacle notion of space must be presupposed—space *contains* ether. The reason for Torrance's neglect of this line of Maxwell's thought can be traced with good reason to his intentions to present Maxwell's idea of space in a certain, carefully reflected light. As was shown above, he explicitly stresses that there are no hints of Newtonian notions of receptacle space in Maxwell's literary production.[14] In this respect there remains a further point to be made.

When defining the electromagnetic field Maxwell maintains that it is "that *part of space which contains and surrounds bodies* in electric or magnetic conditions." A little later he continues: "We have therefore some reasons to believe, from the phenomena of light and heat, that there is an aethereal medium *filling space* and permeating bodies."[15] These two fragmentary examples are not sufficient to give reason to conclude that Torrance is totally wrong in his interpretation of Maxwell's understanding of space. The scientist's care in formulating his idea of ether as filling space should be well noted.[16] These passages offer, nevertheless, first-hand evidence for the claim that Torrance wants to see Maxwell more independent of what is understood as Newtonian influence than he actually was. For Maxwell, then, space is something that contains bodies, something that can be *filled* by ether. At least in this respect Maxwell's space is a receptacle or a container.[17]

The second point in Maxwell's and Einstein's scientific thinking, and especially in the field theory, which Torrance has found fertile for his efforts to restore ontology, is the idea of *relationality*. The problem of the ultimate nature of space described above indicated Torrance's eagerness to contend that Maxwell's space was not absolute like a container but rather relational. While the relational notion of space has been discussed earlier under the heading "Space and Time," there is no need to repeat the insights presented there. The relevant issue at this stage is the bearing of relationality upon ontology along the lines drawn by Torrance from Maxwell's thought. From this perspective it is important to note that the scientist represented such a field theory that was thoroughly relational and not mechanical. Torrance criticizes the latter by blaming it for superficiality and for inability to take seriously what Torrance calls "real connections" in nature.[18] Torrance thus implies the difference between relational and mechanical interpretations of events in the universe by suggesting that relational thinking takes *real* connections between entities seriously and that it is not just a working model but something essentially deeper. A more detailed discussion of Torrance's conception of relationality is necessary, and it will be provided later after an evaluation of Torrance's reading of Einstein.

As has been noted earlier, Torrance finds special pleasure in holding to Albert Einstein's view of space which resulted from his general theory of relativity. There is one point in Einstein's theory that makes Torrance apply it readily in his efforts to reconstruct a sound ontology. Torrance does maintain that Einstein succeeded in invalidating the notion of absolute space as entertained in Newtonian science. But the absoluteness thus removed is not seen primarily as nonrelationality but rather as adherence to a very static understanding of natural phenomena, replacing the idea of absolute space as an inertial frame of reference.[19] In this respect Torrance's reading of Einstein seems to be adequate, for the physicist held that, on the one hand, the structure of space as revealed by the general theory of relativity remains indefinable but that, on the other hand, the nonabsoluteness of space depends, strictly speaking, not on its relational character but on *physical factors* that affect the very nature of space or the field.[20] Thereby Torrance can use Einstein's view of the field as evidence to back his claim that the ultimate character of space must be considered dynamic rather than static.[21]

The above discussion has indicated that understanding space as a field has led Torrance to maintain that the character of space is both relational in the sense that relations are constitutive for its structure, and dynamical, full of continuous events and acts. Space cannot be detached into isolation from realities as if they were mere accidental features of it. But perhaps the most important implication that he wishes to draw from modern physics and its concept of field is considering it a case for contingency.[22] Torrance maintains that Maxwell's view of the universe involved the idea of an inherent intelligibility of its own, which a human inquirer cannot create or project onto reality. This is a manifestation of the contingency of the universe so that it exists under no necessary condition but freely and dynamically.[23] Torrance carefully constructs his argumentation, the aim of which, from the standpoint of the idea of reminder, is clearly seen at this stage: Newtonian science operated with necessities and hard causalism or determinism, whereas Maxwell could restore contingency in the ontology of the universe through his field theory. The revolutionary character of Maxwell's science lies, then, in its ability to discard determinism and initiate a new era in physics, resulting in the indeterminism of the quantum world.[24]

The actual defect in Torrance's argumentation can be found in his conviction that the novelty and revolutionary nature of Maxwell's theory of the electromagnetic field lies precisely in its indeterminism. In order for this claim to be correct, however, strict causality should not be applicable to the concept of field. But in fact it can. This can be argued on two grounds. First, from a mathematical standpoint, the concept of field and its dynamic development employs partial differential equations in which Newton's and Descartes's principles of causality are brought together.[25] This means that the notion of determinism cannot be separated from the concept of field as strictly as Torrance would like it to be. After the initial conditions have been met, we can solve for each of the relevant variables on the basis of partial differential equations in a thoroughly deterministic way. Second, from a philosophical standpoint, the concept of field cannot be regarded as free from determinism on the basis that the causality entertained in it neglects the idea of probability, an essential ingredient of modern quantum mechanics and its indeterminism.[26]

From the first or mathematical point of view, Torrance's argument does not sound too convincing. Although he admits that differential equations of classical Newtonian mechanics have the tendency to confine nature in a formalized, static, and deterministic straitjacket, he does not give a similar account of *partial* differential equations used in the field theory.[27] Here his point is most vulnerable, but in the second case, as it comes to the more philosophical type of criticism, Torrance can defend himself at least in one respect, namely in that his own definition of indeterminism does not need to operate with the idea of probability. In other words, Torrance does not regard the merits of Maxwell's theory of the field and its alleged indetermism from the standpoint of probability but rather from that of relationality. Torrance states: "classical causal relations are replaced by a dynamic inherent relatedness in the universe, in which space and time are included within the internal connections of all empirical realities and processes and are inseparable from them as space-time."[28]

The absence of the idea of probability in this definition of indeterminism should be carefully noted. What Torrance offers in its stead turns out to be most illuminating with regard to his view of the ultimately contingent nature of ontology. Relationships return to the stage here, so that what he calls *inherent relatedness* in the universe forms the constitution of contingency and indeterminism. Torrance prefers to use the term *onto-relation* in this connection, a phrase with which he denotes "the kind of relation subsisting between things which is an essential constituent of their being, and without which they would not be what they are. It is a being-constituting relation."[29] The ultimate nature of ontology consists, then, not primarily of substances but rather of relations without which no being could be what it is. Reality can be viewed as a field based on a network of ontologically constitutive relations between entities. Thus, ontology is basically relational. But to understand this and his somewhat loosely applied concept of *inherent relations*, a question needs to be asked concerning the ultimate nature of these relations in his thinking. Why does he call them *inherent*? Nowhere does he provide an explicit answer, but the context of his usage of the term can provide an implicit one.

In the passage quoted above, Torrance contrasts *classical causal relations*, on the one hand, with *dynamic inherent relatedness*, on the other. This seems to imply that causal relations, as applied in classical physics, are in a sense *external*, precisely because of their causal character. In other words, relations between things can be perceived only as they are connected mechanically or causally, thus rendering the relations *external* with respect to the thing itself. Causal relationality, then, does not even tangentially touch the *being* of a thing. It is part of Torrance's aim to prove that what he calls *dynamical inherent relatedness* meets this requirement: they are not external in the sense of causality but internal in the sense of noncausality or indeterminism or, better expressed, of contingency. The notion of inherence gives the impression that relatedness cannot be considered something external which could then be attached to a thing as an accidental feature. Rather, relatedness belongs to its inherent or *internal* nature, so that being itself is regarded as essentially relational.[30]

This confirms, from another angle, the assertion that Torrance sees Einstein's theories of relativity as a reminder to theology, as well as other disciplines, of the indeterministic and relational nature of ontology in the manner that relations constitute what being is. One could also argue that Torrance is implicitly making the assertion that the theory of relativity might possibly be renamed that of relationality. But whether Einstein's theories of relativity form a sound basis on which to build an argument in favor of indeterminism, or in other words, whether it is possible to view Einstein's approach as really indeterministic, is the question that remains to be answered.

Einstein's relation to the so-called Copenhagen School of quantum mechanics was discussed earlier within the context of the problem of realism. Their differing approaches were noted, as well as Torrance's conscious adherence to the opposition to the Copenhagen interpretation as represented by Einstein. As to the ontological nature of the universe and its alleged indeterminism, the case is very similar. Now the question is whether Einstein was a determinist in his interpretation of quantum physics or not. In opposition to the majority of scholars, Torrance takes the definite view that Einstein was not a determinist.[31] The debate between Einstein and other eminent physicists, such as the Dane Niels Bohr, was due to the varying interpretations of the nature of the behavior of quantum particles. Bohr, Werner Heisenberg, and other members of the Copenhagen School adopted the view that chance is an essential feature in the particle world. Einstein, however, was unable to accept this view because he held that ultimately chance cannot be the way nature behaves, and therefore he wanted to conceive of it, also on the level of quantum physics, as rational and ordered, claiming that any irregularities or disorders to be found must be due to the methods and inadequate instruments used in experimental situations. Reality itself has to be what it is, independent of human investigation, and therefore Einstein believed that a more comprehensive theory of objective reality can be achieved.[32]

Torrance's reading of Einstein includes a strong emphasis on his claims for the indeterminacy and independence of the events in the quantum world. Torrance's own contribution in this connection is his way of identifying Einstein's reliance on the nonarbitrariness of the universe with the concept of contingency. He wants to construct, as it were, a third option between pure chance and causal determinism by claiming that there are two aspects to the concept of contingency, one in which it is dependent upon God as the transcendent Ground of all being and one in which it posits a freedom of its own, restricted only by the freedom of God.[33] This is both a challenging and in some respects hard-to-apply view because of its inevitable metaphysical implications. This approach certainly guarantees that the natural sciences pursued in the contingent, free world remain independent of God as the transcendent Source of the universe, but this freedom is far from absolute because it places limits on what Torrance calls "the freedom and rationality of God." His insistence on the referential character of the rationality and freedom of the universe also strongly remind us that there has to be a target or aim of this reference, namely God. This kind of theological natural science, if such a phrase may be allowed, can hardly be welcomed in scientific communities, and this is the main defect in Torrance's insight concerning the concept of contingency and its applicability. Nevertheless, it is this very kind of contingency as the third option between arbitrariness and necessity that Torrance claims to have found in Einstein's thought.

An important parallel with the theology of the Early Church is to be found in Torrance's argumentation. It should be recalled how he accredited the Greek Fathers with the introduction of a novel understanding of contingency into the corpus of theology and into subsequent Western thinking alike, as has been discussed earlier in our analysis of Torrance's idea of contribution. It is not difficult to see his actual point in the emphasis that modern science, as it was inaugurated by Maxwell and Einstein, rerevealed contingency after a long mechanistic and deterministic period of neglect and even contempt. Thus nothing essentially new was invented, but contingency was only dug up and purified from dualistic and deterministic accretions. To be convincing

Torrance's argument needs to be able to prove that Einstein really held the notion of contingency in the sense described by Torrance, but it is precisely here that difficulties emerge.

In this regard, it should be noted that in Einstein's works there is no evidence that he used this actual concept to describe his idea of the ultimate nature of reality. If contingency is really held to be identical with nonnecessity and indeterminism but not with arbitrariness, then it can, at least to some extent, be justifiably used to express Einstein's ideas. On the one hand, Einstein was quite unwilling totally to reject causality, which he saw as a necessary means to describe regularity in nature.[34] But, on the other hand, Einstein, with a certain degree of caution, took his distance from determinism as an explanatory model, at least as far as quantum mechanics or microcosmic scales are concerned. Instead, he wished to transcend the whole problematic issue of determinism versus indeterminism and suggested that causality holds true only within theoretical explanations of nature but not in reality itself.[35]

The question of Einstein's alleged determinism seems to remain unresolved, resulting, in part, from the fact that the meaning of the term *determinism* in this connection is far from clear. Can it be identified with causality? Are there different forms and degrees of determinism, and to what extent is it only a feature of theories invented by human beings or an essential character of nature itself?[36] These problems have pressed themselves upon Torrance's understanding, too, and as has been emphasized, his position is basically Einsteinian. His stance is not, however, understandable on that basis alone and therefore there are reasons to put forward the claim that Torrance interprets Einstein's statements concerning ontology in a way that is more congruous with his overall theological system than with the scientist's own intentions. It can also be concluded that defending Einstein as the pioneer of a new scientific era and as a representative of indeterminism and ontological contingency is an important detail in Torrance's thinking for another reason. It is significant that otherwise Torrance seems to be rather uninterested in the advances made in quantum physics, which for several other scholars offer a basis for discussing the relations between the natural sciences and theology.[37] For example, he does not see any benefit in developing his thought on the basis of probability, a concept that plays a central role in quantum-theory accounts of the behavior of particles and that, from another angle, would shed light on the idea of indeterminism and unpredictability.[38] Torrance's reluctance toward this kind of application of quantum-theory models can be explained as resulting from his firm belief in a single unifying order which would not allow one to think of it in probabilistic terms or as connected with chance or arbitrariness. And additionally, it would diminish the explanatory power of a theory with probability as its axiom.[39] In this basic motivation Torrance can be seen to maintain a truly Einsteinian attitude.

In spite of the ambiguous and not always convincing way in which Torrance uses both Maxwell and Einstein to support his own position, it should not be overlooked how he wishes to underscore the positive significance of modern physics from the standpoint of ontology. Torrance's reading of the works of the two scientists must be understood not primarily as a faithful exposition of their thinking but rather as a selective effort to find elements in modern science which could be used in an argumentative manner as a manifesto, the aim of which is to lead theology back to its alleged scientific basis. One cannot avoid the impression that his discussion would gain more credibility if Torrance did not lean so much toward the two scientists but presented the insights more as his own modifications of some important issues in modern physics. Why, then, has Torrance not chosen this approach but has sought support from the two physicists?

Torrance's appreciation of Maxwell and Einstein can be explained by reminding oneself that the theories of the electromagnetic field and relativity are mainly concerned with the problems of space and time, and that for Torrance the question of space is of utmost importance. The reason for this is not difficult to apprehend. The doctrine of the Incarnation unites the ontological aspects of both the ideas of contribution and reminder. In other words, Maxwell's and Einstein's views of ontology and their alleged indeterminism offer Torrance not only scientific tools to try to convince his readers of the contingency of reality, which, according to Torrance, was the direct impact of the doctrine of the Incarnation with its concept of *homoousion*. This ontology brings forth space and time as contingent features of the universe, thus highlighting Torrance's conviction that they have no absolute status in the Newtonian sense but are themselves a creation of God.[40] To express this in a more condensed form, the question of *homoousion* is best understood in the general framework of space, that is, how God in his Being can enter his own limited creation and its space-time conditions, while problems in quantum physics do not offer such a direct overlap for science and this specific theological doctrine.[41] This implies as well that Torrance has assigned to both Maxwell and Einstein the role of scientists who have struggled to find justification for taking the Incarnation seriously. There is hardly any other theologically relevant reason for his continuous reference to these men.

One significant point to note is that the idea of reminder in modern physical science confirms something that was originated a long time ago in the era of the Early Church. Repetition of the insight presented at the beginning of this chapter is not trivial but highlights an important cornerstone of Torrance's entire argument. Namely, modern science has not created anything theologically significant but it has only cleansed science from false notions and reintroduced more adequate ones, which had originally arisen within the Christian Church. As has been shown, the main points in Torrance's idea of reminder concentrate on the notions of relationality and contingency or indeterminism as expressed in the view of space, but before we touch upon their implications for theology, the epistemological aspect of the idea of reminder deserves an account of its own.

Epistemology

In Torrance's opinion, the restoration of ontology is not the only contribution that modern science and physics in particular have made to our understanding of the world, but epistemology, too, is involved.[42] This aspect of the "Einsteinian cosmology" becomes clearly discernible in the following quotation: "That is to say, here we have a radical reorientation in knowledge in which structure and matter, form and being are inseparably fused together, spelling the end of the analytical era in science. This involves the restoration of a genuine ontology, the replacing of the mechanistic universe with a dynamic universe conceived in onto-relational terms, and the replacing of the old forms of causal connection and natural law with field-structure and field-laws."[43]

The primacy of ontology is clearly underscored, and Torrance relates onto-relationality, a view of the dynamic universe as well as appropriate epistemology with the idea of the field. In order to comprehend Torrance's point as to the revolutionary nature of the new epistemology as revealed by modern physics, it is now necessary to find out the background against which Torrance wishes to construct his argument. The above passage refers to Newton and his scientific heritage with its inherent determinism and mechanism, but it is important to note that the picture would not be complete without another important factor which, in Torrance's opinion, contributed substantially to the rise and success of dualistic modes of thought. The philosophy of Immanuel Kant becomes the main target of Torrance's criticism, where he makes the claim that Kant, attempting to unite the empirical and theoretical aspects of scientific inquiry, constructed, in fact, a damaging combination of necessity and determinism into the foundation of knowledge.[44] It is suggested that what Newton achieved in the natural sciences was likewise achieved by Kant in philosophy and especially in epistemology. It is worth noting that Torrance consciously draws a continuous line from Newton to Kant, as if the work of the philosopher was a response to that of the scientist.[45] Torrance's critical position can be claimed to be based on two premises, both being indications of alleged dualism in the philosopher's thought.

The first criticism is aimed at the distorted relationship between epistemology and ontology in Kant's philosophy. Torrance interprets him by stating that, in his thought, "'given' and 'objective' do not refer to any character of things in themselves but to the synthetic and necessary form which appearances take in our cognition of them."[46] This means that we cannot gain knowledge of things in themselves but only of their external appearances, leaving objectivity and intelligibility to the subjective side of an epistemological process.[47] Torrance's own view of appropriate epistemology could not make such a claim concerning the unattainability of "the thing in itself" and its firm anchoring in the subjective mind of human beings. His reluctance to accept Kant's scheme can be understood primarily as an effort to reject a view that cannot take ontology seriously but stops short at mere external appearances.

The second point in Torrance's criticism is closely related to the first one, actually being its special case. Interestingly enough, space and time, which formed the crucial area of debate in Torrance's discussion of Newton's dualism, enter the stage once again. When it is asked in what way Kant could so seriously undermine the primacy of ontology in the epistemological process, when denying the possibility of knowledge of things in themselves, Torrance directs attention to the philosopher's conception of the synthetic a priori. Thereby, Torrance insists, Kant united sense experience and structures of the consciousness so that with this tool the human knower could place conceptual order on everything he or she perceived. But how did this shift come to lay emphasis on the subjective side of knowing? Torrance's answer is both bold and simple. Referring to Newton's view of space and time, he claims that Kant now transferred absolute time and space from the mind of God to the mind of the human knower. In the human mind they still have their absolute role in not being affected by experience.[48] While Newton attached dualism to the ontological nature of the universe, Kant did the same with the human mind and epistemology as well.

This claim is masterful in its simplicity, since it combines the conceptual apparatuses of two eminent figures of the history of Western intellectual history in a straightforward way, giving the impression that fundamentally the question is again that of the ultimate nature of space and time, and that there is an unbroken continuum between the two persons and their ideas. Faithful to his basic interpretation of the history of science, Torrance's reading can be seen as a simplification of an issue which is as complex as it is difficult. To evaluate Torrance's claim, it is now argued that this simplification is, on the one hand, driven too far but, on the other, it expresses something essential in Kant's thought.

Torrance's claim is based on the assertion that Newton identified his absolute space and time with the mind of God. In a previous chapter, however, this insight was argued to represent quite a daring reading of Newton's writings. Especially Torrance's unhesitating way of translating Newton's *sensorium* by *mind* was demonstrated to be problematical. This means that the connection he makes between Newton and Kant and especially between the notions of absolute time and space was closer and more continuous than the evidence warrants. Therefore there are grounds for claiming that what is at issue is not a historical reception of Newton's concepts by Kant but rather Torrance's own emphasis on the crucial nature of Kant's epistemological thinking and its alleged indebtedness to Newton.

In another respect, too, Torrance's exposition of the relation between the notions of absolute space and time in Newton's and Kant's thought is implausible. As has been described, Newton held the view of absolute space and time in the sense that through it he rejected their relational conception: space and time are like a container or receptacle holding diverse events and things. Space and time are substantially real, and they cannot be reduced to a mere network of relations as Leibniz had insisted. On the one hand, Kant felt critical toward both notions of space and time, denying their existence outside of the human mind either substantially or relationally.[49] But on the other hand, it has been proved that Kant gave some preference to the absolute notion of space and time because it was closer to his own construction of the basis of human knowledge.[50] It should be underscored now that Kant's space and time were absolute in the sense that nothing external could affect them because of their a priori status. At the same time it is necessary to point out that scholars' opinions vary remarkably as to Kant's philosophy of space and time and that Kant himself wavered between different solutions during his philosophical career.[51]

The confusing feature of Torrance's argument becomes apparent when it is recalled that Kant did not work with the dichotomy absolute–*relative*, as Torrance strongly suggests, but only with the distinction absolute–*relational*. In other words, Newton's alleged identification of God's mind with absolute space and time could take place only with regard to *relative* space and time, not as a feature of absolute space and time in the container or receptacle sense as opposed to the relational notion. In his own discussion, then, Kant treated space and time as absolute, not as the complementary side of relative space and time but only as exclusive of the relational view.[52]

In spite of the fact that the link Torrance draws between Newton and Kant is exaggerated and that there is a certain confusion in his exposition of the reception of absolute space and time from Newton to Kant, he certainly does not miss the mark in his evaluation of the centrality of the human mind in the philosopher's thinking. For Kant, then, space and time are a priori categories of the mind, categories by which we give meaning to our perceptions. "Space is not an empirical concept which has been derived from outward experiences." "Space then is a necessary representation *a priori*, which serves for the foundation of all external intuitions." "Space is no discursive, or as we say, general concept of the relations of things, but a pure intuition."[53] The nature of time is described in similar terms. "Time is not an empirical concept." "Time is a necessary representation, lying at the foundation of all our intuition." "Time is not a discursive, or as it is called, general concept, but a pure form of sensible intuition."[54] Kant summarizes his view illuminatingly in his claim that "space and time, as the necessary conditions of all our external and internal experience, are merely subjective conditions of all appearances, and not things in themselves, presented to us in this particular manner."[55]

Although in this respect Torrance's analysis of Kant's views on the absoluteness of space and time is valid, it is not comprehensive. Kant spoke of space and time in two differing senses, as a priori modes of perception and as metaphysical entities.[56] Torrance's analysis takes into primary account only the first aspect in Kant's discussion. With regard to the second aspect, Torrance seems to suggest that space and time are real for Kant only in the human mind as a priori modes of perception. The problem is, however, more subtle because Kant holds that space and time "are empirically real but transcendentally ideal. They are empirically real in the sense that what is given in experience is in space (if it is an object of the external senses) and in time. . . . But space and time are transcendentally ideal in the sense that the sphere of phenomena is the only sphere of their validity, and that they do not apply to things-in-themselves, considered apart from their appearance to us."[57]

In spite of this restricted point of view, Torrance is able to highlight the absolute nature of Kant's idea of space and time, but only in the sense that nothing empirical can affect them, as they are a priori categories of the human mind. From this standpoint, Torrance is undeniably right when he regards Kant as holding to the absolute view of space and time. This also reveals why Torrance cannot feel at home with Kant's epistemology, for in it objectivity and the primacy of ontology are neglected. This can also be expressed by claiming that Torrance cannot agree with Kant because the thought of the philosopher does not allow the compulsion of reality to exercise its force upon the human knower but reverses its direction so that now it is the human mind that exercises coercive power upon reality.[58] For Torrance, then, this is an untenable example of a perversion in epistemology where there is no hope of getting in touch with reality itself in its ontological depth.

The concept of *homoousion* can convincingly explain Torrance's standpoint and his criticism of Kant. The possibility of the Incarnation and the ultimate nature of Jesus of Nazareth are at stake. If space and time are absolute, and if they are categories only in our mind without any independent objectivity of their own, any idea of God entering this space and time remains inconceivable. And further, if this epistemology prevents us from gaining knowledge from reality itself and can only touch upon appearances or phenomena, then there is no possibility of taking seriously the *homoousion* nature of the Son, or the Deity of Christ. Only the historical phenomenon of Jesus provides an object for theological inquiry. Torrance's reluctance to agree with Kant's epistemological views can be, then, maintained as being based primarily on theological grounds.[59]

The actual content of Torrance's idea of reminder in its epistemological aspects can be fully evaluated only against the Newtonian-Kantian background thus described. While it was pointed out that the inherent dualism in both Newton's and Kant's thought prevented the idea of compulsion from working, it is appropriate, as well, to claim that the bearing of the idea of reminder upon human knowledge has essentially to do with realism in general, as analyzed in the fourth chapter. Theologically reasoned and seen through the Reformed doctrine of election, a correct epistemology, according to Torrance, must submit itself modestly to the constraints of reality as it is, independently of the human mind or, to put it differently, a truly scientific epistemology has no alternative but to be realistic under the demands of reality. There is no need to penetrate once more to the nature of realism in Torrance's thought here, but instead, what is relevant is to discern the actual reminding elements in the epistemology of modern science as expounded by Torrance. In this respect two particular elements or themes should be investigated, elements through which Torrance sees his realism in its epistemological aspects functioning in the foundations of knowledge and which therefore provide the reminding force in modern physics.

First, Torrance claims that the dualistic separation of form and being is now past, so that an era of integration has emerged along with modern physics.[60] The Kantian dichotomy between the noumenal and the phenomenal, the Cartesian distinction between subject and object as well as the division between structure and substance are other implications of dualistic defects that are under reconstruction toward an integrated whole in the basis of knowledge.[61] Torrance especially appreciates the field theory for this advance but he does not forget Einstein's theory of relativity, either.[62] This integration and reorientation in epistemology makes Torrance express his appreciation to Kant because his notion of the synthetic a priori recognized that empirical and theoretical elements work together in all human knowledge.[63]

But what actually is at issue in Torrance's insistence on the integrative character of modern physics with its field, relativity, and quantum theories? A new combination of theoretical and empirical factors is surely one of the points he consistently stresses. Using concepts deriving from ancient Greek philosophy, he holds it to be an implication of the novel unity of form and being or structure and matter, a unity in which distorting dualism is overcome. But it seems, too, that Torrance does not find an ultimate balance between the two complementary poles in the process of integration.[64] His eagerness to stress independent objectivity makes him inevitably assign absolute primacy to objectivity over subjectivity, matter over form or structure and the empirical over the theoretical. This implies that the integration as described by Torrance is not meant to unite the two poles in the integrative process into a balanced whole after a disastrous dualistic separation but rather to combine them into such a unity where the nonsubjective or nonformal side retains its authority to dictate what is allowed to be thought about it. Both relativity and quantum theory, and in particular Torrance's interpretation of the latter, illuminate this feature.

Torrance follows Einstein faithfully in the scientist's view of an independent external world, a belief that is claimed to form the basis for all natural science. Whatever a scientist has found out in his investigations he must be willing to transform his understanding according to the further disclosure of this reality.[65] Space and time as integrated into a four-dimensional space-time implied by the general theory of relativity are no longer separate entities but form a reality that has to be taken for granted.[66] This kind of integration of the objective side of reality reinforces the status of reality independent of an observer to the extent that this objective reality and the subjective human knower cannot be viewed as bearing an equal weight of importance. The role of an observer remains, as noted earlier when discussing Torrance's realism, problematic in spite of the fact that Torrance constantly emphasizes the active role of the subject in the epistemological process.[67] We do have a role, even a crucial one, but we can exercise it only in one way, through our active and appropriate questioning.[68] This kind of stress upon the role of human agents undeniably gives the impression that the subjective side has been taken sufficiently into account, but it is not able to remove doubts concerning the thoroughness of this reflection.[69]

Torrance's interpretation of quantum theory gives an even clearer example of this "unbalanced" integration. Aligning himself with Einstein in his debate with representatives of the Copenhagen interpretation of quantum theory, Torrance admits that quantum physics presents a challenge to traditional ways of conceiving the relationship between theory and experimental results.[70] Despite this, he maintains that quantum theory has not been able to reject dualism entirely.[71] Torrance is of the opinion that although the subjective aspect in a scientific inquiry within quantum physics is taken into account, the ontological or objective side tends to be consistently dropped out. This means that the actual object of quantum physics is not reality as independent of ourselves but only those phenomena that become apparent in the act of observation. Torrance claims that nature itself is not disclosed through this approach but only statistical probabilities, a conceptual tool used in the interpretation of quantum phenomena. The intrusion of a measuring device in the investigation of the atomic world introduces an additional subjective element which, in Torrance's view, gains too significant a role at the expense of assigning the independent reality the central position it deserves.[72] This suggests that Torrance sees, for instance, the so-called *uncertainty principle*, indicating the impossibility of accurately defining both the position and the momentum of a single particle simultaneously, as a product of dualistic thinking which should be overcome.[73] This leads further to the claim that for Torrance any hint of contradictory features in the results of quantum physics are merely due to our inadequate and imperfect human knowledge, not to inherent features of nature itself. Therefore, Torrance's exposition of quantum theory reveals his favoring of integration in the structure of human knowledge, an integration where empirical and theoretical, subject and object, form and being are united. At the same time, however, it proves the correctness of the argument presented here that this integration is not aimed to attain equality or complete balance but only to bring the two elements together so that what is theoretical, subjective, or form would no longer be functioning alone and self-sufficiently but would be transferred to the control of what is empirical, objective, and being itself. Only in this manner is the fluent working of the idea of compulsion guaranteed.

But why is such an integration necessary, in Torrance's opinion? If the Kantian dichotomy between the phenomenal world and the thing in itself or his constructivist notion of the role of space and time were retained, human knowledge could not attain reality, as it is independent of any observer. To state this otherwise, the principle of *homoousion* could not have any positive significance if the human mind were allowed to function on its own terms without taking ontological objectivity into account, that is, what is observed could not be identified with reality itself but only with some constructs of the human autonomous reason. Torrance maintains that restoring the *homoousian* relation is precisely what modern physics has been able to do.[74] The notion of counter-dualistic integration is, therefore, a central theme in Torrance's idea of reminder.

The second element or theme in Torrance's idea of reminder is not without connection to the notion of integration described above, as will become apparent. Torrance makes a distinction between two kinds of openness, namely an ontological and an epistemological one. Applying Einstein, he concludes that the amazing detail in this double openness is that both aspects are integrated or open toward each other.[75] The main concern, however, in the context of the present discussion must be his idea of epistemological openness, for this is the point where, in Torrance's opinion, the reminding feature of modern science is plainly apparent.

Torrance makes use of Einstein's conception of the stratified structure of the scientific system in which any epistemological process is constructed through different levels, aiming at as simple a body of concepts as possible to describe the object of inquiry.[76] The first layer consists of our everyday experience and primary concepts related to it. Intuition and sense experience give rise to our basic concepts. On the second level or the first scientific level a new set of concepts is introduced to collect the meaning and purpose of concepts used on the first level. They are connected with empirical reality but in an indirect way.[77] A deeper abstraction is met on the third or meta-scientific level, which is still poorer in concepts and relations but nevertheless succinctly collects the knowledge with greater precision than is the case on a lower level.[78] When knowledge increases, this kind of stratification becomes inevitable for the new information to be comprehensible.[79] But what is the relation between these layers or levels? What is required for penetration into the next sphere of scientific concepts and relations?

Torrance claims that this hierarchical structure of epistemological levels of inquiry are "open upwards into wider and more comprehensive systems of knowledge but are not reducible downwards."[80] To note the direction of this openness is of primary importance: it is always upward, toward levels of greater theoretical value and conceptual unity. To gain a firmer grasp of this argument, it is useful to refer to Torrance's conviction that the eminent example of a higher-order theory is Einstein's theory of relativity in the sense that it collects the bearing of earlier physical theories into a grand whole so that even Newtonian science appears to be a limiting case within the realm of relativity theory.[81] Openness has therefore a close connection to indeterminism, meaning and signification, intelligibility, and a wide coherence with other theories touching upon the same field.[82] This idea of stratification inevitably reveals Torrance's belief in the possibility of reaching a theory that could explain, in a unified manner and to a considerable extent, different aspects of reality. Like Einstein, however, he is agnostic with regard to the thought that an ultimate theory could be attained, a feature which reminds one of the similarities between Torrance's thought and Karl Popper's idea of *verisimilitude*.

The crucial point to be made is that the idea of the openness of a layer toward a higher one essentially serves Torrance's general view of the submissive role of epistemology. His insistence that a level should be open upward and not downward toward empirical reality or to the first level of everyday experience can be interpreted only as meaning that scientific concepts on whichever layer are closed downward so that they may not be used to manipulate reality but that open reality would be capable of having an impact upon these subsequent layers. Openness upward, correspondingly, is maintained in order to guarantee that new explanations and theories with stronger unifying force are possible in the first place.[83]

The aim of this argument is not hard to find: its purpose is to safeguard the independence of reality from our knowing of it so that through this notion of openness the idea of compulsion might be maintained. According to this line of thought, if the levels were open downward it would lead to assigning primacy to our knowledge over reality as it is, or in other words, it would lead human knowledge back to a mode of Kantian constructivism. But there is another side to Torrance's notion of openness. Whereas the functioning of the idea of compulsion is safeguarded through the notion of openness upward on the different levels in the stratified structure of knowledge, closing their direction downward toward empirical reality, single scientific concepts are seen as open toward this concrete reality.[84] This implies that Torrance uses the theme of openness in two separate manners which, nevertheless, have the same significance. In the latter case openness means readiness to be altered if circumstances in scientific inquiry demand any modification of them. They are not self-subsisting or self-defining concepts, but essentially ones that need to be brought into fuller precision from outside, from empirical reality. In the previous case openness is used to underscore the constant need of revision and thorough consideration in more comprehensive scientific theories when they are related to a higher level of theories with a larger unifying power. While the former is open upward, toward another layer of theoretical unity and coherence, the latter is open downward, toward the hard facts of empirical reality. Whatever the level, a close relation to empirical reality should be maintained all the time.[85]

This epistemological side of the idea of reminder, consisting of the two elements or themes of integration and openness, are seen as a confirmation of Torrance's ambitious aspiration to construct a unifying view of an epistemology which is controlled by ontology and which allows one to view empirical phenomena as *homoousioi* with reality itself. Integration refers directly to this submissive status of epistemology, and openness indicates the receptive role of the human knower, who cannot but consent to the demands of reality. Stressing these features in the reminding character of modern physics reveals, however, that Torrance makes careful selections from the epistemologically tangled field of modern physics and especially quantum theory. The fact that physics, as a whole, is in a state of ferment, due to widely differing opinions, for example, concerning the correct interpretation of physical phenomena in the crossfire of varying theories, makes Torrance's efforts to construct a unifying epistemological basis, largely dependent upon Einstein, quite a strenuous task. Whether his attempts are totally hopeless or not is a question that remains beyond the scope of this discussion. What is relevant here is simply to note those tendencies and features in modern physics which Torrance considers crucial for theology as well. Therefore in the following pages a critical survey will be offered with regard to the implications that the idea of reminder has for theology.

Implications for Theology

The ontological themes of relationality and indeterminism, as well as the epistemological themes of integration and openness, in Torrance's thought coincide both methodologically and substantially in the crucial concept of *homoousion*. Thereby the idea of compulsion is allowed to influence theological thinking. According to Torrance, theology has a realistic lesson to learn from the modern natural sciences, especially from physics, so that an observation is not cut off from its underlying ontological reality to the status of a mere phenomenon and that God's revelation in his incarnate Son is taken with all seriousness as the basic substance of theological inquiry. Only in this way can a sound theological "healing" take place subsequent to an era of dualistic modes of thinking, governed by disintegration, determinism, and static rigidity. But what kind of theology is Torrance willing to construct according to his idea of reminder? This is the question to be asked in this final section before the concluding chapter. Two areas of interest gain special attention, namely the basics of theology, including the problem of Bible interpretation and the issue of ecumenism, where Torrance's standpoint, deriving from his interests in the natural sciences, proves to be, to say the least, highly original.

Basics of Theology

One of Torrance's intentions is to contribute to the theological discussion of the role of the doctrines of the Person of Christ and the Trinity, which are commonly seen to have been undervalued during the past two centuries.[86] New insightful efforts have been made to restore the role of these doctrines in the center of Christian understanding and doing theology.[87] Torrance's originality in this respect lies precisely in his attempt to draw parallels from the recent development of the natural sciences and the philosophy of science influenced by it. In other words, Torrance tries to convince us that the restoration of Christological and Trinitarian doctrines is the inevitable task of theology if it entertains any scientific ambitions and intentions. Without proper Christology and a coherent view of the Trinity, theology would remain devoid of any relevant content and unscientific as well. To fulfill this criterion, it has to attach itself firmly to the only "place" where theology, the science of God, can be pursued, namely God's self-revelation in Jesus Christ, who is *homoousios* with the Father. This scientific character is realized in theology on the two conditions implied in the preceding pages. Methodologically and epistemologically, theology as a science has to submit itself to the compulsion of its Object as he has revealed himself in the incarnate Jesus Christ. Substantially and ontologically, theology has to start from Christology and the doctrine of the Trinity, the actual knowledge of God, and deviate from any attempt to reflect on God beyond his self-revelation in the Incarnation in the manner of an independent natural theology. Torrance is aware of the "Christomonistic" tendency in his thinking, and therefore he admits that all knowledge of God cannot be reduced to mere Christology but rather needs to be made coherent with it, having its main criterion in it.[88] But what does his argument consist of, and how can it be evaluated?

As indicated above, Torrance insists on the desperate need for a scientific theology, which needs to be reestablished on the ruins of dualism, that is gradually being overcome by the modern natural sciences. The era of disintegration of form from being and phenomena from their own ontology has come to an end. Now, the compulsion of reality again exercising its influence upon human understanding and removing artificial anthropocentric modes that manipulated reality in the Kantian fashion, a genuine realism has become possible in theology, too. But what does Torrance mean by his manifesto, derived analogously from the realm of modern science? He maintains that an advantage "of this closer readjustment of scientific method to nature has been to show us something of the damage which dichotomous ways of thinking have done to our knowledge of God, and not least to our knowledge of God in Jesus Christ."[89] In this connection Torrance describes as dualistic those theological tendencies where Jesus is detached from God and from Israel and where Christianity is detached from Christ. The crucial point, then, is disclosed as lying in Christology and especially in the identity of Christ. This further suggests that scientific theology cannot start outside this basic point of God's self-revelation in the Incarnation of his Son. The substance of the concept of *homoousion* is thereby observed, once again, to form the basis of all theological activity according to Torrance.[90] But how can Torrance give reasons for his argument for scientific theology on the basis of the contemporary development in the natural sciences? Torrance's contribution to the discussion concerning the role of Christology in modern theology is highly dependent on the success of this argument.

Torrance's enchantment with the rigor of the natural sciences has had an apparent effect on the way he characterizes the nature of theology as science.[91] Faith involved in theological enterprise is not something irrational but the rational movement of human beings to God's self-revelation in his Word.[92] Torrance wants to reject the conception that faith is essentially contradictory to reason, and this he establishes by making faith a subform of reason and rationality. He claims that it would not be rational to see faith and reason as mutually exclusive because through faith, participating in rationality, reason faithfully tries to understand its object.[93] This means that there is actually no science where a kind of faith-assent would not be present, because all rationality contains belief which "has to do with the elemental interaction between persons and realities other than themselves, entailing a recognition of their independent reality and truth."[94] Belief, then, confirms the objectivity and reality, the otherness, of a thing observed. If there were no faith, there would be no truly objective science, either.[95]

Torrance wants to argue, first of all and not without apologetical intentions, that in every science, through the idea of belief, there is an element of "theology" involved without which it would not be able to remain faithful to its object. But, second, Torrance suggests that when using the concept of faith in its own inquiries theology cannot be blamed for any unscientific mode of thinking because faith belongs to the very core of all rationality. But this uniting bridge does not consist only of rationality but also of two other important ingredients, which highlight Torrance's inclination to see theology as a rigorously pursued science in the manner of the exact natural sciences. Theology consists of positive reflection, basing itself securely on the ground of empirical reality so that it represents essentially a posteriori thinking.[96]

Torrance's effort to characterize theology as a rigorous science in the manner of the natural sciences is based, first, on his insistence that theology has nothing to do with speculation but only with precise, scientific thinking which strictly conforms to the nature of reality under investigation. This means that knowledge must be based on reality, in what it actually and ontologically is, and not on any human construction or convention.[97] Theology takes its Object–God–seriously and submits itself to him. It does not entertain any a priori arguments, loose speculations that do not have any direct contact with the reality under investigation. In other words, theology is not mythology, which, in Torrance's definition, has its center in the human mind and thereby neglects the full weight of the object under investigation.[98] Torrance holds that theology cannot be rigorously pursued if the coercive force of the object is not permitted to influence human knowledge but, instead, the human mind is elevated to have the power to dictate what to think about reality. In Torrance's understanding, genuine, scientifically pursued theology and the idea of compulsion are inseparable, and this makes Torrance claim that theology is a science of an a posteriori nature.[99] It has its roots firmly in the reality which it wishes to obey.

The critical question to be presented in this connection is whether there really is such an a posteriori feature whereby scientific theology could be pursued in a manner similar to natural sciences. Torrance gives a determinedly positive answer, for instance, when he describes theological statements as results from an actual a posteriori encounter with reality.[100] Above, it was disclosed that for Torrance this a posteriori ground for theological knowledge or the objective reality to be encountered is the "Word and Act of God," a phrase with a direct reference to God's self-revelation in the Incarnation. The incarnated Christ as God's unveiling of himself is, then, the a posteriori of scientific theology. An encounter with him provides us with such a knowledge that is not mythological, speculative, or a priori in character.

It is certainly true that theology can offer the incarnate Son of God as the ground for its enterprises, but it remains questionable whether this approach can be named as an example of theological a posteriori thought. The biblical accounts of Christ do offer historically grounded literary evidence upon which Christian theology is based. No doubt theology has an a posteriori nature in this sense. Human thought always needs "material" upon which to build conceptions and theories, and whether this material is called a posteriori is a matter of convention. But Torrance remarkably broadens the area within which the term a posteriori can be applied when he gives the impression that it is our direct encounter with the incarnate Word that forms the ground for theological knowledge. The question is not only of Jesus of Nazareth of the four gospels or the other New Testament books, although the necessity of truthfulness to biblical witness is never questioned, but primarily of the Lord with whom an individual can be in face-to-face contact here and now, in this space and time. Torrance's effort to construct scientific theology by drawing a parallelism of a posteriori thought from the natural sciences is most vulnerable in precisely this respect: Torrance cannot convince his reader that a devotional religious experience could be considered a truly a posteriori ground for theological knowledge and that this character could be acknowledged from a general scientific point of view, such as by the natural sciences or the philosophy of science. Wolfhart Pannenberg has succinctly expressed the problem, reminding us that, in our times, no one "has an experience of him [Jesus] as risen and exalted, at least not an experience that could be distinguished with certainty from illusion."[101]

The problem is even more apparent when the second element in the scientific nature of theology, as presented by Torrance, is considered. He claims that theology is not only grounded upon a posteriori thought but that it is also an *empirical* science. In this respect Torrance feels it necessary to give a more precise interpretation of the word that better suits his purposes. "What is meant here by the 'empirical' is not just what is sensible and tangible so much as the *experienced imperceptible and intangible*."[102] It should be well noted that Torrance makes a conscious decision to present all human experience as empirical, not only experience based on the senses. The idea of a religious experience as an empirical fact itself is well known in the science of religion, where it is considered an object of inquiry. This approach, however, could not possibly fit into Torrance's system because it would mean that our understanding and our experience were made the actual objects of a scientific enterprise, not God as he has made himself known. Mythology would replace theology, centering on human beings would substitute for a center in God. For Torrance, human experience is never the object or target of inquiry but the empirical basis from which the proper object is researched.[103]

This implies further that what cannot be seen, smelled, tasted, heard, or touched falls altogether into the realm of human experience, and this is enough to make theology an empirical science, claims Torrance.[104] As in the case of the concept of *a posteriori*, here, too, Torrance has significantly to enlarge the area within which the concept of *empirical* can be applied, a necessary procedure to make theology look more like an empirical science. At the same time, however, the meaning of what is truly empirical vanishes significantly because this approach makes it justified to accredit the qualification *empirical* to whatever mode of human thinking, based on human experience. Its variety and complexity, versatility and problems, if not based on generally acknowledged criteria of sense experience, can hardly contribute to the rigorousness of empirical science as Torrance would undoubtedly like to see. If experience in general is claimed to provide the basis for theological knowledge, one can justifiably ask, which experience of God, out of the vast number of differing experiences, is the correct one to provide the basis for theology? In a multifaith and pluralistic world, this question is far from irrelevant, but Torrance's approach is not able to cope with it seriously.[105]

Torrance articulately states that his motivation for making such comparisons is not to pursue apologetics in order to safeguard for theology an acknowledged position, with similar terms, among the sciences.[106] This would suggest that Torrance has not intended his claims concerning the a posteriori and empirical nature of theology to be taken strictly at their face value but only as heuristic phrases with which he tries to find an area of common interest for theology and the natural sciences. It has to be admitted that this modest interpretation is certainly attractive and can contribute significantly to the discussion between the disciplines. But Torrance's position, as has been argued above, involves something more. He stresses that the comparison between theology and the natural sciences, also with regard to their a posteriori and empirical natures, helps theologians to guide others to assume a theological position where the reality of the Divine is met. This intellectual witness or persuasion takes place when theologians promote the need to submit one's rational assent to the logic and reality as they are expressed by theology.[107]

After all, then, theology has an apologetic function to fulfill. Its method is to persuade scientists to acknowledge the rationality of theology. But what kind of rationality? A rationality that can be recognized by natural scientists who use empirical methods, a rationality linked closely with *empiria* or experiental basis, so that theological rationality, even in its distinctiveness, would be conceived as based on empirical reality. And further, referring to critical observations with which the truth of a theological statement is evaluated, Torrance points out that these observations are valid in the realm of empirical science as well because in science statements refer to real existents and processes, not to ideas. Torrance draws the conclusion that whenever critical observations are used to cause damage to theology they actually cause damage to empirical science, too. This can only result in skepticism.[108] In other words, denying the truth value of theological statements would inevitably result in the denial of the statements in empirical sciences. They have a common fate. This suggests strongly that Torrance's claims concerning the a posteriori and empirical natures of theology are not mere rhetoric but should be taken quite literally. It should be remembered, though, that Torrance does not regard theology and the natural sciences as totally identical but, together with their similarities, he emphasizes their differences. But it should be noted as well that the differences are connected with the nature of objectivity and subjectivity in theology, not in the empirical or a posteriori nature of theology.[109]

Bypassing problems of the evidential nature of religious experience just mentioned, Torrance clings to his conviction that in scientific theology its empirical character means "thinking out of real experience of God determined by God."[110] The reality of an experience is not, then, defined by psychological factors in the human mind, however genuine an experience might be. The determining force making an experience real, that is, submitted to the compulsion of reality, is God alone, who through his self-revelation in the incarnate Christ, being *homoousios* with the Father, makes himself known to us. Jesus Christ as the incarnate Son of the Father is the *empiria,* the experiental ground and the concrete basis of theology. This conviction suggests that Torrance does have an answer to the critical question posed above, as to who has the correct experience of God. Referring to the Christological decision made at Nicea, Torrance maintains that it was the Christian Church which could interpret truth and the ultimate nature of God's empirical appearance in space and time in the incarnate Jesus Christ.[111] Torrance can use the phrase *the mind of the Church* to explain how the *homoousion* came to be applied and God's revelation interpreted in rational terms.[112] The compulsion of objective reality cannot function without a target, and this target is always a rational entity, the mind. It is the Christian Church which has the ability to interpret and mediate truth about God's self-revelation, and in this respect only the theological experience of the Church is relevant.

This approach is certainly endogenous and immune to criticism, which would prefer to demand external criteria for verification of this kind of theological claims.[113] For instance, the disruption in the Western Church that occurred during the time of the Reformation poses the problem of judging, if Torrance's argument is followed, which party actually had the true "mind of the Church" and on which criteria such a question could be resolved. That in this special case it was the Reformation churches and in particular Calvin's theology which, in Torrance's opinion, represented a correct attitude in "the mind of the Church" is not a matter of surprise. Therefore we should note well that Torrance has made a conscious choice to present his argument in a genuinely Barthian way, whereby God's self-revelation is considered a self-explanatory event which cannot be verified by external criteria alien to the way God has made himself known.[114] As a consequence, only the Christian Church can have such a correct experience of God and a right interpretation of it. This conviction makes theology primarily a function of the Church, a feature which, in the final analysis, belongs essentially to the scientific nature of theology as understood by Torrance. Including the explicit view of theology as a science, it can be claimed that, with regard to theology as a function of the Church, to know Barth's thought is to know Torrance's thought as well, and therefore numerous references to the similarities in their thinking are unnecessary.[115] In this respect one crucial point of criticism, from the standpoint of Torrance's own intentions, can be repeated. We have seen how Torrance does not consider the natural sciences to be self-explanatory or to have reasons for their own satisfactory explanation, as implied by contingency, which Torrance claims belong essentially to the object of the natural sciences. He readily refers to the so-called Gödel theorems involving the view that no logical system can be fully consistent in itself and contain reasons for its own explanation but that these must be sought from outside, in another system.[116] The previous discussion proves, however, that in this connection Torrance does not apply this argument within theology, which contains its own justification and verification. Torrance cannot escape this problem even if he attempted to insist that it is true that theology does not contain its own explanation but for this purpose one has to refer to God as its final judge.[117] Purely theological issues, such as Christology or ecclesiology, certainly must be regarded as matters of theology alone, and in their case no justification "from outside" is necessary. But the question of the scientific nature of theology in general, particularly if it is placed on the same level with the natural sciences, cannot be successfully made a purely internal affair within the boundaries of theology alone and on its own terms.

The crucial issue at this point is whether or not Torrance is able to give a convincing reply to those who ask for further evidence for accepting the Christian Church as the realm within which theological enterprise finds its justification. A parallel from the world of sciences is again used when the Church is seen as the "scientific community" of theology.[118] Torrance applies his phrase "the social coefficient of knowledge," by which he refers to "the capacity of a society or a community to be affected and modified through its advance in knowledge of what is independently real."[119] This coefficient is the social bearer of meaning and significance, which are not to be found in themselves but refer beyond themselves to their objective source.[120] This process takes place when "God reveals himself to us in such a personal way that he establishes relations of reciprocity between us and himself and calls forth a community of reciprocity—the people of God or the Church—which he assumes into covenanted partnership with himself."[121] For Torrance, this implies that a theologian should be committed both to active participation in the experience of his Christian community and to the community of theologians, in which the experience gained in the worshiping community becomes articulate.[122] In both cases the objectivity of God exercises its compelling power and dictates the content of our knowledge and experience of God, for even the world wide scientific community of theologians is considered a servant of the Church.[123] In the way described, the *mind of the Church* is affected by continuous religious and devotional encounter with God and God is believed to exercise his compelling power upon this collective mind. Torrance does not hesitate to apply concepts derived from Christian worship and devotional life to explain his view of how a theologian must behave under the compulsion of his Object. To scientific theology belong *reconciliation* and *renewal*,[124] as well as *repentance* and *conversion*.[125]

In spite of the interesting parallelism Torrance thus draws between the communities of Christianity and science, he does not provide any further evidence to convince the reader that it is the Christian Church, with its long tradition and willingness to submit itself to the commands of its Lord, which alone can have a genuine and true experience of God. From the standpoint of Christianity and its self-understanding, this approach is a challenge to view the role of the community of faith in scientific terms, but it certainly is not sufficient to make it more scientific in the sense that this would, without causing problems, transfer theology into the same category of rigorous science as the natural sciences, which seems to be Torrance's implicit intention.

But his use of the phrase "social coefficient of knowledge" has another relevance which is more theologically reasoned. Torrance states that the social coefficient arises within the communion that God gives to the Church through the Holy Spirit. From the human side it can be viewed as an adaptation to God's revelation and a kind of human response to God's gracious act of establishing this communion.[126] The Holy Spirit is, then, viewed as the Constitutor of the Christian communion, that is, as its ontological foundation in the Godhead. But his task is not only to call forth a community of believers and sustain it in his power but also to serve as an epistemological factor that grants knowledge of God by showing the way to the Son. This, however, can take place only when we are moved by the Spirit under the compelling power of the divine Object of theology.[127] The epistemological role of the Holy Spirit, then, is to act within the Church so that the compulsion, true objectivity, and consideration of an independent reality might become real. Through him God produces in us the ability to apprehend him, an idea implying the exclusion of all other means of obtaining knowledge of God.[128] Therefore natural theology cannot be viewed as scientific or objective; it does not allow the Spirit to influence us and create a reciprocal relation between ourselves and God. This approach can be considered a sophisticated reformulation of the traditional view of the Spirit as the guarantor of the truth of the interpretation of Christian faith entertained within the Church, but materially it can hardly introduce novelties to reinforce evidentially the claim that the Church can be compared to a scientific community in the realm of theology. But Torrance's insistence on the epistemological significance of the Holy Spirit can also be viewed as an effort, within the bounds of theology on the terms presented by Torrance, to make an approach to a *Trinitarian* understanding of the structure of the basics of theology. The overtly Christological nature of Torrance's theology sooner or later raises the question of whether the doctrine of Trinity is taken into account at all and how he is able to relate it to his view of scientific theology.

We must firmly acknowledge that Torrance has taken the doctrine of the Holy Trinity with the utmost seriousness, especially in his later writings. This feature is very significant because it is not uncommon to meet opinions in which Barthian theology is blamed for neglecting the doctrine of the Trinity behind its Christological intentions.[129] On the other hand, theologians have acknowledged that Barth's actual contribution to Trinitarian theology is to be found in the way he commences his colossal *Church Dogmatics* with the doctrine of the Trinity, thus affirming its central and formative position in his thinking.[130] Torrance readily uses this feature in the thought of his former teacher, although not uncritically.[131] With deep appreciation he claims that it is Karl Barth who in modern times has been able to restore theology back to its proper basis in the doctrines of the Incarnation and the Holy Trinity.[132] In other words, Barth has shown the way back to a truly scientific theology, scientific in the sense that the Object is allowed to determine what one is to think about it. This means that his significance is not to be detected only in the reinstatement of the doctrine of the Incarnation as the self-revelation of the Triune God. This effort can also be directly compared to recent revolutionary discoveries in physics in which, as Torrance would express, dualistic modes of thinking have been overcome and realism with the inherent idea of *homoousion* is restored.[133] In this respect Torrance strongly suggests that Barth has achieved something comparable to Athanasius, especially in that both rejected an independent natural theology as the proper means by which to gain proper knowledge of God. In Torrance's view, both

Athanasius and Barth were fighting against various implications of dualism, both epistemologically and ontologically.[134] Torrance draws even further conclusions from this parallelism by claiming that, in fact, Karl Barth is the Athanasius of our time.[135]

But Torrance's appreciation of Barth is not based on the comparison with Athanasius alone but also on one with Albert Einstein. Torrance holds that the rise of four-dimensional geometry, which was later found consistent with Einstein's theories of relativity, meant the end of an independent Euclidean geometry in its idealized form detached from experience of empirical reality.[136] Torrance holds that this traditional Euclidean geometry had been treated as a self-contained conceptual system and as a theoretical construction through which reality had been interpreted. Therefore, in Torrance's opinion, this abstract approach to reality, not firmly grounded in reality itself but seen as an a priori discipline, cannot mediate a trustworthy view of the world, which, after all, cannot be understood solely in two or even three dimensions.[137] To apply Torrance's argument, this can also be expressed as meaning that an independent Euclidean geometry is not *homoousios* with reality but only a construction of the human mind. Einstein, then, is accredited with integrating Euclidean geometry into the larger body of physical science so that it could no longer be used as an independent branch, able to be pursued only from fixed axioms without any direct contact with empirical reality. What Einstein in our times did in physics was achieved by Barth in theology—this is Torrance's grand argument—when he rejected an independent natural theology and integrated it into the corpus of scientific theology, based on God's self-revelation in his incarnate Son who is *homoousios* with the Father.

Torrance's appreciation of Barth and his view of his achievements as products of an eminent pioneer of what he understands as scientific theology is shown to be based on the authority of two great figures in theology and physics, Athanasius and Einstein. These kinds of comparisons are always more or less suggestive and tentative no matter how extensive the evidence one can submit in support of them. Solely the fact that the temporal distance between Barth and Athanasius is immense and that the distance between Barth and Einstein with regard to the object of study is not easily shortened makes all efforts to draw direct links between them problematic. It seems that both Athanasius and Einstein are used by Torrance as authorities with which the relevance of Barth's theological program can be justified and its alleged scientific character underscored. In this respect Torrance cannot avoid giving the impression of representing confessional preference in which grand figures in the history of science and theology are employed to give credibility to his own position and background as a Reformed and particularly Barthian theologian. Great geniuses are always open to a wide variety of interpretations, even contradictory ones, and Torrance's approach certainly presents only one of them. But this is not meant to indicate that Torrance does not have anything important to say by his appreciative interpretation of Barth with the help of Athanasius and Einstein. Rather, his presentation highlights certain significant questions in modern theology, especially in the fundamentals of theology, which have been brought to light by Karl Barth, such as the question of the starting point of theological inquiry, the problem of realism, and justification of theological claims. Although the comparison of Barth with Athanasius and Einstein, as done by Torrance, does not provide convincing evidence to accredit Barth as *the* scientific theologian of our age, it nevertheless offers tempting insights and suggestions that are worthy of further consideration.[138]

Leaving Torrance's personal enchantment with Barth's theology aside, an example of what scientific thinking in theology could mean can be found in the second point to be made in his appreciation of his former teacher. Torrance implies that only Christological and Trinitarian theology can be viewed as scientific.[139] God in his Triunity is the precondition of all theological reflection, its first and ultimate Object from which every other branch must proceed.[140] As has been noted throughout the present work, Christological themes are determinative for Torrance, and this raises the question as to what, in his opinion, is the relation between Christological and Trinitarian doctrines. On the one hand, the crucial role of Christology is emphasized, but on the other hand, an appeal to Trinitarian theology is strongly expressed. Which one, then, is actually the real starting point of theology? Does not Torrance's constant reference to the *homoousion* as the linchpin of theology, reminded of by modern physical science, underscore Christology as the proper field in which a theological inquiry should begin?[141]

Torrance is quite determined in his efforts to integrate Christology and the Trinity with each other in a coherent way, although, as a general remark, it could be stated that explicitly Christological overtones are stressed in Torrance's earlier works, whereas in his mature years he has made efforts to relate them effectively to the doctrine of the Trinity.[142] Nevertheless, his starting point seems to remain thoroughly Christological. "It is, then, this two-fold ground in God and in Christ Jesus, in the Father and the Son, that together with the Communion of the Spirit forms the framework within which the Holy Trinity is to be known, understood and interpreted."[143] The ground for the knowledge of the Trinity is two-fold, not three-fold, as clearly implied in the quotation. The relation between the Father and the Son is, in this sense, absolutely primary and constitutive. The Holy Spirit enters this formula only secondarily, suggesting that the Spirit does not belong to the constitutive ground of theological knowledge. This would be easy to interpret as an indication of an underestimation or even neglect of the full weight of pneumatology but before such a conclusion is reached another remark of Torrance's should be remembered.[144] The reason why precisely *this* relation of the Father and the Son forms the epistemological basis of theology is that "the Incarnation constitutes the one actual source and the one controlling centre of the Christian doctrine of God."[145] It is the Incarnation, God's assuming of human form in Jesus Christ, still *homoousios* with the Father in spite of his humanity, that provides the only acceptable way to know God as he himself has willed. Only here can the way to God be found in this created reality. The ground for theological knowledge is thus established in the "binitary" structure of the Father-Son relationship. Yet one is bothered by the question of what position is reserved for the Spirit in this system.

Torrance provides an answer when he states that it is "the two-way relation between the Father and the Son illuminated for us in the Holy Spirit that provides the frame both for our knowledge of God in his inner trinitarian relations and our knowledge of the Son in his inner hypostatic relations as God and man in one Person."[146] The Holy Spirit has the role of bringing human beings to a relation with God in which this God is known in his Triunity. This observation affirms the point made earlier that the Holy Spirit has an epistemological role in elevating us to participate in God's self-knowledge.[147] Torrance would not express this point as indicating that the Spirit is only a secondary instant not belonging to the constitutive basis of theological knowledge grounded on the Father-Son relation. One is inclined to receive this impression, however, when reading his writings because the event of the Incarnation is accredited such a crucial and epistemologically important position. The Son was incarnated, not the Spirit. The Son entered space and time, not the Spirit.[148] One might pay attention to the fact that the concepts of space and time, drawn from the philosophy of physics, determine the primary relevance of the incarnate Son and the minor epistemological relevance of the Spirit in Torrance's thinking, a useful feature for emphasizing the significance of the Christological approach. The Reformed doctrine of accommodation is apparent as well, since it was only in Jesus Christ that God descended to earth and took human form upon himself, accommodating himself to our mundane situation.

In spite of the apparent primacy given to Christology, there is yet another aspect. We should remember that according to Torrance we could never enter a relationship with the incarnate Christ who is *homoousios* with the Father unless the Spirit draw us into such a relationship.[149] In other words, since there is no proper knowledge of God the Father without his self-revelation in the Incarnation of God the Son, we have first to enter into a relationship with Christ. But this is something that remains impossible unless God the Spirit give us the opportunity for this relationship, participation in the knowledge and love of the Father and the Son. In fact, the role of the Holy Spirit proves to be more crucial than one might expect at first sight. It belongs to the very nature of the activity of the Spirit that he, as it were, hides himself in order to let the light of the triune God shine through.[150] To use this conclusion as an answer to the main question of the relation between Christological and pneumatological aspects in Torrance's theological epistemology, it can now be concluded that he accredits explicit primacy to Christology in the sense of *objectivity*, with regard to the particular event of the Incarnation in Jesus Christ, and to pneumatology in the sense of *subjectivity*, with regard to the action of the Spirit as the remover of human beings to a theologically "scientific" epistemological position before God. This inference would also satisfyingly explain the explicitly Christological character of Torrance's argumentation in his earlier as well as his later writings. Since he is keenly interested in safeguarding objectivity, the natural inclination from this standpoint is to rely on Christology, in which the objectivity of God as revealed in Christ through the Incarnation is highlighted. But on the other hand, on the basis of the preceding analysis, one could also argue that strictly speaking Torrance's starting point in Trinitarian theology does not lie in Christology but in pneumatology, albeit in rather intellectual terms.

This conclusion implies, nevertheless, that Torrance takes the Trinity into account. He does not speak of the Spirit without reference to the Son, who alone can give access to the Father. All three Persons in the Godhead–the Father, the Son, and the Holy Spirit–are *homoousioi* with each other, confirming that the *homoousion*, then, does not apply only to the Son but to the Spirit as well.[151] The concept of *homoousion* must be seen as an inner bond securing the consubstantiality of the three Persons. In the final analysis, then, the *homoousion* is not only a Christological term but essentially a Trinitarian one. But in what way should theology begin from the doctrine of Trinity in order to be scientific? What does it mean to start doing theology from the scientific basis of the Trinity?

The influence of physical science and particularly Einstein's thought could not be more apparent elsewhere than it is in this connection. The stratified structure of knowledge provides Torrance with a tool with which to describe not only the intellectual birth of the concept of *homoousion* but also an analogously applicable way of introducing new concepts into theological inquiry. As was pointed out earlier, this approach presupposes at least three levels of scientific concepts, the first of which starts with everyday experience. Concepts are then organized into a greater unity and coherence on the second or scientific level, and a still more consistent as well as succinct formulation is reached on the third or meta-scientific level. Torrance renames the three levels as the basic level of religious experience and worship, the theological level, and the higher theological level.[152] This is another example of how Torrance makes efforts to give an impression of theology as a rigorously pursued science in which knowledge is strictly achieved and arranged. The interesting point, however, is that Torrance uses this stratification to justify his insistence on Trinitarian theology as the only scientifically acceptable way of doing theology. But in order that this argument (the Trinity is the starting point of theology) be consistent, it is crucially important that the Trinity be taken into account already on the first level and not as late as the third one. In an earlier work Torrance gives the impression that the Trinity of consubstantial relations does not enter the discussion until the second level, where the *homoousion* relation between the Son and the Father is realized. The first level consists of the relation between the human being and Christ, the second level of the relation between the divine Persons in the Economic Trinity, and the third level explicates the consubstantial relation between the Father, the Son, and the Spirit in terms of the Ontological Trinity.[153] (For the sake of clarity, it is appropriate to point out that in theological thought the Economic aspect refers to the activity and the work of the Trinity as they are known to us within the "economy of salvation." The Ontological approach attempts "to formulate the Godhead outside the limiting conditions of time and space," the essential being of the Trinity.)[154]

In a later work, however, Torrance presupposes that the Trinity does not wait until the third level for its realization but is *intuitively* apprehended already on the evangelical and doxological level of everyday religious experience. The fact that the Christian lives in a community where the New Testament witness to Christ is proclaimed takes the Trinity into account in an *implicit* way, although it has to wait for its intellectual explication until the next levels.[155] This feature does not indicate only the progress of Torrance's thinking toward a consistently Trinitarian theology but also the role he has reserved for intuition in the corpus of human knowledge and particularly in what is inexplicable in it. But the validity of Torrance's main argument, that all theology must have Trinitarian "initial conditions," stands or falls with the two premises involved. The first premise holds the Christian Church to be the only environment in which genuine knowledge and experience of God can be attained, through religious life and worship. This claim relates to the view of theology as a function of the Church. The second premise is attached to the concept of intuition as a mode of inarticulate, inexplicable knowledge. Whenever one accepts one or both of these premises, one can undoubtedly agree with Torrance. But whenever such an acceptance is not found possible, Torrance's way of reasoning the Trinitarian initial conditions of all theological knowledge is beset by problems which, in the final analysis, can be seen as a result of the Barthian context of Torrance's thought.

In Torrance's exposition of Trinitarian theology there is one detail in which his invitation to theology to learn a lesson from modern physics seems particularly sound. In the previous section Torrance's view of the nature of ontology as revealed by modern physics was analyzed, and relationality was found to be its most important distinctive feature. This view has direct Trinitarian implications for the patristic reflections on the hypostatic relations of the three Divine Persons and includes the conviction that no Person can be what it actually is without being in relation to the other Divine Persons.[156] Not only the ultimate nature of being in the natural sciences is relational in the sense of internal, real relations, but in Trinitarian theology, too, this understanding is crucial. The relation between these two relationalities in Torrance's thinking becomes explicable on two grounds. First of all, Torrance suggests that because God as the ultimate Being is relational himself, the ontological nature of his contingent creation is also analogously relational. Substances in the created realm exist only insofar as they have a relation to each other.[157] This outlook implies that God's relationality is seen as the cause of the relationality of the ontology of creatures, a view which well fits into Torrance's conception of contingency of the created reality as a reference beyond its own boundaries to its divine Creator. The second point to explain this analogy made by Torrance can be seen to lie in his willingness to prove that it is not scientifically inconsistent to think of real beings as constituted by relations. Or in other words, realizing onto-relations in nature makes it easier to view God's Being as constituted of relations. Modern science, therefore, can remove intellectual obstacles to apprehending the mystery of the Trinity. This analogy is undeniably quite powerful and consistently reasoned, presenting perhaps the strongest point in Torrance's argumentation concerning science and the Trinity.[158]

There is still another area in the basic structure of theology in which Torrance makes efforts to apply his view of scientific theology, namely biblical hermeneutics. Particularly the theme of integration detected in modern science by Torrance comes to the fore, a theme in which object and subject, being and form, were seen as integrated in a union where object is given primacy over subject and being over form, as analyzed earlier. This integration leads Torrance to maintain that "theological exegesis and interpretation of the Holy Scriptures would go along with and be controlled by theological understanding of the divine truths and realities they mediate to us, but theological understanding of those truths and realities would themselves be controlled by constant theological exegesis and interpretation of biblical texts."[159]

Integration and a victory over dualistic thinking in which the Jesus of the Bible is detached from his "natural inherent relations" in Israel and in God's Being means here, in other words, that exegetical and dogmatic theology are closely coordinated.[160] But what does this "scientific hermeneutics" imply, in Torrance's opinion?[161] It should be noted, first of all, that Torrance underscores the referential relations between the Bible and the realities it bears witness to. In other words, the Bible is not itself the reality from which theology and Christian life are derived but only bears a witness to it. Therefore Torrance can state that the realities signified in the Bible remain independent of the Bible itself. The Scriptures, then, have the task of bringing their readers into a relationship with the realities revealed by the Bible, to bring them into relationship with God. This view makes Torrance abandon any fundamentalist standpoint which cannot appreciate the consubstantial relation between God's free self-revelation and its actual content as expressed particularly in the Bible.[162] But this approach makes it impossible, too, to concentrate only on the "liberal" manner of biblical exegesis in which the consubstantial or *homoousion* relation between the Son and the Father is neglected. In both cases "scientific hermeneutics," as understood by Torrance, implies the dismissal of what is seen as dualism.[163] The belief in *homoousion*, then, is detected as offering the hermeneutical presupposition for Bible interpretation.

This leads to the second point worthy of note, namely the nature of the relation between Bible interpretation and "theological understanding" which controls exegesis. This confirms once again the basic structure and aim of Torrance's notion of integration: it is the object which has the primacy over anything thought about it.[164] This explains why, in Torrance's thinking, the idea of compulsion belongs essentially to hermeneutical work where what is called theological understanding must take seriously the *homoousion* relation between the Father and the Son. Of course, Torrance makes a reservation when he states that the results of exegesis should have a controlling influence upon theological understanding, but the overall impression he gives is that this control-relation is only secondary to the first one, where theological understanding is the presupposition for a proper Bible interpretation.

Torrance seems to take seriously the fact that no interpretation can take place in a vacuum as if there were no preunderstanding at all. In theology based on the message of the Scriptures, this preunderstanding is determined by the experience of the Church, which, as has been shown, provides the only reliable context for any relationship to God, whether cognitive or emotional. Torrance thus anchors biblical exegesis and hermeneutics strictly in the domain of the Church. At the same time he suggests that a "neutral" or "objective" interpretation is an impossibility, due alone to the fact that to be truly *objective* means obedience to the compulsion of the object under investigation, rejecting detachment, impartiality, and indifference toward the object.[165] One cannot avoid the feeling that the grand Barthian theme of objectivity is active once again in the form of the idea of compulsion. This implies further that, in Torrance's opinion, all theology, biblical hermeneutics included, can be scientific only when pursued within the worshiping fellowship of the Church.

Torrance has again something very important to say concerning the present situation in modern theology, where the need for a closer relation between biblical exegesis and systematic theology is deeply felt. Torrance undoubtedly offers one alternative for such an integration, but the actual problem is that his conception of the Church as the only proper realm of theological activity seems to draw its limits quite narrowly. We have seen how the Church as a worshiping community forms the basis for scientifically pursued theology, as understood by Torrance, a view that does not give due attention to theological work in academic circles, the scientific nature of which becomes at least questionable if we follow his argumentation. If the Church in this connection were defined along broader lines, not explicitly as the community of worship, Torrance's point would be easier to apprehend. As such, however, Torrance's scientific theology, biblical interpretation included, retains its strong character as a "theology of believers," reserved only for those who can accommodate their religious experience, if there is any, to the common experience of the Church, indicated in its credal formulations.[166] This is, of course, not to deny the value of such a theology. On the contrary, theology that is keenly conscious of its religious presuppositions and aims can have only a refreshing contribution to make. The present discussion is merely to make the point that to view this "theology of believers" as the sole representative of what is conceived as scientific theology can turn out to be strictly apologetical and incapable of entering into serious dialogue with other schools of theological interpretation.[167]

Torrance's view of theology as science has been demonstrated as relying primarily on the Subject-object relation in which the former is submitted to the latter, that is, in which the idea of compulsion operates. It is not so much a certain intellectually and consistently pursued rigorous methodology that determines the scientific nature of theology but rather a submissive attitude, in which obedience toward the Object has to exercise its compelling power upon the human knower and upon his methodology as well. This is what Torrance sees as happening both epistemologically and ontologically when the concept of *homoousion* with all its implications is applied. Theology is not, then, pursued "behind God," that is, detached from him, but before and in conscious relation to him on the terms laid down by him. This takes concrete form in participation in the life of the Church and in living Christian experience within it, with the Trinity as the starting point of theology as well as of Biblical interpretation. Torrance's approach has gained appreciation especially because of its apologetical function. In the same connection it has been criticized for not taking seriously enough the fact that reality as the object of any scientific enterprise as well as human understanding of it are far from uniform and unambiguous.[168] This point is certainly a sound one but it dismisses another problem posed by Torrance's intention of viewing theology as science. This detail has not so much to do with the rich variety and complexity of both reality and human understanding of it but rather with the concept of science itself, which is far from simple and settled. This is, however, what Torrance consistently supposes in his insistence that it is the *homoousion* which makes a science realistic and objective.[169]

Ecumenism

As a theologian deeply involved in ecumenical activities, Torrance has made use of his scientific insights, as expressed in his idea of reminder, in the field of interchurch relations. When this side of his thought is considered, one of the most interesting features in his descriptions and analyses of the causes of the divisions in Christianity is his direct application of concepts, phrases, and patterns of thought that have become familiar to the reader in the course of the preceding discussion. In fact, Torrance's ecumenical maxim is expressed succinctly in a single quotation: "Ecumenical activity takes place within the space and time of this world."[170] Space and time form the stage on which the Christian churches seek a closer unity, a detail that is here claimed to belong to the essence of Torrance's thinking on ecumenical issues. The idea of reminder, then, asserts that modern science has revealed something essential, both ontologically and epistemologically, which can make a notable contribution to a deeper understanding and mutual approval among the churches by showing the way back to a theological basis upon which closer communion can be constructed. The notions of integration and openness have a place in Torrance's ecumenical thinking, too. The space-time reality of this created world does not, however, provide a mere stage for ecumenical activities but is closely bound up with the subject matter of ecumenical interests.

At first sight this connection between a special branch of theology and church life, on the one hand, and the fundamental concepts of the general theory of relativity and the philosophy of physics, on the other, can seem rather strange to most readers of Torrance.[171] It might be useful, therefore, to recall how he views the development of Western scientific thinking by claiming, first of all, that it was the theology of the Early

Church with its emphasis on the Incarnation and the *homoousion* which contributed to the birth and application of the relational notion of space. Because of an emerging dualism, the absolute notion gained priority from the early Middle Ages on and was hardened further still by the deterministic science of Newton and the subjectivist philosophy of Kant. It was not until this century that a revolution took place when Einstein's theory of relativity forced us to think of the ultimate nature of space and time in a new way.

With this background in mind, our attention can now be directed to the straightforward claim made by Torrance as to the reasons for the state of disunion among Christians. He namely holds that reasons for disruptions and disagreements between the churches are nontheological rather than theological.[172] One cannot avoid asking what exactly, in Torrance's opinion, are those nontheological ingredients in Western thought that have given birth to the lamentable circumstances under which the churches have felt compelled to take distance from each other. He maintains that the primary reason for impaired relations between the churches are due to the absolute or container notion of space and time![173] Ecclesiology as well as the theology of the sacraments and ministry have been gravely affected by this view, where space is seen as a closed receptacle confining events and individual things within its tight borders. Torrance criticizes the Roman Catholic conception of the church as an ark of salvation, bringing to the fore the idea of a closed vessel, resulting in the conviction that catholicity was identified with the institutionalized Roman Church as expressed in the acclamation *extra ecclesia nulla salus*.[174] On the other hand, and more generally, Torrance blames dualism for ossifying institutional structures in the Roman Catholic Church.[175]

Torrance argues further that the receptacle notion was a remarkable exponent of dualism that influenced medieval sacramental theology by grounding its premises on the Aristotelian ideas of space with their dichotomy between the earthly center of absolute rest and the immutable, celestial unmoved Mover.[176] In the main, God was thought of in spatial terms without giving temporal aspects any further consideration. This led medieval theologians in the West to think of Christ's real presence in the Eucharist similarly in spatial terms, resulting in "artificial constructions" by which Christ's body was conceived of as being present "in the whole host, in each part of it, and in a multitude of hosts at the same time, without any commensurable relation between His body and the space of the place in which it is contained."[177]

Torrance claims that Lutheran theology became a direct heir of medieval notions of Christ's real presence when it uncritically accepted the receptacle notion of space.[178] He reflects on the problem of space with regard to the doctrines of the Incarnation and Christ's real presence through the controversy known as *extra Calvinisticum* that took place between the Lutherans and the Reformed party shortly after the time of the Reformation. He maintains that when the Reformed theologians adhered to the conviction that the Son of God really descended from heaven in the Incarnation, without actually leaving heaven and his supreme lordship over the whole creation, they were, in fact, in line with the patristic understanding of the issue. But the Lutherans, as Torrance points out, having adopted the receptacle notion of space, had no alternative but to interpret this conviction as leading one to suppose that the Son or Word of God, entering this receptacle, had to leave something of himself outside, *extra se*. Thus the notion of the *Calvinist extra* was introduced into the debate that took place between the two main branches of the Reformation.[179]

Thus Torrance strongly implies that if one holds on to the receptacle notion of space when discussing the possibility of the Incarnation, it inevitably results in a conviction where the impossibility of the Incarnation is easier to confirm than its possibility. Therefore, as is the case in this particular argument cited above, those theologies that work on the basis of the receptacle notion are more vulnerable, and their development will more likely lead to difficult problems. Such a problem really may arise, as Torrance argues, which may be called *extra Calvinisticum*, but at the same time he wants to show how this conclusion is based on distorted premises, having their source in a less compatible view of space. If the relational notion of space were adopted, as was done by the Reformed theologians, in Torrance's view, there would be no need to construct lines of thought that could be described as something *extra*. That is to say, the possibility of the Incarnation must be discussed on the basis of the relational notion of space, where speculations about the possibility of the divine being united with the human would turn out to seem less of a problem. Torrance implies that there is really not any greater difficulty in adhering to the presence of the Son of God incarnate both in heaven and on earth if only we leave behind any distorted view of space as a container, a limited place with strict boundaries, and instead give preference to the relational view.[180]

From the Lutheran point of view, the case is quite different. It has been stated that it was the Reformed party, both Zwingli and Calvin, who adhered to the view of God's spatial relation to the world, with the result that heaven and the phrase the "right hand of God" were conceived as "a place far removed from our world, to which Christ was removed at his ascension and where he will remain until the Last Day."[181] It has also been pointed out that it was Luther who managed to leave behind not only the Scholastic but also the Reformed point of view when he criticized those imaginations in medieval theology that localized heaven and the relation between immanence and transcendence. In addition, the claim has been made that the Reformed party, seemingly modern with all its emphasis on the Spirit, could not, after all, get rid of either the old Aristotelian view of heaven as the *coelum empyreum* (the highest part of that which exists) behind the stars or material and localizing thinking.[182] Accordingly, Torrance fails to take into account the variety of modes in which Luther, applying scholastic distinctions, spoke of three ways of being situated somewhere. *Localiter* or *circumscriptive* means to be located within the limits of the space a being occupies. This is the way in which bodies, human beings included, exist. To be in a place *diffinitive* is restricted to angels and spirits, and also to Christ after his resurrection, whose being cannot be understood in terms of bodily existence as assuming a certain portion of space. The third mode of being in a place is *repletive* or supernaturally, so that being can be omnipresent. According to Luther, this mode of being belongs to God alone.[183] Torrance's criticism of Luther is based on the assumption that the Reformer applied only the receptacle notion of space, which in this case is equivalent to the *localiter* or *circumscriptive* notion. Although the articulated notion of relational space appears later in the history of Western thought, it is evident that its essential content as an exclusive alternative of the absolute or receptacle notion is already anticipated in the *diffinitive* and *repletive* modes of being in space. It can also be suggested that this threefold distinction, as applied by Luther, can fulfill the same function as Torrance's twofold distinction between the receptacle and relational notions of space, namely to underline the possibility and actuality of the divine presence in space through the incarnate Christ. One of the further advantages of this "Luther alternative" is that its origins lie deep in theological thinking and not primarily in the philosophy of physics.

Many of Torrance's arguments seem to rise from confessional motivation, and in this context there is hardly any reason to make final judgments concerning the correctness of either of the parties. The important point to be noted is, however, that on the Lutheran side, too, the problem of spatial concepts has been realized and that these concepts are used in analyzing the Reformer's thought. However, despite apparent one-sided tones, Torrance has highlighted in his analysis quite an important insight related to the Eucharist and Christ's real presence, a central area of discussion in interchurch relations.

But what is the actual motivation behind Torrance's efforts to approach ecumenical problems from the standpoint of the concept of space? The answer is not too difficult to find. It has only to be recalled how Torrance accuses Isaac Newton's scientific heritage of distorting a more plausible notion of space because of its inherent dualism. This leads Torrance to conclude that the Christian Church, too, "has allowed the divisions of the world to penetrate back into itself so that its own unity in mind and body has been damaged, and its mission of reconciliation in the world has been seriously impaired."[184] Precisely here lies the significance of modern science, which has been able to take remarkable distance from both ontological and epistemological dualisms, thus creating a new scientific context which challenges the churches to overcome their divisions and separations based on extratheological dualistic trends.[185] The crusade against dualism is thereby transferred to ecumenical discussions.[186] Torrance's insightful analysis falls, then, fully and coherently into the wider field of his argumentation in which dualism is seen as dangerous primarily because it produces divisions which do not let the compulsive power of reality influence the human mind. His standpoint is consistently reasoned, giving a sound and harmonious impression of the causes of separation as well as exclusive suggestions of how a remedy could be discovered. The full value of this approach can be evaluated only after a look is taken at the consequences this kind of ecumenical theology could possibly have.

For this purpose it is useful to refer to Torrance's own characterization of the ecumenical process, and in this case his account of Reformed-Orthodox relations provides ample reference. At the beginning of his description Torrance reminds us that both churches "continue to seek the realization of that unity in space and time,"[187] a phrase referring to the common scene of every human effort, be it science or ecumenical theology. One of the crucial issues in this ecumenical dialogue is the doctrine of the Trinity. Torrance contends that the churches are aware of the differing views and interpretations entertained in the East and West with respect to the relation between God's Unity and Trinity, "but do not regard any one of them as an explanation of the Mystery of God's Triunity which is more to be adored than expressed."[188] This discloses Torrance's conviction concerning the role of confessional formulas in general. They are not seen as capable of precise definition of the content of faith but, at best, only as good approximations, in the Popperian sense, to the Truth which ultimately remains outside the scope of human expressions.[189] One of the remarks made by Torrance in this respect is certainly a direct result of this approach: "No one may boast of his own orthodoxy."[190] This means further that human formulations are themselves called into question by the very same revelation, the content of which they strive to explicate and which forms their actual basis. Truth does not belong, then, to any single church or denomination, but churches can only participate in it.

The message of the idea of reminder, according to Torrance, is that, while the mechanistic and deterministic understanding of reality has been removed by modern natural science, theology should apply this same approach in its enterprise in the field of ecumenical discussions, too. Strict and obstinate adherence to rigid confessional formulas can bear testimony only to a dualistic and deterministic view, comparable to the rigidity of natural science in the Newtonian context. Therefore, both the theories of relativity and quantum physics remind ecumenical theology of its own roots in relational, nondeterministic thinking to which it should now return in order to make an essential and relevant contribution to interchurch relations.

At first sight, this argument seems appealing precisely because it appears to guarantee significant flexibility. According to it, one might achieve a considerable agreement on doctrinal issues when the incomplete nature of confessional formulae is acknowledged.[191] No one can have the ultimate Truth, but each party has to admit the inadequacy of its own apprehension. But the problem with this approach lies in its inability to find proper criteria for such a common agreement which could form a unifying ground for two differing confessions. This problem, resulting from the Reformed interpretation of doctrinal statements, has become apparent in the so-called *Leuenberg Process*, involving several series of negotiations between the Reformed and Lutheran Churches, in which an implicit distinction is made between the ground and the expression of doctrinal formulations. This feature can be detected also in Torrance's view of the solution to ecumenical problems. If it is acknowledged that all contemporary churches committed to the Niceno-Constantinopolitan Creed participate in Truth, the solution suggested by Torrance must ignore quite significant differences in formulations made upon this common ground. In other words, the relation between the common ground of the churches and its articulate expression in their credal formulas remains problematic and even, one might argue, indifferent toward the divergence and dissimilarity to be found in church confessions. Therefore it remains quite questionable whether ecumenical reconciliation can be achieved by applying this method, for in order to achieve a comprehensive insight of what an ecumenical agreement implies, there certainly should be as precise an articulation of the common ground between churches as possible.[192] Torrance's ecumenical methodology would suggest, then, that an interchurch agreement could be achieved by making a separation between the ground and its expression, not necessarily in a dualistic manner, but in a way that reminds one more of an imprecise mystical attitude rather than of paying serious attention to problems emerging on a cognitive and articulate level.

One of the appealing characteristics of this approach, at least in Torrance's mind, seems to lie in the implicit use of the concept of *homoousion* in this connection. True reconciliation between the Christian churches is possible, in Torrance's opinion, if the participatory nature of different confessions in the one Divine Truth is acknowledged, leading to the conviction, as was stated, that no one can boast of his own orthodoxy. Also the referential nature of the truth claims in credal formulas was noted, implying that no confession can have truth in itself but only insofar as it refers to the Truth it signifies. It seems evident that, for Torrance, agreement upon crucial issues or upon the ground is sufficient for ecumenical consensus. This can also be expressed in such a way that, for reconciliation in Christendom, it is sufficient to recognize and acknowledge that, ultimately, the churches are *homoousioi*, of the same being with each other. Unity cannot be found, in the final analysis, in the coherence of credal formulas, however profound it may be, but exclusively in God himself who has called the Church forth and sustains it by his Spirit as the Body of Christ. This implies further that the unity, the *homoousion*, already exists so that any ecumenical activity cannot create it but only detect it within the divergence of credal formulations.

In addition to this implicit use of the *homoousion* in reasoning arguments in ecumenical theology, Torrance also applies it in a more explicit way. A good example is the *filioque* problem, which, as is well known, constitutes one of the most obstinate ecumenical problems between the churches of the East and the West. Torrance blames the Roman Church for introducing the clause unecumenically into the Nicene Creed, suggesting that the Greek Church could not but reject it, not only because of its alleged heterodoxy but also because its appearance in the creed was based on the authority of the bishop of Rome.[193] It is now argued that Torrance blames the Roman Church only for a technical mistake. The actual or substantial reason lies deeper. Torrance sees there a misinterpretation which evoked the whole *filioque* problem, and he makes efforts to explain it away by referring to the absolute equality of the divine Persons, based on the *homoousion*, so that there are not two distinct sources of the Spirit but that "the Spirit proceeds from the Father through the Son, and leads us through the Son to the Father."[194] In his opinion, the difficulty arose when the Cappadocians tended to see the procession of the Spirit as starting from the *Person* of the Father, a view that attempted to secure a proper monarchy in the Godhead. In the West, however, this solution was seen to imply that there is a hierarchy in the Godhead, subordinating both the Son and the Spirit to the Person of the Father, a fact that made Western theologians add the clause *filioque*.[195]

The reason for the problem, in Torrance's suggestion, is to be found in the theology of the Cappadocians, where a sharp distinction was made between "the οὐσία as a generic concept and ὑπόστασις as a particular concept," resulting in the view that the Person of the Father is viewed as the primary Cause or Principle or Source of Deity.[196] The undeniable advantages of the Cappadocian theology, as Torrance maintains, were achieved "at the cost of cutting out the real meaning of ὀυσία as *being in its internal relations*, and of robbing οὐσία of its profound personal sense which was so prominent at Nicaea."[197] This claim leads us to discern the heart of Torrance's argument: the fatal disruption in the history of the Christian Church was due to a deviation from the Athanasian doctrine of the *homoousion*, in which the onto-relational consubstantiality of the Persons of the Godhead was considered a self-evident matter, securing the full equality of the three Persons. With regard to the problem of *filioque*, this resulted in the view that the Spirit proceeds not from the Person of the Father but from the *Being* of the Father, which is, at the same time, the Being of the Son as well.[198] There cannot be, therefore, two distinct Sources of the Spirit but only one, namely the Divine Being or οὐσία himself, consisting of *homoousial* relations between the Persons.[199]

If the deviation from the Athanasian understanding of the *homoousion* as the constitutive feature of the relations between the Persons was the cause of the *filioque* problem, Torrance suggests a similarly simple remedy: "The problem of the *filioque* falls away if the doctrine of the Trinity is set back again on the Athanasian (and Cyrilian) basis."[200]

This means further that, on the level of explication, the Western version of the creed is condemned, and in this Torrance joins in the criticism presented by the Eastern church. This does not imply, however, that Torrance would agree with the content of the Eastern interpretation of the debate, at least in its Cappadocian form. Instead, he makes his own proposal, based on the concept of *homoousion* and derived from his interpretation of Athanasian theology. His proposal is certainly inspiring and insightful, based on a thorough study of patristic sources. But are there realistic possibilities for his suggestion to be taken seriously in the final test of every ecumenical discussion, namely statements of mutual agreement? It is, of course, impossible to answer the question definitively, but there is a test case which can have some relevance in evaluating the possibilities of Torrance's proposal.

A study of the *Agreed Statement on the Holy Trinity*,[201] a paper in which the achievements in the Reformed-Orthodox dialogue have been collected, reveals first that an effort to base this discussion on the theology of Athanasius is evident: most of the references to patristic sources are to the Church Father's writings. Second, it seems that an apparent consensus concerning the *filioque* question has been attained, although expressed in a short and succinct phrase: "Thus the order inherent in the Trinitarian relations is grounded on the fact that the Son is begotten of the Father and the Spirit proceeds from the Father."[202] The *filioque* is omitted, and no further reference to the problem or any mention of its rejection is made. There is no doubt about the fact that Torrance's own theological work has helped the Reformed party to accept this formulation in which the characteristic tones of Western Christendom are now significantly mitigated. In a commentary reflection on the statement formulated by a group of theologians from the two churches and headed by Torrance, it is stated that "further study in depth of this procession might help us to find ways of cutting behind the division between the East and the West over the so-called '*filioque*', for it does not allow any idea of the procession of the Spirit from two ultimate principles or ἀρχαί."[203] In spite of the significance of the expression concerning the procession of the Spirit, the absence of any articulate mention of the rejection of the *filioque* question suggests that the problem must be viewed as only partly solved. In the same direction refer the words in the commentary reflection, in which the need for further discussion on the issue is pointed out. As such, however, the subject offers an outstanding example of Torrance's ecumenical approach, in which he consciously desires to adhere to the *homoousion* as the structuring principle of interchurch relations both methodologically (the churches are *homoousioi*) and substantially (the *homoousion* is the ground upon which reconciliation is possible). As has been stated, the former tends to present more problems than the latter, in which Torrance's authoritative knowledge of patristic documents provides fertile standpoints and interesting insights worthy of further consideration.

The preceding discussion of ecumenism as a part of Torrance's idea of reminder has disclosed that he makes constant attempts to set the efforts to ecumenical consensus in a wider context of the paradigm shift in the Western cultural climate, inaugurated by the development in the modern natural sciences. The *homoousion* remains the "linchpin" of this process both implicitly and explicitly. Implicitly, differences between the churches are suggested to present only a phenomenal problem due to divergent interpretations of Christian doctrines. The objective reality behind them is common to all Christian churches, and the task of ecumenical dialogue as well as all of theological activity is, according to the maxim of the idea of compulsion, to let this reality dictate what one is to think about it. Integration of form and being, subjectivity and objectivity does not mean in this case the unification of different credal formulations into a more coherent and wider system but, rather, an integration with God as the final judge of confessional formulations. The aim of this procedure is to allow God to exercise his compulsion over theologians involved in ecumenical discussions. Only thus can the alleged dualism, introduced into theology from outside, be overcome and the problems stemming from it, for instance the so-called *extra Calvinisticum*, be solved. Explicitly, on the other hand, the *homoousion* in the ecumenical context of the idea of reminder forms the doctrinal basis upon which a closer unity between the churches can be built. A full appreciation of the consubstantiality of the Father, the Son, and the Spirit should result in taking seriously the heritage of patristic theology as a product of the undivided Church and especially of Athanasius's insights. From this point it is justified to conclude that the *homoousion* is in the center of Torrance's ecumenical methodology and its subject matter itself, a feature which is certainly proper to theology but which now has been revealed once more by the development of modern natural science.

As an overall evaluation of Torrance's approach, it can be stated that in Trinitarian theology, particularly, his arguments are weighty and well reasoned, backed by exhaustive patristic evidence. But linking this line to the wider context of Torrance's argumentation, in favor of realism and opposed to what is called dualism and reasoned through the development in physics, raises problems that characterize his thinking in general. Describing certain features in the history of thought as dualistic excludes a great deal of serious thinking without which a coherent and justified picture of the past cannot be achieved. The effort to apply the philosophy of physics or, better expressed, that of space, in the interpretation of ecumenical problems can be regarded as heuristic and suggestive, containing hints to be taken note of, but this approach can hardly offer a definitive solution to problems that are theological rather than physical. Also, suggestions that the theory of relativity with its new understanding of space, making a relational view more preferable, should inevitably lead to a new understanding of space in theological thinking, sound doubtful at least as far as no further and more convincing evidence of the absolute necessity of the relational view is offered. The fact that this problem plays such a central role in Torrance's argumentation can be explained as resulting only from his conviction that the Incarnation, the descent of the Son of God, who is *homoousios* with the Father, into the space and time of our reality as well as its possibility, forms the structuring principle for all of Torrance's thinking.

6

Conclusion

The aim of this study has been to investigate Thomas F. Torrance's view of the significance of the doctrine of the Incarnation, involving the patristic term *homoousion*, for both theology and the natural sciences and, accordingly, for the relationship between them. Torrance's approach is unique in two respects in particular. First, he makes efforts to build bridges between theology and the natural sciences within the context of Barthian theology, which is usually viewed as indifferent toward issues of nature and the sciences. Second, the scheme presented by Torrance does not start from the doctrine of creation, as has generally been the case on the theological side of interdisciplinary discussion, but from the doctrine of the Incarnation and especially the concept of *homoousion*. In both respects Torrance's enterprise is strongly Christological, a fact that makes his insights and suggestions of great interest. He does not attempt to introduce theological viewpoints by referring to a general conception of God, a supreme Being with certain qualities such as omnipotence or eternity. Instead, his proposal brings the fundamental tenets of Christian faith into the discussion. His attempts can be considered a "Christianizing" of the quite neutral conception of God entertained in purely philosophical speculations, and a search for fruitful ingredients for theology-science relations on the basis of special revelation and Christology.

The hypothesis on which the structure of this work has been constructed is that Torrance's view of the relevance of the doctrine of the Incarnation for theology and the natural sciences is characterized by the three ideas detectable in his thought. The idea of contribution asserts that the incarnational theology of the Early Church provided Western thinking with intellectual substance which made the rise of empirical science possible. The idea of compulsion explicates Torrance's conviction that the doctrine of the Incarnation was both a result and, later, a channel of realism. Through his idea of reminder Torrance claims that now, the natural sciences having made significant progress, particularly in the field of physics, theology has something important to learn from them, so that it is able to return to its original foundation, which was severely obscured by dualism throughout the centuries.

Several academic studies of Torrance's thought analyze his view of the relationship between the natural sciences and theology. In this regard their defects can be summarized under five points. First, they have not critically evaluated Torrance's view of the fruitful role of the incarnational theology of the Early Church in the rise of modern science, as expressed in his idea of contribution. Second, they fail to detect the deep roots of Torrance's realism in his theological position as a Reformed and Barthian theologian. Third, they have taken Torrance's interpretation of some historical figures of physics for granted, which has made them incapable of properly evaluating the relevance of Torrance's arguments. The same also applies to their uncritical acceptance of his view of dualism. The fourth defect is that Torrance's proposals with regard to the reminding force of the natural sciences have not been studied adequately. In Torrance studies, references to the analogous relation between theology and the natural sciences are abundant, but what these studies fail to see is that, in Torrance's case, the relation does not consist of mere analogies but, rather, of connections. The fifth and last defect in earlier works on Torrance's views on theology and science is that, in spite of their attentiveness to the centrality of the Incarnation and the *homoousion* in Torrance's thought, the implications of this insight have not been taken far enough. It needs to be recalled that the concept of *homoousion*, as used in the doctrine of the Incarnation, is of crucial importance to Torrance. According to him, it is not only "the linchpin" of all theology. Physics and the other natural sciences have their own *homoousion*, too.

The present work has attempted to take these points into account in its own systematic analysis, concentrating on the context of Torrance's thought and evaluating problems inherent in his solutions. Torrance's own historical framework, within the boundaries of which he reflects on the issues of theology and science, is based on a rather harsh periodization of the history of Western science. The conception of the three eras of Ptolemaic, Newtonian, and Einsteinian cosmologies certainly gives ready entrance to some major trends, persons, and currents in the history of science, but, in Torrance's case, its failure to pay adequate attention to other important details in the development of Western thought and to both scientific and theological thinking reveals his preferential reading of the past. The principal criterion for these preferences is the phenomenon of dualism, a concept used by Torrance in a specific sense as an "anti-incarnational" frame of mind.

Torrance's idea of contribution consists of the argument that it was the Christian theology of the Early Church that injected such fertile insights, concepts, and forms of thought into Western thinking as have made the rise of modern empirical science possible. The important point in Torrance's claim is that it was not the doctrine of creation alone that contributed to the rise of science, as held by some other scholars, but the doctrine of creation interpreted through the doctrine of the Incarnation. This insight has been called Torrance's *modified creation argument*. To investigate the justification of this claim, Torrance's idea of the contributive role of the Incarnation and the concept of *homoousion* was analyzed in two respects, epistemologically and ontologically.

The epistemological side of the concept of *homoousion* in the idea of contribution was seen to include two roles, a negative and a positive one. The negative role expresses Torrance's conviction that the incarnational theology of the Early Church excluded every other means of receiving knowledge of God except the Incarnation of the Son of God. Torrance claims, therefore, that the *homoousion* denies any natural theology pursued independently of God's revelation. He points out that natural theology is a movement from us as human beings toward God, a human effort to comprehend the incomprehensible, and therefore it can be viewed as mythology. Quite the opposite is the case in theology, based on revelation, as the movement from God toward us humans. Here God takes the initiative and authoritatively determines what we can know of him. Torrance's aim is to prove that true knowledge is never based on human speculations but on the disclosure of the object. Torrance makes a sharp distinction between two "frames of mind," a Hellenic and a Hebraic one. The former represents an epistemological attitude originating in the human mind, thus being speculative mythology, whereas the latter starts from the mind of God, being genuine theology.

The problem in Torrance's argument was seen to be twofold. First, Torrance applies his preferential reading of history so that the distinction between the Hellenic and Hebraic frames of mind become two separate and mutually exclusive categories. Historical studies, however, prove that the relation between Hellenism and the Judeo-Christian religion was far more subtle and complex than Torrance suggests. The second and also the most central argument in Torrance's treatise was shown to be that, according to him, Athanasius used the concept of *homoousion* epistemologically to reject any independent natural theology, or mythological way of thinking. Comparison with Athanasius's own writings, however, revealed an ambiguity in Torrance's interpretation. On the one hand, he seems to interpret Athanasius correctly in that he discerns the epistemological side in his usage of the term *homoousion*. But on the other hand, his reading is one-sided in that he considers the Church Father as having categorically abandoned any appropriate knowledge of God through the creation strictly, unconditionally, and in principle. Two conclusions were reached. First, Torrance's interpretation reveals an anachronism in that he would like to see Athanasius as having also struggled against natural theology. However, the problem of natural theology, understood according to Torrance's nuances, was not acute until later times. This discovery reveals the second point, namely that the scheme presented by Torrance is Barthian rather than genuinely Athanasian in that he uses the Church Father's theology to provide support for his own theological position. To follow Athanasius, the two modes of theology, natural and revealed, cannot be contradictory simply because both of them are based on the activity of the Word. This point means that what he sees as natural theology can be understood as revealed, too, and that it is not necessary to think of it as centered on human beings because, in Athanasius's opinion, the creation is always Word-centered.

With regard to the positive epistemological role of the contribution of the Incarnation, and the concept of *homoousion*, Torrance's understanding of disclosure models was taken as the starting point. In his opinion, a disclosure model helps the human mind in scientific activity to grasp a reality which transcends it, a reality beyond its finite capabilities for total comprehension. A disclosure model unites the human subjective pole and the transcendent objective pole of an epistemological process so that scientific knowledge becomes possible. It was pointed out that the term *disclosure* in this connection refers to the theological notion of revelation where the transcendent God discloses himself to human beings. This implies that there is a link between disclosure models, as presented by Torrance, and the doctrine of the Incarnation. If the concept of *homoousion* is a disclosure model, it would make Torrance's claims about the "*homoousion* of the natural sciences" more understandable, underscoring the contribution of the incarnational theology of the Early Church to the empirical sciences.

An apparent development in Torrance's thought was detected with regard to the relation between the *homoousion* and the disclosure models in science. In his earlier writings Torrance makes no suggestion of identifying them with each other, but in his later writings such a connection becomes more apparent, although not explicit at all. One of the features given to disclosure models by Torrance was seen to be the "referential" and "open" relation it has to reality, thus giving reality the possibility of determining what we actually know of it. As such a disclosure model is never finite and determined but rather a flexible and "transparent" medium through which reality in its depth is grasped. To the question of whether the *homoousion* can be seen as a disclosure model in the realms of theology and the natural sciences, in the sense that it is open and flexible to constant revision, a twofold answer was given. On the one hand, Torrance suggests that the *homoousion* itself must be open for revision and new definition. Therefore it could be called a disclosure model. But then, on the other hand, the problem in his thinking is that the *homoousion* remains for him the fundamental dogma that is normative for all theology and therefore not open for reconsideration. This ambiguity in Torrance's thought gave reason to conclude that, for him, the *homoousion* is not a disclosure model but rather an authoritative and normative methodological tool, applicable both in theology and the sciences.

The concepts of norm and authority were noted as referring to another problem in Torrance's view of the positive epistemological role of the contribution of the incarnational theology of the Early Church, the position of the human knower in the epistemological process. The question is that of the nature of faith and its relation to reason. Torrance maintains that the concept of *homoousion*, as well as all other elements in the theological work of the Church Fathers, originated under certain epistemological conditions where faith played a crucial role. In fact, faith is a guarantee for the objectivity of knowledge because it places the knowing subject under the compelling power of the object. Thereby true scientific thinking and faith are closely interrelated, according to Torrance. It was shown that Torrance has made efforts to prove that faith is actually an essential part of the contribution that theology has made to the development of the sciences. Torrance's intentions with regard to faith were seen as problematical. Does he refer to a general epistemic attitude, also able to be entertained in a secular context of the natural sciences, or is faith in this context explicitly Christian? Our analysis revealed that Torrance does not refer to some secular intellectual attitude in general but to religious and explicitly Judeo-Christian theism. His way of presenting the views of James Clerk Maxwell and the relation between his deep Christian conviction and his scientific work was seen to disclose Torrance's conviction that, even in modern times, Christian faith can make a positive contribution to the advance of science. The most problematic part of Torrance's argument, however, was shown to be in his insistence that theistic belief was the motivating factor in Albert Einstein's science, too. It was noted that he was aware of the difficulties of this stance, mainly because of the explicit appreciation of Spinoza's nontheistic theology by Einstein, but in spite of this awareness he was seen to insist on the close interaction between theistic belief and scientific thinking in Einstein's thought. The conclusion was that Torrance has to interpret Einstein in this way because Torrance's system requires such an approach.

The idea of contribution also includes an ontological side. Torrance was proved to hold that the ontologies of God and the universe are interconnected through the doctrine of the Incarnation and the concept of *homoousion*, indicating thereby his insistence upon the insufficiency of the doctrine of creation to form the basis of the ontology of the universe. The aim of our analysis was to find out whether Torrance is able to provide convincing evidence for his modified creation argument—the idea that it was not primarily the doctrine of creation but that of the Incarnation that contributed to the rise of modern science. On the basis of his authoritative knowledge of patristic sources, Torrance makes the claim that the doctrine of creation, interpreted through that of the Incarnation, gave birth to certain fundamental ideas concerning the ontological nature of the universe, without which the development of empirical science would not have been possible. According to Torrance, there are three ideas contributed by Early Church theology to the rise of modern science, namely the rationality, freedom, and contingency of the universe. Comparison of his earlier and later writings revealed a development in his thought whereby the notion of contingency has attained a more central and independent role.

Torrance was seen to develop his conception of contingency as a Christian antithesis to the Greek notion which identified contingency with change, chance, and randomness and, therefore, with irrationality. However, it was not before the Christian conception of contingency had emerged that the advance of scientific thinking became possible, in Torrance's opinion. Two questions were then asked, one concerning the relation of contingency to necessity in Torrance's thinking, and another about the justification of the way in which he attempts to give credit to the Incarnation theology for the new idea of contingency. Torrance was seen to hold that while Greek thinking gave primacy to necessity over contingency, early Christian theology reversed the order, making contingency the principal feature of the universe. It was pointed out, however, that this was not yet such a radical or synchronic contingency as emerged as late as the Middle Ages, for instance, in the work of John Duns Scotus, who held that a thing is contingent if its negation is possible at the same time. In this scheme, contingency is based on God's free will and his possibility of continuously choosing another alternative.

The question of the justifiability of Torrance's insistence upon the primacy of the doctrine of the Incarnation, in the contribution of Early Church theology to the rise of science, appeared problematical. First of all, this emphasis is the only conceivable one for a theologian committed to underscoring the significance of an articulate Christian understanding. The doctrine of *creatio ex nihilo* is certainly biblical but also originally a characteristic of the Hebrew religion. The doctrine of the Incarnation, however, is genuinely and exclusively Christian, which is the reason for its centrality in Torrance's modified creation argument. It was discovered that Torrance cannot, after all, provide very strong evidence for his argument, a feature revealed, too, in his choice of words such as "radicalizing" or "deepening" to describe the special contribution of the Incarnation to the idea of contingency. The conclusion was reached that although the *Logos*'s entering the created reality as the Incarnate One implied the affirmation of the reality of the universe even for the Creator himself, it remains unresolved as to what the connection is between this reaffirmation and the idea of contingency or its radicalizing and deepening.

The second contributive element provided by the incarnational theology of the Early Church to the view of the ontology of the universe, as claimed by Torrance, is the rationality of the universe. His appreciation of the rational harmony of the creation was noted, as well as its basis in the *Logos* Christology of the Early Church. For Torrance this implies that there is only one, all-pervading order that excludes any other modes of rationality than the existing one, conceivable by human beings. This rational monistic order, according to him, is divided into two aspects, God's uncreated rationality and the created rationality of the universe. This led to the question, what is the relation between these two aspects of rationality? Because Torrance denies an analogical correspondence between the two rationalities and because he, nevertheless, has to link them together in order to safeguard the idea of the creatureliness of the universe, he makes use of the very vague term *coordination*. A further study of the concept disclosed Torrance's thought in this respect. The two rationalities are coordinated so that there is a resemblance between them, in that they tend to avoid chaos and disorder. The two rationalities, then, can be viewed as being connected by a kind of umbilical cord through which rationality and order is channeled from the uncreated to the created side. This connection is essential; without it the created would lack any meaning, according to Torrance. Their relation is referential and open, not strictly defined as in a one-to-one correspondence.

Torrance's exposition of the rationality of the universe as a contribution of Early Church theology to the rise of science was seen to have one apparent problem, similar to the one detected in his appreciation of the concept of contingency. The actual contribution of the doctrine of the Incarnation, distinct from that of creation, remains obscure in his thinking. Torrance wishes to link the doctrines of the Incarnation and creation so closely together, in order to emphasize the contribution to the view of the ontological rationality of the cosmos, that because of the lack of evidence he uses the concept of confirmation. This cannot but imply that the doctrine of the Incarnation, compared to that of the creation, could not contribute anything strikingly original in this case, either. The conclusion was reached that, implicitly, Torrance has to admit that while the *Logos* was already active in the creation, the rationality of the universe can be claimed to have its sufficient reason in the creation. This means that in this respect, too, his modified creation argument remains unverified.

The third contribution of the incarnational theology of the Early Church to the view of the ontology of the created world, according to Torrance, is the freedom of the universe. Closely linked with the notion of contingency, the freedom of the universe is for Torrance another indication of nondeterminism and of nonarbitrariness as well. In other words, the freedom of the universe means that it is neither necessary nor a product of pure chance. It was noted that Torrance maintained, on the one hand, that, due to the idea of the freedom of the world, the natural sciences can pursue their investigations of nature without having to fear any intrusion into the territory of divine powers, and, on the other, that the cosmos itself, in its freedom, cannot be interpreted as an overall net of necessary causes and effects. As was seen to be the case in rationality, here, too, Torrance makes the distinction of uncreated-created, indicating that the former is the ground and reason for the latter. In its distinctness, created freedom is conditioned and limited by the uncreated freedom of God. A development in Torrance's thought was detected. In some earlier writings he was found to emphasize the dependency of created freedom on the uncreated one, implying that their relation could be characterized as relative independence. In his later writings, however, the idea of participation becomes the determining feature of the relation of the two modes of freedom: the created is allowed to participate in the uncreated. The patristic understanding of deification (θεοποίησις) was detected as the underlying conceptual framework for Torrance's exposition.

The critical question of the originality of the doctrine of the Incarnation, compared to that of creation, with respect to the idea of the concept of freedom was then asked.

While the doctrine of creation asserts that God's preexistent Word was already active in the event of creation, giving the universe its ontological status as a contingent, rational, and free cosmos, it was claimed that one cannot easily find a legitimate place for the doctrine of the incarnate Word as having made an essential contribution in this respect. This means that in all the three points analyzed—the contingency, rationality, and freedom of the universe—Torrance's modified creation argument cannot successfully meet the requirements imposed on it by Torrance himself, namely to prove that the doctrine of the Incarnation contributed something original to the view of the ontology of the universe. Another strand of thinking was detected in Torrance's thought which was claimed to offer more fruitful ingredients for a positive evaluation of the contributive role of the doctrine of the Incarnation for the view of the ontology of the universe and, through it, for the rise of modern science. His views of the Incarnation as the expression of God's inherent love toward his creation and as the reaffirmation of the reality of the universe for God himself were considered more plausible insights than the three "masterful ideas" for demonstrating the contributive impact of the incarnational theology of the Early Church on the view of the ontological nature of the universe. This theologically reasoned aspect of Torrance's thinking, in spite of its plausibility and advantages, was discerned as being overshadowed by the more philosophical one.

One of the most original characteristics of Torrance's idea of contribution was claimed to be his view of a novel conception of space and time as a product of the incarnational theology of the Early Church. Torrance's argument is that the Incarnation of the Son of God compelled the Early Church Fathers to find an adequate expression for it and, for that purpose, to transform the static notions of space and time entertained in the intellectual world of Antiquity. Space and time are important for Torrance not only as an aspect of the doctrine of the Incarnation, articulating God's relation to created space and time, but also because, according to him, space and time are the actual scene common to both theology and the natural sciences, the scene in which they attempt to penetrate deeper into the field of their investigation. Precisely due to their createdness, space and time are, in the final analysis, theological entities. And furthermore, it is part of the "mission" of human beings as members of the creation with its space-time structure to give articulate expression to the creation.

Torrance's theological position becomes very apparent when he enters into discussion on the ultimate nature of space and time, a subject commonly reflected on in the philosophy of physics. Torrance holds that the incarnational theology of the Early Church radically changed the Greek conception of space and time. In our analysis the role of space turned out to be crucial. Torrance was seen to make use of the dichotomy receptacle (absolute or container) versus relational notions of space. Roughly defined, the former refers to space as a sphere with strict boundaries and containing independent particles. The relational notion, however, regards relations between existents as constituting the structure of space. This means that the relational notion remains more open and flexible than the absolute one.

Torrance's point is that it was the incarnational theology of the Early Church, having been compelled to reflect on God's relation to space, that contributed to the birth of the relational notion and remarkably anticipated the view of space expressed in the theories of relativity of Albert Einstein. The characteristic feature of Torrance's understanding of the relational notion of space was then asked, compared to the traditional definition given by Leibniz. It was noted that while Leibniz considers horizontal relations between existents as the constituting factor of the structure of space, Torrance replaces it with a vertical relation between God and space as his creation. With regard to his claim that the view of space entertained in the incarnational theology of the Early Church actually belongs in the same continuum with the view of space presupposed by Einstein's theory of relativity, the critical remark was made that there is no inherent necessity in the theories of relativity for adhering to the relational notion alone. In the philosophy of physics, interpretations adhering to the absolute notion are possible as well. It was concluded that the connection Torrance makes between the views of space entertained in the theology of the Early Church and Einstein's theories of relativity is insightful but daring and in desperate need of further reflection and evidence. The reason for Torrance's determinate support for the relational notion of space is apparent: openness and flexibility make it easier intellectually to accept the idea of the Incarnation than would be the case if the absolute notion were acknowledged. Therefore, Torrance's preference is strongly reasoned theologically. Similar remarks resulted from our analysis of Torrance's understanding of the relational notion of time.

The approach applied in the study of Torrance's idea of compulsion was twofold. First, his view of realism, especially its very essence and its theological context, was examined, and second, his view of the threat posed by dualism to all realistic science was analyzed. Both points of views were considered as being of utmost importance because, so far, no single study has paid sufficient attention to the theological context of Torrance's realism or his idea of dualism.

Torrance was seen to maintain that the concept of *homoousion* is both a result of realistic thinking and a guarantee of its proper functioning. What is observed is "of the same being" as reality in its ontological depth. Torrance's view of realism was first examined from the standpoint of diverse definitions of realism. Due to the varying results of some earlier studies of Torrance's realism, efforts to define it on a purely philosophical basis were found to be inadequate. The concept of *homoousion* represents, then, the key to understanding Torrance's realism, emphasizing the conviction that since the reality under observation, independent of human beings as its observers, must be allowed to determine what is the content of our knowledge, there is no other option than to acknowledge the supremacy of reality over human conceptions of it, the supremacy of ontology over epistemology. Torrance holds that reality compels us to know it according to its true nature. He insists that the compelling force of reality can be effective only if the *homoousion* relation between observations and reality is acknowledged.

The critical question was asked, then, what kinds of possibilities there are for Torrance's realism, characterized primarily by the idea of compulsion, to succeed in the realm of the natural sciences. His remarks on the so-called Copenhagen interpretation of quantum physics were evaluated from this point of view. It was noted that Torrance is in favor of Einstein's criticism of the Copenhagen interpretation, which is commonly held as rejecting traditional forms of realism. This means that Torrance cannot accept the thought that knowledge of quantum phenomena does not reach reality itself as it is independently of ourselves but is restricted to appearances and phenomena. It was discovered that there is a difference in the way in which realism is used by some quantum physicists and by Torrance. For instance, Werner Heisenberg holds that realism inevitably presupposes strict correspondence between scientific theories and observations. This means that a realistic approach in science demands knowledge of reality before a comparison between it and a test result could be performed. Torrance's answer to this was seen to consist of the conviction that the distinction between correspondence and coherence theories of truth is unfortunate. Our analysis showed that Torrance prefers a referential relation to a correspondence relation between reality and scientific theories, so that the vagueness of human conceptions and the absolute supremacy of reality are emphasized. The reality observed is therefore always beyond; it is transcendent. This view reinforces the point that we as humans have no ability to define reality because it always remains transcendent and totally independent of us. This leaves a knowing observer with only one choice, to submit to the compelling power of this transcendent reality where he can rely on the *homoousion* character of his observations: reality forces one to obtain genuine knowledge of reality through the phenomena he observes. Reality cannot be defined beforehand but is actually disclosed in scientific investigations, and, accordingly, the power of a scientific theory does not lie in human definitions or theories themselves but in reality as their ultimate judge. The conclusion was reached that Torrance's program, based on the idea of *homoousion*, might be possible for an application in quantum physics if the realist position were accepted in principle, an approach that is not, after all, totally unknown within its field.

Torrance's strong conviction of the capability of the human mind to apprehend reality in its independence and depth was noted as suggesting that there are no limits to human understanding. An analysis of how Torrance uses the concepts of *symbol* and *sign* disclosed that the referential relation between reality and human understanding, implied by the *homoousion*, does not allow any human construction the right to claim to own the totality of truth. Human knowledge can never be comprehension but, rather, apprehension, a detail in Torrance's thought that brings him close to Karl Popper's notion of *verisimilitude*.

Torrance thus maintains that reality never loses its sovereignty over human conceptions but exercises its compelling power, submitting the knowing subject to its force. This raises one special problem, namely the way in which Torrance relates the knowing subject to the independent object in an epistemological process. He suggests that to be a knowing subject means to be subject to the compelling power of reality. Torrance presents Michael Polanyi's concept of "personal knowledge" as a relevant epistemological approach in which the problem would diminish. Torrance's views of intuition and logic were also examined, and the conclusion was drawn that, on the one hand, our role as human subjects is receptive in that our submission to the coercive force of reality is demanded. On the other hand, however, this role is not totally passive because even in our submission we remain active in our relentless questioning. Accordingly, the *homoousion* must not be seen as an indication of the objective side of the process only but as that of our subjective participation as well. It was also pointed out that Torrance's reliance upon the self-evident compelling power of independent reality cannot avoid certain features of a circular argument: we know that our knowledge accords with reality if we allow its compelling power to affect our understanding, and we know that the coercive force acts upon us when we adapt our knowledge to reality. Torrance's approach was evaluated as resulting in a situation where the problematic relation of subject and object remains both strict and imprecise at the same time: the difference between them is emphasized relentlessly and rigorously, but the difficulty lies ultimately in the inaccuracy of the demarcation line between them.

Some earlier studies of Torrance's writings usually insisted on pointing out the problem of the subject-object relation in his thought. In this study, therefore, an explicit effort was made to relate to each other the question of realism, subject-object relations, the idea of compulsion, and the concept of *homoousion* in Torrance's own theological context. The Reformed doctrine of election proved most fruitful for this purpose. It was first discovered that in Torrance's thinking divine election and compulsion are deeply connected, almost identical. Torrance's interpretation of Calvin proved this conclusion to be correct. He emphasizes the Reformer's view that what God has elected in his sovereignty cannot be resisted or altered by human efforts. Election, then, discards all human projections toward God and confirms the direction from God to human beings. We can only acknowledge God's decision through the election in its coercive force.

Torrance's interpretation of Calvin raised the question of the relation between the Incarnation and election in the Reformer's thought. If the Reformer's account of election is taken at its face value as presented in his *Institutes*, one cannot avoid the impression that Christ and his atoning work have only a secondary role in his theology. In other words, Christ's centrality is based on his ability to give believers assurance of being elected and that Christ's atoning sacrifice does not add anything essentially new to God's eternal election. For Calvin, then, election is not primarily a Christological act of God. Torrance, however, gives the impression that Calvin actually thought of election as centered in Christ. The reason for Torrance's reluctance to accept the interpretation of predestination entertained in and after the Reformed Orthodox era, one that left no specific room for Christ in the act of predestination, can be seen to lie in its *necessitarian* character. Torrance thus maintains that individual predestination, based on God's decree issued before the beginning of time, implies that we can find ourselves only in a fatalistic, strictly predetermined relation with God. Therefore he defines election in another manner: the Incarnation of the Son of God is the divine election. The direct implication of this is that the idea of compulsion, too, becomes essentially Christocentric.

Torrance's interpretation of Calvin, then, takes on familiar Barthian tones according to which God's election has its only target in Jesus Christ, the one elected and also rejected, so that the election does not concern individual persons directly, on the basis of God's eternal decree, but only indirectly, through the person and act of Christ. Torrance's idea of compulsion was noted as correlating with Barth's conception that God's election in Christ cannot be frustrated by human unbelief or resistance. It can only be accepted in obedient submission. After these remarks the problem of the relation of election to the Incarnation, and its implications for scientific thinking in general, was stated. The shift from theology to the realm of the natural sciences and philosophy does not produce insuperable problems for Torrance since, in his opinion, election forms the basis for the coercive force of reality in that field, as well. The connecting link is his conception of rationality. In the natural sciences we meet a rationality of the universe that transcends our comprehension and simultaneously exercises its compelling power over us. Ultimately, the commanding force is not the rationality of the universe as such but God who is the Source of created, contingent rationality. In other words, God's compulsion meets us through the rationality of the universe. God has chosen to endow his creation with rationality, and therefore rationality is compelling to the extent that there is no other way to know reality.

Torrance was found to have introduced a remarkable addition to the traditional Reformed doctrine that only rational beings—human beings and angels—can be elected. By widening the range of rationality from humans and angels to the whole universe, Torrance maintains that God has elected his creation to bear rationality and knowledge and, additionally, that this election does not take place outside Christ but in essential connection with him, for he is the *Logos* through whom the universe was created and who was elected to be incarnate under its temporal conditions. Barth's Christocentric election is, therefore, applied by Torrance to the whole universe and, implicitly, in the field of the natural sciences. It was concluded that, in his opinion, the idea of compulsion and the realism based upon it are direct results of God's election: the world we inhabit is the only one realized out of innumerable possibilities and so now we must accept this and yield to the coercive power of reality, a power that is not self-explanatory but is, in the final analysis, identical with the power of God, the divine and sovereign Elector. On this basis any natural theology pursued apart from the Incarnation and the *homoousion* is conclusively excluded: the creation reveals not only God in the general sense of a Divinity or Creator but explicitly God who is manifest in the Word or *Logos* made flesh. In the incarnate Word the ultimate nature of the universe is revealed.

It was pointed out that Torrance's argumentation could be considered theologically reasoned realism. As a somewhat daring but evident interpretation of Torrance's position, it was claimed that he suggests a unifying theory of all human knowledge using the concept of *homoousion*, applicable in every field of investigation. There is, then, no other way of gaining trustworthy knowledge. The remark was also made that Torrance's strong theological position and his appreciation of God's revelation in his interpretation of how nature discloses its secrets in natural scientific investigations make it justified to call his stance *revelatory realism*, a phrase combining the idea of compulsion, the concept of *homoousion*, based on the revelatory act of the Incarnation, and the doctrine of election.

Torrance maintains that, in scientific thinking, a realistic attitude, based on the concept of *homoousion*, has been seriously impaired by its inimical counterforce, dualism. To express briefly the relation between realism, the idea of compulsion, the concept of *homoousion*, and dualism, it can be stated that, in Torrance's opinion, dualism prevents any application of the idea of compulsion and the idea of *homoousion*, resulting in the rejection of a realist approach. It was pointed out that, according to Torrance, Arianism, viewed as a distinctively dangerous heresy by Athanasius, constructed its Christology on dualistic notions which had to be overcome so that the correct way of obtaining knowledge of God could be safeguarded. The actual problem, however, is that, in spite of the fact that dualism occupies such a central position in Torrance's theology not only in its patristic emphases but also throughout his thinking, he has not given adequate attention to properly defining it. The same defect applies, as well, to earlier studies of Torrance. Therefore, in the first place, an effort was made to define Torrance's use of the complex concept of dualism, on the basis of expositions by some other scholars and of his criticism of Arian theology. The analysis showed, in general terms, that the application of the concept of dualism usually reveals three characteristic features defining the concept. First, there is a certain opposition between the two principles involved. Second, there is a certain degree of internalization in the use of the concept, for instance, when it is seen as a mere theoretical framework or as a comprehensive worldview. And third, the use of the concept of dualism usually raises certain value judgments concerning itself so that it is considered either a good or a bad thing.

The most comprehensive definition given by Torrance himself is the view that dualism expresses the tendency of the human mind to see reality through dichotomies. This means that dualism cannot be seen as a property of reality but rather as one of the human mind. Also, his classification of dualism into three categories—ontological, epistemological, and anthropological—was noted. This definition by Torrance himself was, however, considered insufficient due to its philosophical overtones. Consequently, it was assumed that, for him, dualism is theologically reasoned. In order to verify this assumption, Torrance's exposition of the Arian controversy was analyzed. Several critical remarks on Torrance's reading of the conflict were made, such as his unreserved acceptance of Athanasius's accounts of the Arian position as well as the historically problematical claim that Athanasius, among other defenders of the Nicene formulation, represented Hebraic rather than Hellenistic "habits" of mind. Torrance's dualism was then proved to be Christologically based: in the context of the Arian controversy over the *homoousion* of the Son with the Father, Christological dualism is for him an imprecise mixture of divine and human features coexisting in an unbalanced way in the person of Christ. Torrance was seen to express critical reservations as to the possibilities of logic to provide adequate means for understanding the Incarnation of the Son, thus implying a close link between dualism and logic.

The analysis showed that the existence of two poles in a distinction is not sufficient to render it dualistic. From his Christologically oriented premises Torrance's dualism is constituted by the relation prevailing between the poles. Finally, Torrance's idea of dualism was defined as a paradigmatic, most often unconsciously applied but deeply internalized, way of perceiving reality which is seen as consisting of two principles. In the final analysis the poles are God and the world, or the Creator and the creation, between which there is no real interaction of dynamic relation as indication of the ontological and epistemological openness of God and the universe toward each other, thus offering a negation of the *homoousion*.

In Torrance's periodization, dualism was seen to play a central role. In the time of the Early Church it was Ptolemaic cosmology that introduced dualistic modes into the heart of Western thinking. And when the natural sciences emerged in the seventeenth century, dualism was once again involved. Torrance's view of Newton's scientific thought was analyzed by asking the question, How do Newton, his physical thinking, and theological thinking, on the one hand, and Arianism, its Christology, and its understanding of reality, on the other, relate to each other? The crucial point of convergence was found in the concept of space. For Torrance, Newton's distinction between absolute and relative space and time is another indication of dualistic modes of thinking. Accordingly, he criticizes Newton for holding to the receptacle or container notion of space, leaving no room for its relational understanding.

The actual point of Torrance's criticism of Newton is the latter's alleged identification of absolute space with the mind of God. Comparison with Newton's writings and studies made of his life and work disclosed a major problem in Torrance's reading of Newton. Torrance's argument is based on a single and very short passage in one of

Newton's works, so that it cannot be seen as a reliable indication of his true thoughts. It was then concluded that Torrance needs this interpretation to give an explanation for his idea of the interconnection between dualism, scientific thinking, and the view of the impossibility of God's incarnation. He was seen to maintain that the Newtonian dualism between absolute and relative space and time has the effect of closing God and the world off from each other so totally that any "active interaction" between them becomes not just weaker but plainly impossible. If God's mind is identical with absolute space, he cannot become incarnate within it. Torrance presents Newton as a man of strict logic who cannot escape from the receptacle notion of space, to which his notion of absolute space, according to him, inevitably leads. His emphasis on Newton's appreciation of Arius and his theology was confirmed to be historically correct. He was seen to suggest that there is at least a parallel in the thought of Arius and Newton in the sense that both men rejected Athanasius's *homoousion* and supported the receptacle notion of space. For both of them the *homoousion* represented an unworthy way of thinking of God as if God were divided into several individual persons or as if something created were included in his substance. Accordingly, seeing space as a receptacle made the Incarnation logically impossible for both of them. In short, Torrance holds that a dualistic understanding of space made both Arius and Newton disregard the Incarnation and the full Deity of Christ, presupposed by the *homoousion*, suggesting that a definite preconceived presumption concerning a physical issue, such as space, had certain inevitable consequences in the religious understanding of the two men.

Another problem in Torrance's reading of Newton is his suggestion that the scientist was a deist. It was pointed out that the principal criterion for Torrance to call a religious phenomenon deistic seems to be that the reality of the Incarnation and the *homoousion* is denied. He claims implicitly that any denial of the possibility of the Incarnation is a mark of deism. Torrance's definition is not, however, the one usually applied, in which deists are those who believe in God on the basis of natural reason and not any divine revelation. Newton's firm belief in the God of the Bible was then pointed out, and the conclusion was drawn that Torrance's claim with respect to Newton's deism is open to dispute. It was also inferred that Newton, as the eminent exponent of modern physical science and scientific thinking in general, is the target of Torrance's criticism for one basic reason: in his opinion, Newton's dualistic thought inevitably leads to the exclusion of the possibility of the Incarnation and the notion of the consubstantiality of the Son with the Father. A further implication of this would be, according to Torrance, that scientific thinking cannot accept the idea of the Incarnation nor a truly realist position in its investigation.

In spite of the problems inherent in Torrance's reading of Newton, he was shown to have articulated an important area of discussion which could offer new insights for the theology-science dialogue. His analysis of the two roles of God in Newton's system is an effort to provide reasons for the exacerbated relations between theology and science after the Scientific Revolution of the seventeenth century. He holds that, on the one hand, Newton had a clear view of the universe as dependent on God, but on the other hand, the scientist's tendency to bring God's actions into his argument to explain irregularities that he could not comprehend with his own mathematical devices eventually turned out to be unfortunate. The more this view became explicable through scientific investigation, the less was left to be explained by God's actions. This problem was seen to trouble Torrance a great deal because it consists of a contradiction: Newton's view of the universe as dependent on God, as contingent, seems to safeguard its openness and indeterminism, but, on the other hand, God's role as an explanation for otherwise inexplicable irregularities relates to the scientist's tendency to view the world as closed and deterministic. Torrance makes the claim that whenever God is seen simultaneously as both the reference point of the contingent universe and an explanatory principle for irregularities in its mechanistic and necessary structure, there is always the danger that these features are totally identified with each other so that the idea of contingency is inevitably excluded. Related to the doctrine of the Incarnation, Torrance was shown to have formulated a "circle of denial" where rejecting contingency means rejecting the possibility of the Incarnation, and, in turn, rejecting the possibility of the Incarnation means rejecting the contingency. And further, abandoning the idea of contingency means that any ontological interaction of the *homoousion* type between God and the world, in which a phenomenon is regarded as identical with reality, is ruled out. Therefore, there is no possibility that the idea of compulsion might determine what is the ontological bearing of reality in its inherent force upon human knowing.

Torrance's interpretation of Newton was evaluated as not being able to do full justice to Newton's thought nor his own intentions to reinforce the idea of contingency as a basic concept for both theology and the natural sciences. Therefore it was concluded that Torrance's idea of dualism, as the opposite of the profound unitary realism in the scientist's thought, affecting the later development of modern science along mechanistic and deterministic lines, proves to be based on insightful ideas rather than on convincing evidence. However, as such, they were mentioned as offering a fertile resource, asking important questions and suggesting some possible referents for solutions in the relationship between science and theology, as well as in their history of reciprocal influence.

Torrance's idea of reminder contains two elements: first, the view that the modern natural sciences in general and physics in particular have been able to discard dualistic modes of thinking and have returned to the truly scientific and realist position where the compelling power of reality is allowed to function. The second point is that, in this new situation, the natural sciences can remind theology of its own forgotten basis in the Incarnation and the *homoousion.* It should be carefully noted that Torrance does not appreciate modern physics because of its ability to teach theology something essentially new but only because it can remind theology of those basic elements of science that have been neglected for a long time because of all-pervading dualism.

The content of Torrance's idea of reminder was analyzed in two respects, ontologically and epistemologically. Ontologically, the concept of field was detected as an essential element in his argumentation. Torrance holds that James Clerk Maxwell's theory of the electromagnetic field anticipated the Einsteinian revolution, culminating in the theories of relativity. Three relevant aspects were pointed out. First, the centrality of space in Torrance's appreciation of the concept of field emphasizes the difference he sees between the Newtonian and Maxwellian views of space. In his opinion, Maxwell was able to detach himself from Newton's absolute, container notion of space. This conception was claimed to be ambiguous because Maxwell interpreted the problem concerning the influence of bodies on each other from a distance with the idea of an ethereal medium filling the space. Maxwell presupposed implicitly a kind of receptacle notion of space—that is, space contains ether. Also, some other pieces of textual evidence showed that

Maxwell was not, after all, as free from the container notion of space as Torrance would like us to believe.

The second relevant point in Torrance's appreciation of the concept of field is the idea of relationality, which, in his opinion, rejects any strictly mechanistic understanding of the universe. His appreciation of Einstein's theories of relativity was proved to be important for his argumentation, underscoring the role of a dynamic conception of the ontology of the universe. In this respect his reading of Einstein was evaluated as adequate. The third ontologically important point revealed by the concept of the field is the idea of contingency. Torrance was observed to be carefully constructing his argument, the aim of which is to assert that Newtonian science operated with necessities and hard causality or determinism, whereas Maxwell could restore contingency to the ontology of the universe through his field theory. The revolutionary character of Maxwell's science lies, then, in its ability to discard determinism and initiate a new era in physics, resulting in the indeterminism of the quantum world.

A problem was found in Torrance's claim that the novelty and revolutionary nature of Maxwell's theory of the electromagnetic field lies precisely in its indeterminism. This requires, however, that strict causality cannot be applied to the concept of field. The counterargument was then put forward that it can indeed be applied. Both the use of partial differential equations in field calculations and the neglect of probability prove that the concept of field is not as free of determinism as Torrance claims. With regard to the idea of probability, Torrance's point was considered justified if his premise were accepted, a premise according to which indeterminism can be defined not from the standpoint of probability but from that of relationality.

Due to the central role of the idea of relationality, it was further analyzed. Torrance's use of the phrase *inherent relatedness* or *onto-relations* revealed that, for him, the ultimate nature of ontology consists, then, not primarily of substances but rather of relations without which no being could be what it is. Reality can be viewed as a field that is based on a network of ontologically constitutive relations between entities. Thus, ontology is basically relational. It was also shown that their being called *inherent* is understandable only as an exclusive alternative to their being called *external* or causal relations. Causal relationality, however, does not touch even tangentially on the being of a thing. Torrance has made efforts to prove that what he calls the *dynamical inherent relatedness* meets this requirement: these relations are not external in the sense of causality but internal in the sense of noncausality or indeterminism or, better expressed, of contingency.

Torrance's appreciation of indeterminism was stated as falling in line with the philosophical emphasis in modern physics, but his way of giving reasons for it by referring to the ideas of Albert Einstein was found to cause a problem. To the question of whether Einstein was a determinist, Torrance's answer is absolutely negative. On the contrary, he was seen to accuse the Copenhagen School of quantum mechanics of reintroducing determinism into scientific thinking. Torrance's own contribution in this connection was detected in his way of identifying Einstein's reliance on the nonarbitrariness of the universe with the concept of contingency. He wants to construct, as it were, a third option between pure chance and causal determinism by claiming that there are two aspects in the concept of contingency, namely one in which it is dependent upon God as the transcendent Ground of all being and one in which it posits a freedom of its own, restricted only by the freedom of God.

It was then pointed out that, to be convincing, Torrance has to be able to prove that Einstein really had such an understanding of contingency as Torrance claims, contingency as a middle point between determinism and chance. The remark was then made that Einstein did not explicitly use the concept in his works to describe the ultimate ontological nature of the universe. It was concluded that if contingency really is held to be identical with lack of necessity and indeterminism but not with arbitrariness, as Torrance maintains, then it can be, at least to some extent, justifiably used to express Einstein's views. It was then concluded that Torrance's standpoint is not understandable from the perspective of Einstein thoughts on ontology only but primarily in the context of his overall theological system. Belief in God as the Creator and Guarantor of order in the universe was observed to be the principal reason for Torrance's acceptance of Einstein's ideas on the ontological nature of the universe.

On this basis it was suggested that Torrance's discussion would gain more credibility if he did not lean so much toward the two scientists, Maxwell and Einstein, but presented the insights more as his own modifications of some important issues in modern physics. This not being the case, his intentions behind this actual solution were then sought. It was pointed out that Torrance's grand theological theme in the doctrine of the Incarnation and the scientific theories of Maxwell and Einstein coincide in one important respect: both are concerned with the problems of space. It was also claimed that the doctrine of the Incarnation unites the ontological aspects of both the idea of contribution and that of reminder. In other words, Maxwell's and Einstein's views of ontology and their alleged indeterminism offer Torrance not only scientific instruments to try to convince his readers of the contingency of reality which, according to Torrance, was the direct impact of the doctrine of the Incarnation with its concept of *homoousion*. This ontology brings forth space and time as contingent features of the universe, thus highlighting Torrance's conviction that they have no absolute status in the Newtonian sense but are themselves a creation of God. Within this context, the question of *homoousion* is best understandable in the general framework of space, that is, how God in his Being can enter his own limited creation and its space-time conditions, while problems in quantum physics do not offer such a direct overlap for science and this specific theological doctrine. In Torrance's opinion, both Maxwell and Einstein have made the doctrine of the Incarnation intellectually easier to understand.

The epistemological side of the content of Torrance's idea of reminder was then analyzed, beginning from his conception of the relevance of Immanuel Kant's philosophy. Torrance's criticism of Kant concentrates on the philosopher's alleged dualism between the phenomenal and the noumenal as well as on his injection of elements of necessity and determinism into the foundations of knowledge. Kant's conviction that human knowledge cannot reach "the thing in itself" but only phenomena sharply conflicts with Torrance's intentions to bring ontology into the heart of epistemology. A special case of Kant's dualistic epistemology, according to Torrance, is his notion of space and time as the unchangeable, absolute categories of human understanding. Torrance holds that precisely the view of absolute space and time connect Newton and Kant: the scientist identified absolute space and time with God's mind, the philosopher transferred them from God's mind to the human mind. While Newton linked dualism to the ontological nature of the universe, Kant did the same with human mind and epistemology as well.

The connection between the two intellectual geniuses, as presented by Torrance, was then analyzed, and it was concluded that his idea of the transfer of absolute time and space from God's mind, as an original view of Newton, to the human mind, as was allegedly done by Kant, is problematical in two respects. The first premise in this argument presupposes that Newton actually identified God's mind with absolute space and time. But as was demonstrated earlier, this reading of Newton is not based on sufficient evidence. Therefore it was concluded that what is at issue is not a historical reception of Newton's concepts by Kant but rather Torrance's own emphasis on the crucial nature of Kant's epistemological thinking and its alleged indebtedness to Newton. The second problematical point was in reference to a slight confusion of terms. Newton developed his view of space and time on the axis absolute-relative, giving both a justifiable position in his thinking. However, he rejected the *relational* notion of space and time and adhered to the absolute notion, absolute in the receptacle or container sense. In other words, Newton's absolute-relative axis is included in his container notion. It is this absolute space and time as the pair of *relative* space and time with which Newton allegedly identified God's mind, not that absolute space and time which is the exclusive counterpart of the *relational* notion. In order for Torrance's interpretation of Kant to be correct, he would need to prove that Kant operated with absolute space and time as the other pole in the axis absolute-relative. But on the basis of Kant's writings it was shown that this was not the case. In his own discussion, then, Kant treated space and time as absolute, not as the complementary side of relative space and time but only as exclusive of the relational view. In other respects Torrance's reading of Kant and the philosopher's efforts to give the human mind a determining role in an epistemological process was found to be valid.

It was concluded that Torrance's criticism of Kant arises primarily from his esteem for the doctrine of the Incarnation and the concept of *homoousion*. If space and time are absolute, and if they are categories only in our mind without any independent objectivity of their own, any idea of God entering this space and time remains inconceivable. And further, if this epistemology prevents us from gaining knowledge from reality itself and can only touch upon appearances or phenomena, then there is no possibility of taking seriously the *homoousion* nature of the Son, or the Deity of Christ. Only the historical phenomenon of Jesus provides an object for theological inquiry. Torrance's reluctance to agree with Kant in his epistemological views can, then, be maintained as being based primarily on theological grounds.

Torrance's criticism of Kant gave reasons to analyze further the epistemological impact and relevance of the modern natural sciences. Torrance was seen to maintain that dualism has had the effect of disintegrating form and being. The novel situation in physics works in precisely the opposite direction, toward the integration of form and being. For Torrance, integration means bringing together both the subjective and the objective poles in the act of knowing into a coherent synthesis. It was proved, however, that Torrance's procedure does not guarantee a totally balanced coexistence of form and being, subject and object, or epistemology and ontology. In the final analysis, his suggestions lead inevitably to the supremacy of being over form, object over subject and ontology over epistemology. It was pointed out, consequently, that the integration, as described by Torrance, is not aimed at reaching equality or complete balance but only at bringing the two elements together so that what is theoretical, subjective or form might no longer function alone and self-sufficiently but might be transferred to the control of what is empirical, objective, and being itself. Only in this manner is the fluent working of the idea of compulsion guaranteed. It was shown that this approach is necessary for Torrance because the principle of *homoousion* could not have any positive significance if the human mind were allowed to function on its own terms without taking ontological objectivity into account, that is, what is observed could not be identified with reality itself but only with some constructs of the human autonomous reason. Torrance maintains that restoring the *homoousion* relation is precisely what modern physics has been able to do. The counterdualistic integration is, therefore, a central theme in Torrance's idea of reminder.

Another crucial theme in Torrance's idea of reminder is the notion of openness. He was observed to be making use of Einstein's conception of the stratified structure of a scientific system, in which the first layer consists of our everyday experience and concepts related to it. On the second level, that is, the first scientific level, a new set of concepts is introduced to collect the meaning and purpose of concepts used on the first level. A deeper abstraction is met on the third or meta-scientific level, which is still poorer in concepts and relations but nevertheless succinctly collects knowledge with greater precision than is the case on the lower levels. Torrance was seen to hold that these layers are always open upward, toward an ever increasing refinement and precision. The direction of this openness was noted to be important because it was observed as representing Torrance's view that theories must not manipulate reality but that empirical reality, as the ground under all the three layers, should determine the content of knowledge in the hierarchical structure. The conclusion was reached that, with the help of this argument, Torrance wishes to emphasize the role of the concept of *homoousion* as the key term of realism, in the direction of which the modern natural sciences, physics in particular, refer.

After the analysis of the content of Torrance's idea of reminder, its implications for theology, as he understands them, were studied. His notions of contingency, indeterminism, openness, and integration were seen to be united into a realistic program, the aim of which is to emphasize the importance of the doctrine of the Incarnation as the basis of theology. Torrance holds that, without a proper Christology and a coherent view of the Trinity, theology would remain both devoid of any relevant content and, therefore, unscientific. Methodologically and epistemologically, theology as a science has to submit itself to the compulsion of its Object as he has revealed himself in the incarnate Jesus Christ. Substantially and ontologically, theology has to start from the doctrines of Christology and the Trinity, the actual knowledge of God, and separate itself from any attempt to reflect on God beyond his self-revelation in the Incarnation in the manner of an independent natural theology. Taking realism into account, as has been done in the modern natural sciences, means that dualistic habits of thought should be discarded, resulting in a scientific theology.

An analysis of Torrance's programmatic views was then pursued. It was observed that, in his opinion, scientific theology cannot start outside the basic point of God's self-revelation in the Incarnation of his Son. The concept of *homoousion* was confirmed, once again, as forming the basis of all theological activity according to Torrance. It was then asked what kind of evidence Torrance can provide to support his argument for a scientific theology on the basis of the contemporary development in the natural sciences.

It was pointed out that Torrance sees faith and reason as close allies between which there should not be any contradiction. Torrance's notions that belief plays a crucial role in every scientific activity and that it confirms the objectivity and reality, the otherness of an object, were noted. This implies that without faith there would be no truly objective science, either. This claim was considered apologetically motivated in order to show that there is a hint of "theology" in every science. The other implication was shown to be that theology cannot be blamed for applying unscientific devices when it appreciates the role of faith in its own inquiries.

Torrance was seen to have the intention to argue that theology be classified as a rigorous science on its own terms, comparable to the natural sciences. Theology has nothing to do with speculation but only with reality as it is given in its compelling force. Therefore, genuine scientifically pursued theology and the idea of compulsion are inseparable, and this leads Torrance to claim that theology is a science of an a posteriori nature. Torrance's view of the a posteriori nature of theology was then critically examined, and it was concluded that, in order to be able to classify theology as such a science, Torrance substantially broadens the area within which the term *a posteriori* can be applied. He holds that it is our direct encounter with the incarnate Word that forms the ground for theological knowledge. The question is not only of the Jesus of Nazareth of the four gospels or the other New Testament books, although the necessity of faithfulness to the biblical witness is never questioned, but primarily of the Lord with whom an individual can be in face-to-face contact here and now, in this space and time. It was concluded that Torrance cannot convincingly prove to a critical reader of his writings that a devotional religious experience could be considered a truly a posteriori basis for theological knowledge and that this character could be acknowledged from a general scientific point of view, that is, by the natural sciences or the philosophy of science.

The same problem was found in Torrance's claim that theology should be regarded not only as an a posteriori science but as an *empirical* science as well. The central point in Torrance's argument was found in his insistence that the term *empirical* need not be applied to pure sense experience only but also to those experiences which are not sensible or tangible. This proves that Torrance makes efforts to enlarge the applicability of the term *empirical* in order to include theology within its scope. His solution was found interesting, but its main defect was noted as lying in its apparent inability to take seriously the vast variety of differing and contradictory experiences of God in the multireligious and therefore pluralistic world.

Torrance holds that Jesus Christ, as the incarnate Son of the Father, is the *empiria* of theology. This was seen to suggest that Torrance does have an answer to the critical question posed above as to who has the correct experience of God. It was discovered that, in Torrance's opinion, it is the mind of the Church which has the ability to receive and express a theological experience realistically according to the way things are and based particularly on the Incarnation. It was pointed out that Torrance faithfully follows Karl Barth in seeing theology as the function of the Church. This approach was, therefore, regarded as endogenous and self-sufficient, having no need to seek support for its validity from outside the Church. It was concluded that purely theological issues, such as in Christology or ecclesiology, must certainly be regarded as matters of theology alone and in their case no justification "from outside" is necessary. But the question of the scientific nature of theology in general, particularly if it is placed on the same level as the natural sciences, cannot be successfully interpreted as a purely internal affair within the boundaries of theology alone and on its own terms.

In Torrance's argumentation, the phrase *the mind of the Church* as an indication of a social matrix in which reality is interpreted is viewed as a parallel with a scientific community. On the one hand, this identification was evaluated as insufficient to give convincing reasons for calling theology a science in the sense of the natural sciences. On the other hand, this phrase was seen to be more plausible in its purely theological context. Torrance uses it to introduce the action of the Holy Spirit as the Constitutor of the Christian communion and as an epistemological factor who bestows knowledge of God by showing the way to the Son. The epistemological task of the Holy Spirit, then, is to act within the Church so that the compulsion, true objectivity, and consideration of an independent reality might become real. The doctrine of the Trinity is thereby also introduced into this argument.

Torrance's appreciation of Karl Barth as the theologian who restored the doctrine of the Trinity to the heart of theology was noted. This gives him reason to name his mentor as *the* theologian who has shown the way back to a truly scientific theology and to the doctrine of the Incarnation. It was claimed that both Athanasius and Einstein are used by Torrance as authorities by whom the relevance of Barth's theological program can be reasoned and its alleged scientific character underscored. In this respect Torrance cannot avoid giving the impression of a kind of confessional patriotism in which grand figures in the history of science and theology are used to lend credibility to his own position and background as a Reformed and particularly Barthian theologian. Although the comparison between the three figures was deemed daring, its positive contribution was claimed to lie in its ability to highlight certain significant questions in modern theology, especially in the fundamentals of theology.

Torrance was noted as having made relentless efforts to relate Christology and the doctrine of the Trinity with each other. His constant claim has been that the doctrine of the Trinity actually forms the basic grammar of theology. It was pointed out that, on the one hand, Torrance's strongly Christological position makes his demands for the centrality of the Trinity a problem. But the analysis showed, on the other hand, that introducing the Holy Spirit as the One who transfers a human being into such a relation to the Son that knowledge of the Father becomes possible, proves that, especially in his mature writings, the Trinitarian structure and its foundation in the *homoousion* relations between the Persons are well discernible.

Torrance's way of discussing the *homoousion* relations in the Godhead was pointed out to be the theological basis for his view of the relationally ontological nature of reality, as revealed by the theories of field and relativity. Torrance suggests that because God as the ultimate Being is relational himself, the ontological nature of his contingent creation is relational, too. Substances in the created realm exist only insofar as they have a relation to each other. This implies that God's relationality is seen as the cause of the relationality of the ontology of creatures, a view that well fits into Torrance's conception of the contingency of created reality as a reference beyond its own boundaries to its divine Creator. It was further argued that realizing onto-relations in nature makes it easier for Torrance to view God's Being as constituted of relations. Modern science, therefore, can remove intellectual obstacles to apprehending the mystery of the Trinity. This analogy was regarded as comparatively powerful and consistently reasoned, presenting perhaps the strongest point in Torrance's arguments concerning science and the Trinity. This implies that his highlighting of Trinitarian theology is not only the result of his wish to construct a genuinely theological basis for the discipline of divinity, but that his insistence upon the centrality of the *homoousion* relations, revealed in the Incarnation, serves as a tool for apologetical purposes.

The integration of form and being, as occurred in the natural sciences in the way Torrance maintains, has a direct impact on biblical hermeneutics or, as he puts it, scientific hermeneutics. The Bible must not be viewed as a corpus of texts with a one-to-one correspondence with the divine reality but rather with a referential, open and flexible relation to them. The belief in *homoousion* was detected as offering the hermeneutical presupposition for biblical interpretation, because only then can the compulsion of the reality observed exercise its power and function properly. This was seen to confirm the basic structure and aim of Torrance's idea of integration: it is the object which has the primacy over anything thought about it. This explains why, in Torrance's thinking, the idea of compulsion belongs essentially to hermeneutical activity, in which what is called theological understanding must take seriously the *homoousion* relation between the Father and the Son. Torrance thus anchors biblical exegesis and hermeneutics strictly in the domain of the Church, suggesting at the same time that "neutral" or "objective" interpretative activity is an impossibility, due alone to the fact that to be truly objective means obedience to the compulsion of the object under investigation, rejecting detachment, impartiality, and indifference toward the object. Here the grand Barthian theme of objectivity was again detected. It was concluded that it is not so much a certain intellectually and consistently pursued rigorous methodology that determines the scientific nature of theology but rather the submissive attitude in which obedience toward the Object has to exercise its compelling power upon the human knower and his methodology as well. This is what Torrance sees as happening both epistemologically and ontologically when the concept of *homoousion* with its implications is applied.

In the final section of the work, special attention was paid to the implications of the idea of reminder for ecumenism. Space and time were seen to form the crucial stage on which ecumenical activity, in Torrance's opinion, takes place. He was seen to hold that distortions in church divisions result mainly from alien frameworks of thought, disturbing a purely theological approach. He contends that the main reason for impaired interchurch relations is the obstinate persistence in the absolute or container notion of space and time, making the Incarnation a needlessly problematical doctrine and resulting, consequently, in distortions of ecclesiology and of sacramental theology. Special attention was paid to Torrance's interpretation of Luther's view of Christ's real presence in the Eucharist, and its defects were pointed out with the remark that Torrance's argument seems to rise from a confessional standpoint.

The actual motivation behind Torrance's interpretation of ecumenical problems through the concept of space was detected in his view of dualism. Having overcome dualistic habits of thought, the modern natural sciences can now remind theology of its own forgotten basis in the doctrine of the Incarnation and its concept of *homoousion*. This strongly suggests that Torrance has transferred his programmatic crusade against dualism into the field of ecumenism. This became evident when his conception of the role of doctrinal formulae was discussed. He was seen to maintain that confessional formulae are not capable of precise, definite explanation of the tenets of faith but, at best, only as good approximations, in the Popperian sense, to the Truth which ultimately remains outside the scope of human expressions. Truth does not belong, then, to any single church or denomination but they can only participate in it. Strict and obstinate adherence to rigid confessional formulae can bear testimony only to a dualistic and deterministic view, comparable to the rigidity of natural science in the Newtonian context. Therefore, both the theories of relativity and quantum physics remind ecumenical theology of its own roots in relational, indeterministic thinking to which it should now return in order to make an essential and relevant contribution to interchurch relations.

The appealing character of this approach was noted, as well as the problem, namely its inability to find proper criteria for such a common agreement that could form a unifying ground for two differing confessions. It was concluded that Torrance's proposal bears the marks of the distinctive approach assumed by the Reformed party in the so-called Leuenberg discussions between the Reformed and the Lutheran Churches. Torrance's solution implies that the relation between the common basis of the churches and its articulate expression in their credal formulae remains problematic and even, one might argue, indifferent toward the divergence and dissimilarity to be found in church confessions. It was finally argued that Torrance's ecumenical methodology, suggesting that an interchurch agreement could be achieved by making a distinction between the basis and its expression, emphasizes an imprecise devotional attitude rather than paying serious attention to problems emerging on a cognitive and articulate level. The idea of the Christian churches as *homoousioi*, being of the same essence, was seen as a determining feature in Torrance's thought.

According to Torrance, a special ecumenical case in which an implausible interpretation of the doctrine of the Incarnation with its *homoousion* has damaged interchurch relations is the question of the *filioque* clause. With the reminding function of the modern natural sciences in mind, Torrance was seen to blame the Cappadocian theologians for not properly appreciating being in its internal, inherent relations, which later had an impact on the process of the division of the Christian Church. Torrance's proposed remedy is likewise simple: the churches should reflect anew on their theological basis in Trinitarian theology and the *homoousion* relations, but now in the Athanasian and not in the Cappadocian context. With regard to Torrance's view of ecumenism, the conclusion was drawn that the *homoousion* remains the "linchpin" of the ecumenical process both implicitly and explicitly. Implicitly, differences between the churches are suggested as presenting only a phenomenal problem due to divergent interpretations of Christian doctrines. The objective reality behind dissimilarities is common to all the Christian churches, and the task of ecumenical dialogue as well as all theological activity is, according to the maxim of the idea of compulsion, to let this reality dictate what one is to think about. Explicitly, on the other hand, the *homoousion* in the ecumenical context of the idea of reminder forms the doctrinal basis upon which closer unity between the churches can be built. A full appreciation of the consubstantiality of the Father, the Son, and the Spirit should result in taking seriously both the heritage of patristic theology as a product of the undivided Church and especially Athanasius's insights.

This work has proceeded through analyses of some apparent problems in the three ideas detected in Torrance's thought. It was found that the idea of contribution, asserting Torrance's belief in the central contributive role of the incarnational theology of the Early Church to the rise of modern empirical science, is in many respects extremely problematical. Due to the lack of sufficient historical evidence, his view of the contributive character of the doctrine of the Incarnation and the concept of *homousion* remains based on an inspiring and positively provocative vision rather than on historical facts. The idea of compulsion is certainly the core of Torrance's argumentation, emphasizing the realist significance of the *homoousion* for both theology and the natural sciences. The doctrine of election as the heart of the idea of compulsion proves that a distinctive theological doctrine can contribute in a fruitful way to innovative thinking. With regard to the phenomenon of dualism, Torrance's line of thought is open to misinterpretations due to the complex way in which the term is understood. Our analysis showed, however, that a Christologically reasoned definition offers the surest path for apprehending Torrance's intentions, although it should be noted at the same time that his way of applying the concept is bold and historically quite problematical. The idea of reminder is the least well thought-out part of Torrance's discussion. His detailed exposition within the idea of compulsion should be applied here, too, so that the connection he draws between the natural sciences and theology might become more articulate.

The problems detected can be seen as resulting from two features in Torrance's thought. First, his preferential reading of history, based on a certain view of the three periods in the history of Western thought and science, compels him to view the course of time through the realism–dualism dichotomy. Its strict application means that long and significant periods as well as many relevant details are not taken into account. Second, Torrance's Barthian context inevitably raises problems, especially with regard to the relation between the object known and the subject who knows. In this respect, the doctrine of election proved to be the key for understanding. It can be even concluded that Torrance's appreciation of Einstein, a fact which unfortunately is obviously harmful for a wider acceptance of his views in scientific circles, results from his Barthian emphasis on the supremacy of ontology, object and being over epistemology, subject and knowing.

On the other hand, however, it is precisely this Barthian context that makes Torrance's discussion both inspiring and fertile. Torrance has been able to prove that the Barthian theological paradigm is not incapable of entering into serious discussion with the natural sciences. In fact, it can provide some helpful analogies and points of common interest showing its numerous possibilities. Therefore the dialogue between theology and science can substantially benefit from the insights presented by Torrance. His call to reconsider the basis of theological enterprise, as explicated in the idea of reminder, should not be neglected, either.

The three ideas of Torrance's Christological bridge between theology and the natural sciences reveal some further points worthy of further reflection. Torrance's emphasis on the role of explicitly Christological themes instead of purely general theistic considerations does not provide only interesting insights and applications of the role of the *Logos* but undoubtedly helps to bring the Christian Church closer to the theology-science dialogue, suggesting that it is not reserved for academic theologians alone. In our times, the churches are in need of being able to meet the challenge of our Western culture, which is based so much on the achievements of the natural sciences. The future of humanity might turn out to be dependent on fruitful interaction between what is called "religious" and what is called "scientific." In this respect Torrance's program is a most important contribution.

The second significant point in Torrance's program is his insistence on the importance of realism, however great the problems this attitude may contain. What is reality, how should it be interpreted, and what is its relation to observations are questions that continuously make people, and not only philosophers, scientists, and theologians, search for meaning and for tools for understanding. Despite evident problems in Torrance's own conception of realism, his emphasis on the relevance of a "common sense" attitude will perhaps never lose its appeal.

The third important feature of Torrance's Christological bridge between theology and the natural sciences, between the Incarnation and physics, is his conviction that all human knowledge, despite its diverse forms and appearances, is ultimately common to all human intellectual enterprises. His insistence that both theology and the natural sciences share in the same single rationality, though in diverse fields, is most imperative. It is hard to believe that the theology-science dialogue could lead anywhere if this conviction were denied. It can be claimed that, due to their mutual participation in the single rationality, theology and the natural sciences are *homoousioi*–they are of the same being.

Notes

Chapter 1

1. The classical exposure of this theory is Andrew D. White's work, *A History of the Warfare of Science with Theology in Christendom*, originally published in 1896. In spite of its harsh criticism of theological dominance over science, its intentions are rather benign. White, *A History of the Warfare*, 1:xii. See also 2:393–66.

2. Drees, *Beyond the Big Bang*, xiii.

3. For a succinct account of the historical development that resulted in the "Galileo affair," see Shea, "Galileo and the Church," 114–35. See also Brooke, *Science and Religion*, 99–105.

4. Nebelsick, *Circles of God*, xiii.

5. Lindberg, *The Beginnings of Western Science*, 21–27, 39–40, 61–62; Nebelsick, *Circles of God*, 1–41. On the relevance of religious myths in ancient civilizations for their conception of nature, see Alioto, *A History of Western Science*, 5–20.

6. Henry Chadwick, *Early Church*, 23–31.

7. This uncritical approach is followed, for example, by Alioto, *A History of Western Science*, 103–13.

8. Lindberg, "Science and the Early Church," 25–26.

9. Ibid., 28. "Nevertheless, his general attitude toward science moved in the direction of accommodating science to Scripture whenever possible." Nebelsick, *Circles of God*, 104.

10. Lindberg, "Science and the Early Church," 28.

11. Lindberg, *The Beginnings of Western Science*, 226.

12. Aquinas, *Summa theologica*, I:1:5 (pp. 7–8); Owens, "Faith, Ideas, Illumination, and Experience," 453–54.

13. Grant, *The Foundations of Modern Science*, 70–85; Lindberg, *The Beginnings of Western Science*, 236–40.

14. "At the beginning of the sixteenth century everyone that mattered in the Western Church was crying out for reformation." Owen Chadwick, *Reformation*, 11.

15. Deason, "Reformation Theology," 167.

16. Nebelsick, *Circles of God*, 205–6, 237–39, 257–59; Westman, "Copernicans," 81–85.

17. It is important to stress the difference between the concepts of *scientific revolutions* and *the Scientific Revolution*. The former is a philosophical idea expressing the view that significant progress in science takes place in a way that can be characterized as revolutionary, as is well evident in Thomas S. Kuhn's book *The Structure of Scientific Revolutions*, published originally in

1962. The latter concept, however, refers to a particular and relatively short period of time, usually the first decades of the seventeenth century, that signified a radical novelty in the history of Western science. See Cohen, *Scientific Revolution*, 21–24.

18. Scheurer and Debrock, *Newton's Scientific and Philosophical Legacy*, v.

19. Torrance, *The Ground and Grammar of Theology* (GGT), 70.

20. McMullin, "Natural Science and Belief in a Creator," 66.

21. Koyré, *From the Closed World*, 276; McMullin, "Natural Science and Belief in a Creator," 66–67. The significance of Laplace's statement is far reaching: "It was as if Laplace had given the nineteenth century a peremptory command that religious considerations were superfluous for an understanding of nature." Hahn, "Laplace and the Mechanistic Universe," 256.

22. Hunter, *Science and the Shape of Orthodoxy*, 225–44; Jacob, "Christianity and the Newtonian Worldview," 238–49. For a wider context, see Westfall, *Science and Religion*, 1–25, and "The Rise of Science," 218–37. The idea of the mathematization of nature as a precondition for the advance of natural science is the main argument of Edwin A. Burtt in his significant work *The Metaphysical Foundations of Modern Physical Science* (see esp. 74–83, 106–11, 172–77). See also Cohen, *Scientific Revolution*, 88–97.

23. Byrne, *Glory, Jest and Riddle*, 162–78. For the role of some Enlightenment thinkers in the development of modern atheism, see Buckley, *At the Origins of Modern Atheism*, 194–321. See also Brooke, "Science and Theology in the Enlightenment," 7–27.

24. Barbour, "Ways of Relating Science and Theology," 26. See the discussion of Brooke, *Science and Religion*, 275–320. Dupree ("Christianity and the Scientific Community in the Age of Darwin," 351–52) points out that "Darwin and Darwinism came out of a profoundly Christian culture." Among scientists favorable to his views were individuals who saw no contradiction between their Christian convictions and the novelties of Darwin's theory. By the end of the nineteenth century, however, the theory of evolution had achieved a remarkable ideological status which was seen as hostile by orthodox theologians. Gregory, "The Impact of Darwinian Evolution," 383–88. A thorough special study of the relation between theology and Darwinism at the end of the nineteenth century is Moore, *The Post-Darwinian Controversies*. See also his article "Geologists and Interpreters of Genesis," 322–50. For the fate of natural theology after the publication of Darwin's theory, see Clifford, "Darwin's Revolution in *The Origin of Species*," 281–302. For Darwin's own account of the reception of his theory, see *The Origin of Species*, 370–77.

25. For the response of German theology to the challenge of natural sciences, see Gregory, *Nature Lost?*.

26. Gregory, "The Impact of Darwinian Evolution," 386.

27. From the beginning of the twentieth century, the so-called Vienna Circle, entertaining what is called *logical positivism*, directed influential criticism at Christian theology by demanding empirical verification for any claim of truth in religious life. For another theological response to the rigorous program imposed by the Vienna Circle, see Norris, "Mathematics, Physics and Religion," 457–70.

28. For a succinct account of the scientific worldview of the nineteenth century, see Knight, *The Age of Science*, esp. 32–51.

29. Drees, *Beyond the Big Bang*, 1–3. "Thus for me and many other scientists of Christian conviction it is God *and* the big bang; the why and the how." Polkinghorne, *One World*, 66. It should be noted, however, that Einstein's theories, in their novelty, with the new space-time metric as expressed e.g. in the *Minkowski space-time*, also faced severe rejection from theologians. See Jaki, "The Absolute Beneath the Relative," 5–18; and Lamont, *Christ and the World of Thought*, 40–43.

30. Polkinghorne, "The Quantum World," 333–39.

31. The appreciation and criticism of science is aptly described by Mary Midgley (*Science as Salvation*, 3), who states, "There has indeed long been an explicit anti-science strain in our cul-

ture, with impressive ancestors such as Blake, and it has gained some strength lately. But it is still a descant; the main anthem is still one of praise." See also Peacocke, *Theology for a Scientific Age*, 7–8; and Jacques Laeuffer's analysis of *scientism* in his "Scientism and Scientific Knowledge of Things and God," 185–91.

32. See Richardson and Wildman, *Religion and Science*, xvii.

33. Gustafsson, "The Current Scientific World View," 1–9; Stoeger, "Key Developments in Physics," 183.

34. Wildman, "The Quest for Harmony," 53, 57–58. Wildman's analysis of the total context of the discussion is illuminating in that he describes it as a typical indication of the cultural crisis of the Western world in which the period of *criticism* has been replaced by that of *self-criticism* (43–47). With certain cautiousness, one could also call this phenomenon a shift from *modernism* to *postmodernism*. Wildman reminds us that "postmodernity may be best understood as an essentially modern, self-critical rebound of modernity against itself, a necessary step toward a more consistent appropriation of the Enlightenment principle of independent criticism" (46). An effort to reflect on Torrance's thinking concerning theology-science relations as another expression of the postmodern paradigm has been made in Miller, "Beyond Dualism."

35. Pannenberg, *Basic Questions in Theology*, 2:46–64; Pannenberg, *Theology and the Philosophy of Science*; Pannenberg, *Systematic Theology*, 1:73–82, 2:84–102.

36. Rahner, "Weltall—Erde—Mensch," 34–85. "Science is reflection: reflection is preceded by a metaphysical and historical *a priori* which inescapably guides it—always and necessarily. This *a priori* can never be completely caught up with. Therefore the reflection of science cannot be the one and only root of human existence." Rahner, *Theology of the Spiritual Life*, 398.

37. A good indication of this is the central role of Jesuits such as William R. Stoeger and George V. Coyne in the process launched by the Vatican in the 1980s and resulting in a most significant collection of essays entitled *Physics, Philosophy and Theology—A Common Quest for Understanding*, published in 1988. On the role of the Jesuit Order in the advance of modern science, see Ashworth, "Catholicism and Early Modern Science," 153–60.

38. "[T]he *homoousion* is the ontological and epistemological linchpin of Christian theology. With it, everything hangs together; without it, everything ultimately falls apart." GGT, 160–61. For the historical background of the importance of the *homoousion* in Torrance's thought, see McGrath, *Thomas F. Torrance*, 151–58.

39. The concept of *homoousion* is generally used in its neuter form, except when it is placed as an adjectival attribute together with a masculine or feminine noun.

40. GGT, 162–63; "Theological Realism" (TheRe), 191. For the history of Torrance's development on the field of the dialogue between theology and science, see McGrath, *Thomas F. Torrance*, 195–211.

41. Torrance himself considers *Theological Science*, *The Trinitarian Faith*, and *The Christian Doctrine of God* his most important books. McGrath, *Thomas F. Torrance*, 107.

42. Guthridge, *The Christology of T. F. Torrance*.

43. Gray, "Theology as Science."

44. Achtner, *Physik, Mystik und Christentum*.

45. Yeung, "Being and Knowing."

46. Richardson, "Trinitarian Reality."

47. Spjuth, *Creation, Contingency and Divine Presence*.

48. Morrison, "The Self-Given Knowledge of God as He Is in the World."

49. Simmons, "The Semantics of God and Nature."

50. Miller, "Beyond Dualism."

51. Stamps, "The Sacrament of the Word Made Flesh."

52. Meardon, "A Study of Space, Time and Incarnation in the Work of Thomas F. Torrance."

53. W. Jim Neidhardt ("Introduction," xxxiv) expresses Torrance's idea of this two-directional movement between science and theology by stating that "Judeo-Christian theology motivates and gives meaning to natural science, which in turn sharpens and clarifies theology." This notion is certainly quite correct. In the present work this mutual relation of giving and receiving is analyzed on a profounder level, with special reference to its actual content in the doctrine of the Incarnation.

54. As mentioned earlier, McGrath, *Thomas F. Torrance*, is particularly helpful.

Chapter 2

1. The most comprehensive work on Thomas F. Torrance and his theology published so far is McGrath, *Thomas F. Torrance*. Succinct but informative introductions to Torrance's thinking have been provided, for example, by Achtner, *Physik, Mystik und Christentum*, 18–51; Morrison, "The Self-Given Knowledge of God," 1–57; Palma, "Thomas F. Torrance's Reformed Theology," 2–46; and Richardson, "Trinitarian Reality," 1–37. A detailed exposition of Torrance's thought, with an effort to relate it to its wider context, is offered by Gray, "Theology as Science," 1–109. See also Hardy, "Theology through Philosophy," 257–61; and Palma, "Torrance, Thomas F.," 372–73. Hardy makes the interesting claim, relevant also to the present issue, that "Torrance's work is a major attempt to construe the relationship between theology and the natural sciences at the deepest level" (261).

2. McGrath, *Thomas F. Torrance*, 13–14.

3. Ibid., 47–85.

4. Ibid., 87–107.

5. Palma, "Thomas F. Torrance's Reformed Theology," 2.

6. Torrance, *God and Rationality* (*GR*), 31–39; Torrance, *Theological Science* (*TS*), 59. A further discussion will be provided in chapter 3.

7. Torrance, "Introduction," in *The Calvin Treasury* (*CT*), viii-xii; Torrance, *The Hermeneutics of John Calvin* (*HJC*), 161–65.

8. Torrance, *Trinitarian Perspectives* (*TP*), 41–76.

9. *TP*, 50–51.

10. Torrance, *Confession*, 53, 74. (Questions 13 and 97 in the Larger Catechism.)

11. Calvin's Geneva Catechism (1541), the Heidelberg Catechism (1563), and Craig's Catechism (1581).

12. Torrance, "Introduction" in *The School of Faith* (*SF*), xvii.

13. *SF*, xvii–xix.

14. Torrance, *The Mediation of Christ* (MC), 95. Especially illuminating is the chapter "The Word of God and the Response of Man" in *GR*, 137–64. Torrance's idea of the vicarious humanity of Christ has been studied, e.g., by Kang Phee Seng ("The Concept of the Vicarious Humanity of Christ") and Christian D. Kettler ("The Vicarious Humanity of Christ and the Reality of Salvation"). The latter sees this emphasis of Torrance as a critique of any anthropocentric soteriology in its insistence upon the fact that in God's humanity in Christ nothing is left unaffected but that vicarious atonement also involves vicarious humanity (p. 93). See also the chapters by Torrance, "The Christ Who Loves Us" (CLU) and "Preaching Jesus Christ" (PJC), in Dawson and Stein, *Passion for Christ*, 9–32.

15. *SF*, xviii; Torrance, *Scottish Theology* (*ST*), 125–27.

16. Barth, *Church Dogmatics*, 4:1:56. An illuminating survey of the history of Federal theology is provided on pages 54–66.

17. "They (Federal theologians) brought the whole under the concept of the covenant, but they did not read the concept out of this one event (the atonement in Christ). Instead, they

imported the concept into this one event like all the others, as the supposed essence of the varied occurrence at every stage." Barth, *Church Dogmatics*, 4:1:56.

18. See, e.g., Barth, *Church Dogmatics*, 1:1:11–17; and Hunsinger, *Karl Barth*, 107–8.

19. Hunsinger, *Karl Barth*, 35. In his famous study of Barth's theology, Hans Urs von Balthasar has pointed out that in the *Church Dogmatics* the central concept "the Word of God" is replaced in a barely noticeable but nevertheless inevitable way by the concept "Jesus Christ, God and a human being." Balthasar, *Karl Barth*, 124. "Christology in the sense of a reference to Jesus Christ as the object and content and therefore the formative norm of justifying faith may very well be described as the climax of the doctrine of justification." Barth, *Church Dogmatics*, 4:1:637. Sin, too, is defined Christocentrically: "The sin of man is the human action which does not correspond to the divine action in Jesus Christ but contradicts it" (415).

20. Lindroth, *Kyrklig Dogmatik*, 1:76.

21. Tillich, *Systematic Theology*, 1:12.

22. Barth, *Evangelical Theology*, 11.

23. McGrath, *Thomas F. Torrance*, 94–102.

24. For Torrance's role in the British Barth-reception, see McGrath, *Thomas F. Torrance*, 113–45.

25. Morrison, "The Self-Given Knowledge of God," 31. See also McGrath, *Thomas F. Torrance*, 147.

26. Barbour, "Ways of Relating Science and Theology," 28. See also Spjuth, *Creation, Contingency and Divine Presence*, 13; and Yandell, "Protestant Theology and Natural Science," 450–51.

27. Spjuth (*Creation, Contingency and Divine Presence*, 13–14) reminds us aptly that "it seems to be in Scandinavia that Barth's doctrine of creation has been most firmly denounced." His analysis of the ideas of some Nordic critics of Barth is most informative (13–15). See also Sigurdson, *Karl Barth som den andre.*

28. MC, 85–86. That this critical attitude has for a long time, if not always, been a part of Torrance's thought is discernible in another memory, where he recalls that when in 1940 he for the first time made the acquaintance of an English edition of Calvin's *Commentaries*, he realized that "Calvin's own theological position was very different from the hardened system that has long passed under the name of Calvinism." Torrance, *Calvin's Doctrine of Man* (CDM), 7.

29. *ST*, 287–317. See also Tuttle, *So Rich a Soil*, 22. Tuttle reminds us that Campbell, as if anticipating Torrance's efforts, was keenly interested in the relation between religion and science (13).

30. *ST*, 295.

31. Scotsman, "Moderator," 1.

32. Torrance, "Reaffirming the Cornerstone of Faith" (RCF), 4.

33. Torrance, "The Truth of the Virgin Birth" (TVB), 11.

34. Barbour, "Ways of Relating Science and Theology," 21–48; Barbour, *Religion in an Age of Science*, 3–30. See, for another classification of the relationship between theology and science, Peters, "Theology and Science," 323–43. How Peters himself applies one of the modes of the interaction can be seen in his article "On Creating the Cosmos," 273–96.

35. Barbour points out, as his interpretation of the Barthian paradigm, that "Science is based on human observation and reason, while theology is based on divine revelation." He also maintains that Torrance has "developed further some of the distinctions in neo-orthodoxy." Barbour, "Ways of Relating Science and Theology," 28, 30; *Religion in an Age of Science*, 11, 13.

36. Barbour, "Ways of Relating Science and Theology," 34; *Religion in an Age of Science*, 17–18.

37. Torrance is not alone in underscoring the possibilities of Barthian theology for positive theology-science interaction. Herbert Hartwell (*The Theology of Karl Barth*, 56–57) states that

"In a theology as thoroughly christological as that of Barth a final conflict between theology and science is not to be expected; on the contrary, it entails and actually envisages a fruitful co-operation between these two activities of the human mind. In such a theology there is therefore no danger that its teaching will be at variance with the results of scientific discoveries; it has nothing to fear from them." Although underlining cooperation between the two disciplines, Hartwell nevertheless confirms their strict independence. Another effort to evaluate positively Barth's capability of entering interdisciplinary discussion has been made by Günter Howe ("Parallelen," 409–22), who has found both historic and systematic parallels between Barth's theology and contemporary physics.

38. Roland Spjuth (*Creation, Contingency and Divine Presence*, 84, n. 80) totally rejects Barbour's way of setting Torrance in the *independence* category. Torrance has not, however, driven his emancipation from Barth as far as Spjuth suggests, for he wants to guarantee a certain kind of independence for both disciplines. Stephen Simmons ("The Semantics of God," 1–17), after an account of Barbour's classification, interprets Torrance's point of view as a reply to the challenge posed by the *integration* approach.

39. Barth, *Church Dogmatics*, 3:2:12.

40. Ibid., 12–13.

41. *GR*, 29–30; Torrance, *Theology in Reconciliation* (*TRc*), 267.

42. *TRc*, 268; Torrance, *Trinitarian Faith* (*TF*), 47–49.

43. *TRc*, 268–69.

44. GGT, 70.

45. *TRc*, 270.

46. *TF*, 48.

47. See, e.g., Kelly, *Early Christian Doctrines*, 5–6.

48. Meijering, *God Being History*, 17.

49. Torrance draws the dualistic line, marking the end of fertile science-theology relations based on patristic incarnational theology, from Pope Leo's address to the Council of Chalcedon, condemning the "monophysites" and interpreting the term nature in Aristotelian terms, through Augustinian Neoplatonism to the rise of Aristotelian philosophy and its culmination in medieval scholastic theology. GGT, 60–65; *RST*, 8–12. This nondualistic line is eventually quite simple: starting from the Greek Fathers and going through the scientist John Philoponos of Alexandria (d. c. 570) and the theologian John Duns Scotus (c. 1265–1308), both of whom criticized Aristotelian philosophy, it finds its first eminent post-scholastic representative in John Calvin. This line finally terminates in Karl Barth. See Torrance, "Science and Access to God" (SAG), 43–58.

50. Grant, "Science and Theology in the Middle Ages," 49.

51. Torrance's best accounts of the Enlightenment period and its relevance for theology is in *Transformation and Convergence in the Frame of Knowledge* (*TCFK*), 1–59, which is a masterful example of Torrance's approach, determined by his rejection of dualism. His way of commencing the Enlightenment period with Descartes and concluding with Kant is also followed by James Byrne (*Glory, Jest and Riddle*). See also Brooke, *Science and Religion*, 152–91, and "Science and Theology in the Enlightenment," 7–27.

52. Willem B. Drees (*Beyond the Big Bang*, 41–75) points, e.g., to the Big Bang theory of the beginning of the universe as a representative of a quantum cosmology in which quantum theory, in addition to the general theory of relativity, is a constitutive part. See also Stoeger, "Key Developments in Physics," 195–97.

53. Torrance's periodization as such is not a unique feature in the thinking of those who are involved in interdisciplinary reflection. While Torrance draws his distinctions according to the three eminent scientists, Stanley L. Jaki (*The Relevance of Physics*, 3–137), for instance, makes equivalent divisions by designating the prevailing views of the world as those of an *organism*, a *mechanism*, and a *pattern of numbers*.

54. For instance, renaming Torrance's Newtonian cosmology as that of *Copernicus* would not violate history in that it was this Polish canon who was the first to present, in his *De Revolutionibus*, a coherent account of the heliocentric universe. See Nebelsick, *Circles of God*, 200–273; and Westman, "The Copernicans and the Churches," 76–113.

55. *TRc*, 270–71; GGT, 43. Torrance suggests that this time-lag is the reason for the fact that approval of theistic belief is more common among physicists than among those representing these "delayed" sciences. The view of the appearance of theism is well reasoned, but the proposal for its causes in a "time-lag" are more disputable. "We do know, however, that for reasons not entirely clear, physical scientists in the twentieth century have remained more orthodox theologically than their colleagues in the biological, behavioral, and social sciences." Hiebert, "Modern Physics and Christian Faith," 434.

56. Eminent examples of these approaches are Kant's idea of religion "within the limits of pure reason" and Schleiermacher's way of assigning primary value to subjective religious experience. Literature on the theology of the eighteenth and nineteenth centuries is vast, and general surveys can be found in any exposition of the history of theology. For a succinct presentation of nineteenth-century theology by Barth himself, see Barth, *The Humanity of God*, 11–33.

57. See, e.g., *TCFK*, vii.

58. A proof of this is, for instance, the sophisticated argumentation in *Divine and Contingent Order* (*DCO*), a book with the clear intention, implied by his contribution idea, of providing scientists with insights of an originally theological nature.

59. Torrance suggests that "the unified contingent intelligibility of the universe sets the perspective within which we may consider the *transfinite* reference of beauty in science and theology." "Transfinite Significance of Beauty in Science and Theology" (*TSB*), 396. In this essay an interesting link is drawn from Mozart's music to Karl Barth's theology, thus suggesting implications arising from Barth's well-known attraction to this master of music.

60. GGT, 21.

61. Ibid., 43.

62. Willem B. Drees (*Beyond the Big Bang*, 167–69) considers Torrance's effort to unite the spiritual and the material as basically quite a differing mode of the Barthian approach if compared to that of some continental Barthians. Drees considers the latter as more conscious of the problematic nature of any appeal to reality than does Torrance. On the other hand, it has been suggested that the theology-science dialogue itself is a *quest for harmony*, "for a unified, consistent understanding of human nature and the world in which it has arisen." Wildman, "The Quest for Harmony," 60.

63. Peat, *Superstrings*, 19–21, 71–75.

64. One of the most famous contemporary thinkers in this respect is Stephen W. Hawking, who has presented some theologically challenging insights on the basis of his scientific work. "However, if we do discover a complete theory, it should in time be understandable in broad principle by everyone, not just a few scientists. Then we shall all, philosophers, scientists, and just ordinary people, be able to take part in the discussion of the question of why it is that we and the universe exist. If we find the answer to that, it would be the ultimate triumph of human reason—for then we would know the mind of God." Hawking, *A Brief History of Time*, 175. A reply to Hawking's argument is offered in Driessen, "The Question of the Existence of God," 203–13. See also Davies, *The Mind of God*.

65. Barrow, *Theories of Everything*, 7–8.

66. *DCO*, 82; *Preaching Christ Today* (*PCT*), 48–49. Roland Spjuth (*Creation, Contingency and Divine Presence*, 87) points out that although Torrance is searching for a kind of synthesis, he consciously avoids any monistic synthesis "that flattens everything to the same level."

67. *DCO*, 83. Compare this to Rahner, *Theology of the Spiritual Life*, 385–400.

68. *DCO*, 83.

69. Ibid.

70. *PCT*, 60–71. See, e.g., Bowker, "Did God Create This Universe?," 98–122, where the doctrine of the Incarnation is clearly subordinated to that of creation with respect to theology-science relations. See also Kaiser, *Creation and the History of Science*; McMullin, "How Should Cosmology Relate to Theology?," 17–40; and Peacocke, *Theology for a Scientific Age*, 27, 168–77. Peacock's position is extraordinarily interesting in that he seems to have increasingly directed his attention to the doctrine of the Incarnation, a detail anticipated vaguely already in the work mentioned (179–80) but expounded only recently in Peacocke, "The Incarnation of the Informing Self-Expressive Word of God," 323–39. Broadly speaking, it seems that Christology in general and the doctrine of the Incarnation in particular are gradually gaining wider attention by scholars involved in the dialogue. An illuminating and classic example is the idea of the complementarity of two seemingly contradictory ways of describing reality, as the wave-particle complementarity in quantum physics, and its relevance for the doctrine of the two natures in Christ. A succinct history of the theological applications of the concept of complementarity is offered by Kaiser, "Quantum Complementarity," 291–96. See also the authoritative accounts in the same volume by Edward Mackinnon, "Complementarity," 255–70; and Loder and Neidhardt, "Barth, Bohr, and Dialectic," 271–89. The latter essay is of particular interest because it relates Barth's theology to Niels Bohr's view of complementarity.

71. Descartes needed God not primarily to save human beings but to stand for the only reliable ground upon which we can trust in the truthfulness of his thoughts and reason. See Descartes, *Essential Descartes*, 127–33. Newton (*Mathematical Principles*, 443–47) saw God not primarily as saving but as ruling the universe. The rational arguments for the existence of God put forward by Descartes and other thinkers were gravely defeated by Kant (*Critique of Pure Reason*, 402–33, or A581/B609–A640/B668), restricting the relevance of belief in God to only within the limits of morals. See also Ashworth, "Catholicism and Early Modern Science," 139–41; and Byrne, *Glory, Jest and Riddle*, 53–74, 157–60, 203–28.

72. Olding, *Modern Biology and Natural Theology*. See also Dawkins, *The Blind Watchmaker*; and Monod, *Chance and Necessity*. An example of serious attempts to deepen dialogue between theology and biology is Russell, Stoeger, and Ayala, *Evolutionary and Molecular Biology*. See also Gregersen, Drees, and Görman, *The Human Person in Science and Theology*.

73. In physics, Stephen Hawking (*A Brief History of Time*) suggests the redundancy of God in creation and scientific explanation but, unlike in the biological accounts referred to, his approach is not based on mechanical causation but, rather, on his interpretation of time. The reason for the apparent positivistic attitude in the thoughts of the biologists mentioned is an interesting problem which, generally speaking, could explain the apparent difference in tone when compared to physicists. Holmes Rolston III ("Science, Religion, and the Future," 65) sums up this feature, saying: "Outspoken monotheist biologists are as rare as those who think physics is compatible with monotheism are common. Typically, biologists seem to insist that if, from the perspective of science, they find what looks like contingency, then God is eliminated." One explanation for the position of biology in theology-science discussion is undoubtedly the occasionally bitter debate between evolutionists and creationists in North America at the beginning of the twentieth century and revived in the 1960s. See Numbers, "The Creationists," 391–423.

74. Isham, "Creation of the Universe," 405.

75. Polkinghorne has found Torrance's view of realism particularly challenging: "For a scientist with a serious interest in theology, this emphasis on the epistemic primacy of the object of knowledge is one of the most valuable and congenial aspects of Torrance's thought." Polkinghorne, *Science and Christian Belief*, 33. See also pp. 38, 42–43, 140, and 155, as well as his *Reason and Reality*, 1, 5, 16. On p. 93, he expresses his deviation from Torrance in the detail of the concept of the field and its application in theological language. An introduction to real-

ism in the discussion between theology and the natural sciences is provided in McGrath, *The Foundations of Dialogue*, 140–64.

Chapter 3

1. An excellent and impressive study in this field is Cohen, *The Scientific Revolution*.
2. Neher, *Jewish Thought*, e.g., 1–2.
3. Duhem, *Medieval Cosmology*, 4. See also Ariew, "Preface," xxii–xxiii; and Cohen, *Scientific Revolution*, 52–53. Pierre Duhem (1861–1916) was a physical chemist as well as a historian and philosopher of science who, apart from his contribution argument, was of the opinion that metaphysics and religion have practically nothing in common with science, making all claims about their alleged conflicts unjustified. See Hiebert, "Modern Physics and Christian Faith," 438–40.
4. Grant, "The Effect of the Condemnation of 1277," 537–39; "Science and Theology in the Middle Ages," 51–70; Lindberg, *The Beginnings of Western Science*, 234–40.
5. Cohen, *Scientific Revolution*, 53.
6. Merton, *Social Theory*, 628–42. The thesis was originally presented by Merton in 1938 in his dissertation, "Science, Technology and Society in Seventeenth Century England," but it runs as a leading theme throughout Merton's scientific writings. Of the nature and the reception of the Merton thesis, see Cohen, *Scientific Revolution*, 314–21.
7. Merton, *Social Theory*, 637–39; Lindberg and Numbers, *God and Nature*, 4–5. This method, used by several pro-religion historians of science, has been called the "headcounting" strategy. It has been pointed out that "scholarship on the question of Protestantism and science has become hopelessly mired because it started off in the wrong direction, counting heads, collecting Calvinist-sounding quotations, and seeking parallels instead of connections." Ashworth, "Catholicism and Early Modern Science," 136. This criticism is targeted particularly at R. Hooykaas, a Dutch historian of science who has unhesitatingly applied the Merton thesis to his own investigations. Hooykaas, *Religion*, 98–114.
8. Jaki's argument is not quite original because it was formulated in the circles of the Royal Society, e.g., in 1934 by M. B. Foster. Polkinghorne, *One World*, 1, 99. Jaki's significance lies, therefore, in his attempts to use the argument in the present phase of theology-science discussion and to find more support for it in the history of science.
9. Jaki, *Science and Creation*, viii. Hooykaas agrees with Jaki on essential points, e.g., when he states that the biblical worldview gave rise to a *mechanistic* world picture which replaced the earlier *organic* one, a distinction that belongs to the structure of Jaki's history of science. See Hooykaas, *Religion*, 13–16; and Jaki, *Relevance of Physics*, 1–137. Anthony M. Alioto gives quite a contrary picture of the contributive role of Christian theology in the rise of scientific thinking. After giving a rather unsympathetic impression of the contributive role of early Christianity, he refers to the Catholic Church of the Middle Ages as having entertained as its primary goal the salvation of humankind, not scientific enterprises as such. Alioto, *A History of Western Science*, 103–12. An intermediate position is proposed by F. Seitz (*The Science Matrix*, 23), who contends that "[b]y the time Rome adopted Christianity in the 4th century A.D. and formed what might be termed an orthodox Christian church, great changes in values had occurred, none of which gave great stimulus to the advancement of science."
10. Lindberg, "Science and the Early Church," 42. See also Nebelsick, *Circles of God*, xix.
11. See Lindberg and Numbers, *God and Nature*, 5.
12. Lindberg, "Science and the Early Church," 40. See also Lindberg, *The Beginnings of Western Science*, 149–51.
13. *DCO*, 2–5, 23–25; *GGT*, 73; *TF*, 77.
14. Gary B. Deason ("The Protestant Reformation," 221–40) makes a distinction between strong and weak interpretations of the relation between the Reformation and the rise of science,

claiming that exponents of strong interpretations are convinced of the direct impact of a certain theological doctrine or attitude upon the rise of science. The weaker alternative holds that while both remained independent of each other, Protestantism could offer few ingredients for scientific expansion. Deason himself maintains that "the strong interpretation is too strong and that the weak one can be strengthened," and suggests an interpretation based on an indirect causal relation between the Reformation and the rise of science (221–22).

15. *TS*, 59.

16. *GR*, 31–39 (esp. 37).

17. Ibid., 31–32.

18. Ibid., 34. Torrance's analysis is based on the Scholastic understanding of questioning. An illuminating introduction to Thomas Aquinas's idea of questions is provided in Aertsen, *Nature and Creature*, 7–53.

19. Ibid., 37.

20. Ibid., 39.

21. Torrance, *The Christian Frame of Mind* (*CFM*), viii.

22. *RST*, 198–99. Crawford Knox also points strongly in this direction, although with a different tone and emphasis, in *Changing Christian Paradigms*, 11.

23. *DCO*, 91–92.

24. See, e.g., *TRc*, 14.

25. The answer to the problem can be found only partially in the realm of religion, as is shown by C. Floris Cohen (*Scientific Revolution*, 308–77) in his impressive study of the historiography of Western science.

26. *TF*, 7. This kind of idea of contribution has been criticized, e.g., by Alan Olding (*Modern Biology and Natural Theology*, 3–4), who argues against what he calls "purposive explanations." Ian G. Barbour also feels critical of any overestimations of the scientific role of Christian understanding of the creation (Barbour, "Ways of Relating Science and Theology," 34).

27. *TF*, 47.

28. Ibid.

29. Torrance, *Divine Meaning* (*DM*), 289–90; *RST*, 79; *GGT*, 36.

30. *DM*, 289.

31. Bultmann, *Existence and Faith*, 197. See also Robinson, "Barth or Bultmann?," 275–90.

32. See, e.g., Hesse, "Physics, Philosophy, and Myth," 198–99; and Sharpe, *From Science to an Adequate Mythology*, 13–27. Sharpe points out the difficulty in defining the term because of its great variety of usages. Sharpe's summary of the different qualities of a myth is informative. He defines a myth as having eight properties: First, a myth is a story, second, it usually involves "extra-mundane beings" (human-like, animal-like, or object-like), groups, or events. Third, myths usually take place in a different setting from our space and time, in transcendence. Fourth, myths offer an explanation as to the "origin and operation of this world." Fifth, they are "models of behaviour," and, sixth, they "have the peculiar status of being believed true." Finally, they tend to be authoritative and associated with rituals.

33. J. Rebecca Lyman (*Christology and Cosmology*, 129) shows that Athanasius accused the Arians of entertaining "mythology" in their Christology, a feature that has undoubtedly influenced Torrance's way of using the concept.

34. This, in fact, is the approach adopted by Sharpe (*From Science to an Adequate Mythology*, 13): "We need a mythology adequate for our society, and that must be founded on the integration of secular-scientific knowledge and that of the Christian religion in which neither is subordinated to the other." It is worth noting, too, how closely Torrance links myth to a *symbolic* presentation of reality. See TheRe, 172.

35. *TF*, 37–39.

36. J. N. D. Kelly (*Early Christian Doctrines*, 7–8) explicitly says about Alexandria that "in earlier days it had produced the Septuagint version of the Old Testament, and in the Christian period it proved a highly sympathetic channel for introducing Hellenistic culture to the early Church." See also Stead, "Greek Influence on Christian Thought," 175–85.

37. See the authoritative presentation by Christopher Stead, *Divine Substance*, 190–222. See also Kelly, *Early Christian Doctrines*, 235. Torrance is not explicitly interested in the history of the concept, but his main concern is to argue that although the term had been in use before it was the best possible word to express and safeguard what the Nicene theologians had understood. Therefore he neglects the political side of the doctrinal debates and the undeniable role of the emperor as an influential backstage person in the development of the term. (For more historical accounts of the background to the controversy, see, e.g., Chadwick, *Early Church*, 116–51, Christie-Murray, *A History of Heresy*, 45–55, Heick, *A History of Christian Thought*, 1:154–60; and Kelly, *Early Christian Doctrines*, 223–51. A historical account of Athanasius's life is provided, e.g., by Barnes, "The Career of Athananius," 390–401; and Pettersen, *Athanasius*, 1–18.) On the other hand, Athanasius himself describes the efforts of the theologians of the Nicene era as an epistemological process and not totally without psychological considerations. Athanasius, "De Decretis," 18–24 (445–60).

38. For an introduction to research on the relationship between Platonism and the Church Fathers, see Meijering, *God Being History*, 1–18. The view of Athanasius's alleged Hebraic "mode of mind" is discussed in Lyman, *Christology and Cosmology*, 130.

39. *TF*, 51; *GGT*, 108.

40. *TF*, 51.

41. Ibid., 19–20; *DM*, 179–80; McGrath, *Thomas F. Torrance*, 160–63.

42. An illuminating definition of natural theology is given by G. Smith ("Natural Theology," 61): "There are two knowledges or theologies about God. First, there is the knowledge that He has of Himself. This knowledge is communicated in part to men only by faith in divine revelation and is caused only by grace. . . . Secondly, there is the knowledge or theology about God that men derive from their natural knowledge of things. This second theology has come to be called natural theology."

43. *GGT*, 75–87; *RST*, 32–63. It is part of Torrance's theological enterprise to try to integrate natural theology with revealed theology, and thus he allows natural theology a legitimate place but only if submitted to what he calls "revealed theology." Wolfgang Achtner states that, for Torrance, this integration means a double approach to natural theology. It must reflect the rational structures of the creation, on the one hand, and must be included in Christology, on the other. Achtner, *Physik, Mystik und Christentum*, 37. See also Polkinghorne, *Science and Creation*, 13–15. An excellent account of natural theology and its Barthian interpretation in Torrance's thought is presented by McGrath, *Thomas F. Torrance*, 174–94.

44. *TS*, 103.

45. Natural theology, which, according to Torrance, has flourished especially when dualism has prevailed, has tried to build these "logical bridges" in order "to give some rational support for faith." Torrance, *Reality and Evangelical Theology* (*RET*), 32. An implicit reference to traditional apologetics is to be discerned here.

46. Torrance's emphasis on the direction from God to the humans in knowledge of God can be seen as an effort to give a reply to one of the sternest critics of natural theology, the philosopher David Hume (1711–76), Torrance's fellow countryman, who in his *Dialogues Concerning Natural Religion* affirms, through a fictitious role character, that "a person, seasoned with a just sense of the imperfections of natural reason, will fly to revealed truth with greatest avidity." Hume, *Dialogues*, 130. Hume's critique in his *Dialogues* has as its actual target religious reasoning, where a kind of a priori knowledge of God and arguments for his existence, according to the "argument from design," are applied. Christian D. Kettler ("The Vicarious Humanity

of Christ," 124) maintains: "It is obvious that any concept of a religious *a priori* in humanity is anathema to Torrance."

47. John Calvin's epistemological orientation is apparent as early as the very first chapters of his *Institutes*, where he explicitly examines our knowledge of God. Calvin, *Institutio*, I:II:1 (34). For Torrance's own comments on Calvin's epistemology, see *HJC*, 163. E. L. Allen (*The Sovereignty of God*, 12) considers "the assertion of divine sovereignty" to be the "authentic note of Calvinism," quoting Karl Barth (from his book *The Resurrection of the Dead*, p. 103): "What is at stake in Christianity is the rule of God and nothing else."

48. *TF*, 51.

49. *DM*, 188; *GGT*, 76.

50. Muller, "The Barth Legacy," 687.

51. *GGT*, 76.

52. In his most important anti-Arian work, *Contra Arianos*, Athanasius applies the term only once, although, let it be admitted, the idea "from the substance" ἐκ τῆς οὐσίας appears constantly. Stead, *Divine Substance*, 235–36. Stead also contends that Athanasius "uses the term *homoousios* perhaps 150 times in his genuine works; but many of these instances are mere reports of the opinions of others, and less than half of them really illuminate his own usage" (260). For another study of Athanasius's application of the term, see Bienert, "Logos-Christologie," 412–19, where it is maintained that for the Church Father the concept means a frontier between our understanding and God's being and, though not literally biblical, it safeguards the biblical belief in salvation within the reach of human beings.

53. J. Rebecca Lyman's interpretation comes close to Torrance's when she claims that "the eternal and incarnate son, who was the essential image of the Father, was the only guide able to reveal and embody divine truth for fallen humanity." Lyman, *Christology and Cosmology*, 129. Lyman's references to Athanasius's writings do not, however, provide the reader with ample evidence to make this exclusive claim.

54. Athanasius, *De Incarnatione*, 11–12 (158–65).

55. Ibid., 14 (166–69).

56. Alvyn Pettersen (*Athanasius*, 37–52) reminds us that Athanasius saw God as revealing himself through four channels, namely "a human's soul, created in God's image and likeness, a harmonious creation, the Old Testament Scriptures, and Christ" (39).

57. Athanasius, *De Incarnatione*, 16 (172–73).

58. *DM*, 230, 385.

59. For example, Thomas Aquinas, *Summa Theologica*, 1:12:12 (84), points out that our natural cognition can know *whether* God exists but not *who* he is. Increasingly, from the time of Isaac Newton, natural theology became highly rationalistic to the extent that any revealed knowledge of God, as apparent in Athanasius's theology, was considered unnecessary.

60. Pettersen, *Athanasius*, 50. "The principal issue here is not God's self-revelation, but man's appreciation of God's self-revelation" (51).

61. *GGT*, 90–91.

62. An effort to construct Torrance's "integrated natural theology" is made by Achtner, *Physik, Mystik und Christentum*, 34–41. On pages 29–35, Achtner analyzes Torrance's indebtedness to Karl Barth and his Christocentrism in relation to natural theology. He states that Christology holds a key position with regard to natural theology, so that it is not legitimate to search knowledge of God as Creator without Christology (32). Achtner gives no answer to the question concerning the difference between natural and revealed theologies in Torrance's thought: if natural theology is also Christologically based, what is the point in calling it natural theology? See also Barth, *Evangelical Theology*, 24; and Hunsinger, *Karl Barth*, 76–81; as well as *GGT*, 94, where Torrance explains his wish to develop Barthian natural theology in order to give it a positive place in his system as subjected to revealed theology.

63. For the role of natural theology in modern theology-science discussions, see, e.g., W. Norris Clarke, "Is a Natural Theology Still Possible Today?," 103–23; and Gingerich, "Is There a Role for Natural Theology Today?," 29–48. A succinct description of some of the problems related to natural theology is Murphy, "Postmodern Apologetics," 105–20. Discussions on arguments for the existence of God from a natural theological perspective are given in the essay by Emerton, "Arguments for the Existence of God," 72–86; and in the debates in Craig and Smith, *Theism, Atheism, and Big Bang Cosmology*. An ultimate conclusion from the possibilities of natural knowledge of God is reached by Paul Davies (*God and the New Physics*, ix, 229), who holds that "science offers a surer path to God than religion."

64. GGT, 125, 161–62; *GR*, 95; *TCFK*, 274–75; *TS*, 318.

65. GGT, 125.

66. One of the problems Ian Barbour finds in religious disclosure models is that they tend to be *self-authenticating*, resulting in the ambiguous situation where there are no commonly acceptable criteria by which their effect and truthfulness with respect to reality could be evaluated. Barbour, *Myths, Models and Paradigms*, 64. See also Barbour, *Religion in an Age of Science*, 41–51. For the role of models in revelational theology, see Dahlfert, *Theology and Philosophy*, 197–203; and in the interaction between science and theology, see Case-Winters, "The Question of God," 351–75.

67. Spjuth (*Creation, Contingency and Divine Presence*, 101) reflects Torrance's view of the relation between reality and human beings, referring to the contradiction between objectivism and subjectivism, or realism and skepticism, by pointing out that in this discussion Torrance relies on the traditional realist position that emphasizes the role of *intuition* as the bridge between object and subject.

68. *GR*, 95. (Italics mine.) See also *TS*, 318.

69. *TCFK*, 275. The opposite of the disclosure model is what Torrance calls "a picturing model," where strict one-to-one correspondence between a scientific statement and reality is presupposed. *RST*, 162. It is also worth noting that Torrance does not favor any theoretical models which, in his opinion, are products of human a priori conceptions. GGT, 125. See Achtner, *Physik, Mystik und Christentum*, 41–42.

70. GGT, 160.

71. Torrance, "Introduction," in *The Incarnation: Ecumenical Studies in the Nicene-Constantinopolitan Creed A.D. 381* (*INC*), xiii.

72. See GGT, 126–27, where he refers to the Chalcedonian Christological formula as a disclosure model. It has to be admitted that the issue underscored by Torrance deals with the interrelation of the human and divine natures in Christ rather than his being *homoousios* with the Father.

73. *Theology in Reconstruction* (*TRs*), 33; *DM*, 378.

74. GGT, 162–63.

75. *TF*, 19.

76. Ibid. The intellectualism of Torrance's conception of faith has been underscored by Gray, "Theology as Science," 382.

77. *TS*, 33. See also ibid., 132.

78. For example, in ibid., 33, n. 2. Torrance explicitly rejects any antithesis between faith and reason, claiming that faith is a reasonable attitude toward its divine Object. A good glance at contemporary Anglo-Saxon reflection on the issue is Richard Swinburne's book *Faith and Reason*.

79. See, e.g., Calvin, *Institutio*, I:VII:5, and III:VII:2 (60, 506); and Barth, *Evangelical Theology*, 99. Philip Clayton (*Explanation from Physics to Theology*, 9–14) gives a short but informative account of the history of the faith-reason debate, pointing out that in the Early Church it was Tertullian who attempted to make a radical separation between faith and reason and to

assign the primacy to the former. This *fideism* found another influential representative in John Calvin, who submitted reason to faith or rationality to divine revelation. A contrary position is named *rationalism*, where faith is subordinated, as happened in the philosophies of Locke, Kant, and Hegel. Clayton reminds us that the controversy has embodied itself in the view of *faith seeking understanding*, which has a long line of adherents, from Clement of Alexandria, through Augustine as its most explicit exponent, to Anselm of Canterbury.

80. *TF*, 20.

81. Following Torrance's line of thought, Iain Paul (*Science and Theology in Einstein's Perspective*, 16) holds that if "scientific faith motivates research, it does so by fostering trust in the reliability of the universe." Note the phrase *scientific faith*, which is another modification of the idea, entertained strongly by Torrance, e.g., in "The Framework of Belief" (FB), that every scientific activity is based on belief. See also Torrance's article "Scientists Commemorating Third Centenary of Newton's *Principia*" (SC), 105–12.

82. FB, 12. It is, indeed, one of Polanyi's major claims that in scientific activity belief is always involved. See, e.g., Polanyi, *Personal Knowledge*, 160–61. A lengthy discussion of Polanyi and a brief analysis of Torrance's relation to his thought is provided by Simmons, "The Semantics of God," 123–57. A more profound approach to this relation has been achieved by Morrison ("The Self-Given Knowledge of God," 114–20), who says that Torrance uses both Einstein's and Polanyi's thought to show that "the sciences have been throwing off the 'myth of induction' for the scientific method" (114).

83. Torrance seems to use the concepts *belief* and *faith* interchangeably, although a slight difference can be discerned in his application of the terms, since belief refers mostly to the fiducial attitude in the specifically *scientific* sphere, whereas *faith* has mainly to do with theology. The actual content of the phrases remains, nevertheless, identical.

84. *TCFK*, 219.

85. Stanley Jaki (*Science and Creation*, viii), whose argument Torrance modifies in his own idea of contribution, maintains that scientific enterprise could not properly develop until faith in a rational Creator permeated the whole Western culture, as was the case from the start of the Middle Ages. Torrance cannot agree with Jaki in his appreciation of the science of the High Middle Ages as the inaugural phase of the contribution of Christian belief to scientific, creative thinking because for him this phase lies further in the past, in the Early Church and its efforts to formulate its understanding of the Triune God. *TF*, 47–52. It is also worth noting that Torrance draws a parallel between the Jewish nation as the mediator of God's revelation and Jewish scientists as mediators of the new understanding of science. MC, 21.

86. *TCFK*, 219. Note the idea of the knowledge of God *through Jesus Christ*, implying both the negative and positive aspects of his idea of contribution.

87. See also Jaki, "Theological Aspects," 149–66.

88. That the role of religious conviction in Maxwell's scientific thought can be viewed quite dissimilarly is evident when a comparison is made between the scholars Hendry and Tolstoy. See Hendry, *James Clerk Maxwell*, 119; and Tolstoy, *James Clerk Maxwell*, 58–61.

89. Torrance's articulate interpretation of Einstein's religious thought is published in his article "Einstein and God" (EG).

90. Einstein, *Ideas and Opinions*, 36–40. To show the interrelatedness of science and religion in Einstein's thought, a short quotation from him is sufficient: "Science without religion is lame, religion without science is blind." Einstein, *Ideas and Opinions*, 46. Einstein appreciated those religious "geniuses" who were often condemned as heretics (38). For a further discussion of Einstein's attitude toward religion, see Lawhead, "Religion, Relativity, and Common Sense," 37–45. On the Jewish impact upon Einstein's humanism, see Tal, "Jewish and Universal Social Ethics," 297–318.

91. Torrance, *Christian Theology and Scientific Culture* (CTSC), 58–59; EG, 7–8. Einstein's affection for Spinoza is seen, for example, in Einstein, *Ideas and Opinions*, 38, 44–45. Spinoza's metaphysics involves the idea of a pantheistic view of divinity and monism, according to which there is only one substance in its infinity. See Copleston, *A History of Philosophy*, 4: 214–29, 244.

92. *CTSC*, 58.

93. Ibid., 59.

94. Ibid., 59–60.

95. EG, 12.

96. *TF*, 109.

97. Ibid., 76. Athanasius indeed makes a clear distinction between the two attributes of God, preferring the concept of Father to that of Creator: "Οὐκοῦν εὐσεβέστερον καὶ ἀλεθὲς ἂν εἴη μᾶλλον τὸν Θεὸν ἐκ τοῦ Υἱοῦ σημαίνειν καὶ Πατέρα λέγειν, ἢ ἐκ μόνων τῶν ἔργων ὀνομάζειν καὶ λέγειν αὐτὸν ἀγένητον." Athanasius, "Contra Arianos," I:34 (81–82).

98. Athanasius, *De Incarnatione*, 17 (174–77). E. P. Meijering has shown the close link between the *Logos* and the universe by pointing out that Athanasius considered the activity of the incarnate *Logos* in a human body as taking place as if the world, too, were a body. See Meijering, *Orthodoxy and Platonism in Athanasius*, 51.

99. *DCO*, 2–5; GGT, 52–60. See also *TF*, 98–109.

100. *DCO*, 29–30. According to Achtner (*Physik, Mystik und Christentum*, 109), the Greeks viewed contingency not only as the opposite of what is necessary or rational but also as universal and timeless.

101. *DCO*, 30. This Aristotelian distinction was later approved by Catholic Scholasticism, as is evident in Thomas Aquinas's theology: "Sic igitur contingentia, prout sunt contingentia, cognoscuntur directe quidem sensu, indirecte autem ab intellectu: rationes autem universales et necessariae contingentium cognoscuntur per intellectum." Aquinas, *Summa Theologica*, I:86:3 (618).

102. *DCO*, 31–32.

103. Ibid., 43.

104. Ibid., 38. Also worth mentioning is Torrance's view that "chance and random events inevitably yield necessary patterns, for behind all chance there is ineluctable necessity" (31). This means that for Torrance "nothing in God's creation happens by mere chance" (CFM, 5). Another example of how chance and necessity are treated in connection with the theology-science debate is given by John Polkinghorne (*Science and Christian Belief*, 77): "I have suggested that from a theological point of view the roles of chance and necessity should be seen as reflections of the twin gifts of freedom and reliability, bestowed on his creation by One who is both loving and faithful." The dissimilarity is evident: while Torrance wants to reject chance totally, Polkinghorne is ready to give it an accepted position.

105. Knuuttila, *Modalities in Medieval Philosophy*, 1–44; Knuuttila, "Nomic Necessities," 222–30.

106. Knuuttila, *Modalities in Medieval Philosophy*, 138–54. See also Vos Jaczn., Veldhuis, Looman-Graaskamp, Dekker, and den Bok, *Contingency and Freedom*, 23–28, and Lindberg, *The Beginnings of Western Science*, 240–44. For Scotus's role in the wider context of Scottish pre-Reformation philosophy and theology, see Broadie, *The Shadow of Scotus*, 1–34.

107. *DCO*, 32.

108. Ibid., 33; *TF*, 101–2.

109. *DCO*, 33.

110. Torrance's efforts to connect creation and redemption is analyzed by Roland Spjuth, who, surprisingly enough, sees a remarkable overlap in the thought of Torrance and some modern Catholic theologians. Spjuth, *Creation, Contingency and Divine Presence*, 72.

111. *TF*, 102. Torrance refers to Athanasius's writings, in which the reason for the Incarnation is defined along similar lines.

112. GGT, 95.

113. The Stoics had a developed philosophy of *logos* with strong religious nuances. They held that the supreme *Logos*, or God, contained an inferior type of "seminal *logos*" as the active ingredient by which all beings are generated. It was further maintained that in us our soul is our *logos*, which was divided into the 'immanent *logos*' and the 'expressed *logos*,' the former referring to our reason per se and the latter to the expression of his reason. A closer resemblance to the Christian *Logos* doctrine can be discerned in the thought of the Jewish thinker Philo, who took his ideas concerning the *Logos* mainly from the Stoics and regarded this principle of rationality "as intermediary between God and the universe" with a double role as "God's agent in creation" and as "the means by which the mind apprehends God." Kelly, *Early Christian Doctrines*, 10, 18–19.

114. Roland Spjuth (*Creation, Contingency and Divine Presence*, 99) proceeds to make the claim that Torrance's "argumentation becomes highly rationalistic."

115. *DCO*, 3; GGT, 52–53; McGrath, *Thomas F. Torrance*, 220–28.

116. This interpretation is further supported by the fact that Torrance feels disinterested in the challenging insights of those cosmologies which suggest the existence of other possible universes such as "mother" and "child" cosmoses, the first giving birth and an independent existence to the latter with an order and space-time structure of its own. See Davies, *The Mind of God*, 70–72 and Drees, *Beyond the Big Bang*, 46–47.

117. GGT, 6; *TS*, 298.

118. It should, however, be remembered that Torrance himself explicitly rejects any notion of the divine Logos as an immanent cosmological principle. *TF*, 103.

119. *DCO*, vii, 36. It is precisely here that Torrance departs from Aristotle, for example, who also considered the universe to be a rational whole. As was seen earlier, Torrance holds that contingency as the guarantor of the non-self-explanatory character of the cosmos eliminates all kinds of inherent necessities in nature. Aristotelian causes made the world consistent and rational but not unnecessary, since it was regarded as necessarily including its own reasons. This is another repetition of Torrance's argument, according to which the real difference made by early Christian theology in relation to the philosophy of Greek antiquity lies in the idea of contingency. On Aristotle as the First Scientist, see Alioto, *A History of Western Science*, 54–71. For Torrance's interpretation of Aristotle, see *CFM*, 31, 43; *DCO*, 5–6, 30–31, 69; and GGT, 46, 64.

120. GGT, 53–56.

121. *DCO*, 3. See Richardson, "Trinitarian Reality," 8–9.

122. *RST*, 24.

123. Ibid. Willem B. Drees (*Beyond the Big Bang*, 109) points out that there are several cosmologies that try to argue from the intelligibility of the creation toward God, and he comments on them, saying that "the further development of the argument from intelligibility in the world to a divine intellect is certainly beyond science."

124. *TF*, 103.

125. *GR*, 112.

126. An interesting parallel to this difficulty in finding a decisive and fundamental role for the Incarnation is to be found in John B. Zizioulas's analysis of Origen's theology in *Being and Communion*, 77.

127. See Heller, "Scientific Rationality and Christian Logos," 149, where Heller affirms thus very plainly, but not without problems, that the rationality of the world was not truly established until the Word became flesh and gave meaning to the creation by changing his transcendence to immanence, immanence identical with God's rationality.

128. *DCO*, 4.

129. Ibid., 5.

130. John Polkinghorne (*Science and Christian Belief*, 79) formulates a similar idea as follows: "The point is that we are considering possibilities that arise from the intrinsically open character of physical processes, not from transient patches of current human ignorance. . . . God's gift of 'freedom' to his creation is conveyed by his respect for the integrity of these processes." Compare this with Torrance's view that some of the Church Fathers saw that, in its relation to the freedom of God, nature is not perfect or self-sufficient but, rather, bears the characteristics of openness, spontaneity, and freedom. *TF*, 108. It should be remembered that, as was shown, Torrance excludes the possibility of chance, which cannot, therefore, be seen as an indication of the freedom of the universe.

131. *TF*, 105.

132. *DCO*, 4–5.

133. *TF*, 105.

134. See *DCO*, 4–5, as an earlier, and *TF*, 105, as a later, writing.

135. Torrance, *The Christian Doctrine of God* (*CDG*), 217.

136. *CDG*, 95–96; *TF*, 139.

137. *TF*, 109.

138. Along similar lines, Zizioulas (*Being and Communion*, 96–97) explains the transformation of the *Logos* from a cosmological explanation in Origen's theology to the doctrine of the Incarnation, as evident in Maximus the Confessor's thought. On this basis, he considers the significance of the Incarnation more explicitly than does Torrance: "The incarnate Christ is so identical to the ultimate will of God's love, that the meaning of created being and the purpose of history are simply the incarnate Christ" (97).

139. *DCO*, 14.

140. *RET*, 30.

141. Ibid.

142. GGT, 6.

143. Torrance, Space, Time and Incarnation (*STI*), 2.

144. GGT, 5–6. The idea of the "two books" of nature and Scripture has become well known, especially through the ideas of Francis Bacon (1561–1626), whom R. Hooykaas calls "the herald of modern science." Hooykaas, *Religion*, 40, 63. Wolfgang Achtner claims that the appeal to the analogy of the two books occurs especially within the Scottish Reformed tradition. Achtner, *Physik, Mystik und Christentum*, 9. The Scottish roots of some remarkable natural scientists give him reason to present this argument. See also Peacocke, *Creation and the World of Science*, 1–49.

145. GGT, 6. For Torrance, this community of reciprocity refers specifically to Israel as the chosen people of God. See MC, 32, where the stress is given to the Incarnation as the culmination of the reciprocal community between God and Israel. Torrance speaks of the triad God–world–human being as the framework of dialogue between us and God, on the one hand, and nature, on the other. *RET*, 30. Kurt Richardson ("Trinitarian Reality," viii) holds that this triad is one of Torrance's key structures, the parts of which should be constantly kept in mind when one of them is given more precise consideration than the others. It seems obvious that among Torrance scholars this three-part concept leads easily to a temptation to interpret it as some kind of analogy of the Trinity. The two-book view is, however, a more fruitful approach to searching for the background of this concept in Torrance's thought.

146. GGT, 1–2.

147. Einstein, *Ideas and Opinions*, 40.

148. See *DCO*, 83.

149. See Smart, *Problems of Space and Time*, 5. Torrance traces the origins of the receptacle notion back to the ancient Greeks (*STI*, 4–10), as does Max Jammer in his article "The Concept of Space in Antiquity," 27–42.

150. Sklar, *Space, Time, and Spacetime*, 167.

151. *STI*, 14.

152. See Torrance's account of the Reformed-Lutheran controversy with regard to the notion *finitum capax infiniti* in *STI*, 62–63. Torrance's view of the ecumenical relations between the Reformed Church and the Lutherans, seen as a problem of spatial concepts, is analyzed in chapter 5.

153. *STI*, 57–59.

154. Torrance, *Space, Time and Resurrection* (*STR*), 130.

155. Smart, *Problems of Space and Time*, 6. Leibniz himself was of the opinion that these relations exist between indivisible "monads" so that space is only "an order of coexistences." It is also worth noting that Leibniz makes a sharp distinction between space and place. Leibniz, "The Relational Theory," 89–92. Torrance refers briefly to Leibniz, but he finds the original roots of the relational notion as early as Plato's Timaeus. *STI*, 58.

156. *DM*, 371 (italics mine).

157. *DCO*, 79. That such a strongly theological interpretation is possible in the first place results partly from the elusive character of the concept of relational space-time. John Earman (*World Enough and Space-Time*, 12) points out that there are at least two reasons for this: firstly, "there is no relationist counterpart to Newton's Scholium, the *locus classicus* of absolutism," because even Leibniz's definition cannot offer any coherent description of relationality; and secondly, "there are almost as many versions of relationism as there are relationists."

158. *DCO*, 11.

159. *STI*, 58. For the history of the origin of the special theory, see Pais, *"Subtle Is the Lord . . . ,"* 111–37; and of the general theory, see the essays in Howard and Stachel, *Einstein and the History of General Relativity*.

160. A good description of this discussion is given in Smart, *Problems of Space and Time*, 15–17.

161. Sklar, *Space, Time, and Spacetime*, 67. A point of consideration is also provided by John Earman, who quite uncompromisingly reminds us that "relativity theory not only does not vindicate relationism but actually proves to be inimical to relationism" (*World Enough and Space-Time*, 174).

162. One of the most far-reaching implications of the theory of relativity is the idea of the continuous field of space-time, where space and time are considered as belonging essentially and inseparably together so that they cannot be regarded as totally independent elements. Space and time are thus deeply interlocked so that there is no space without time, or vice versa. The German Hermann Minkowski (1864–1909) became famous for his construction of a space-time that was later found to be in line with the theory of special relativity. See Sklar, *Space, Time, and Spacetime*, 56–59. Torrance seems to be well aware of the novel character of space-time to the degree that he readily makes use of it in his theological argumentation, for instance, in *DCO*, 79. This understanding of space-time reveals that Torrance attaches to it the same feature of openness from above as he does to the relational notion of space-time. See especially *DM*, 371; and *STR*, 184–90. But apart from this, Torrance has not given very much further thought to other possible and interesting implications of relativistic space-time. Such insights can be found, for example, in Davies, *God and the New Physics*, 199–217; and Stoeger, "Key Developments in Physics," 189–94.

163. A brief but illuminating discussion of time from the standpoint of theology-science dialogue is in Polkinghorne, *Science and Providence*, 77–99.

164. *TF*, 104.

165. Leibniz, "The Relational Theory," 89. See also Popper, *Realism and the Aim of Science*, 77–78.

166. *STR*, 130. For Torrance's view of the role of time in scientific and historical research, see his article "Time in Scientific and Historical Research" (TSHR), 292–97; and *TS*, 312–77.

167. *CDG*, 241.

168. Ibid.

169. The interesting insight presented by Torrance is that God has his own time. God's life in eternity is thus not timeless, although it does not have any beginning or end. *CDG*, 241. As Torrance does, Augustine also thinks that time is something created. However, the Church Father does not admit that God's eternal being could have anything to do with created time. Augustine, *Confessions*, XI:14 (267). See also Sorabji, *Time, Creation and the Continuum*, 98–136, in which an authoritative analysis is offered of the history of the concept of eternity with its relation to time; Sheldon-Williams, "The Greek Christian Platonist Tradition," 477–83; and *DCO*, 134–42.

170. *DCO*, 134–42.

171. *STI*, 58.

172. "Thus time, as a concept unequivocally determined by phenomena, was first deposed from its high seat. Neither Einstein nor Lorentz made any attack on the concept of space." Minkowski, "Space and Time," 304.

173. An illuminating discussion of time from the perspective of the new physics is provided by Paul Davies, who has reflected on the effect of Einstein's theory of relativity on the understanding of time in his book *About Time*, 32–33, 146–62.

174. For an insightful discussion of time in relation to personality, a subject that comes close to Torrance's interests through the concept of hypostasis, see Emmanuel Levinas, *Time*, 51–55.

Chapter 4

1. Torrance's own language includes the phrase "theological realism," which is both implicitly and explicitly apparent in his works. Kurt Richardson ("Trinitarian Reality," 2–5) has given a short but informative account of the history of the phrase in Torrance's writings.

2. *RET*, 39–40; TheRe, 173–74. See also Richardson, "Trinitarian Reality," 4. In short, what Torrance understands both by observationalism and phenomenalism is best seen in his constant criticism of Kantian notions of the unattainability of reality itself. He wants to confirm that genuine knowledge has to be based on reality, not just observations of it or phenomena occurring within it.

3. *TF*, 37–38.

4. See, e.g., Gray, "Theology as Science," 135–36; McGrath, *Thomas F. Torrance*, 211–20; Miller, "Beyond Dualism," 132–33; Spjuth, *Creation, Contingency and Divine Presence*, 94–101; and Trook, "The Unified Christocentric Field," 6.

5. Runzo, *Is God Real?*, xiii–xiv; Hick, "Religious Realism and Non-Realism," 4–5. See also Brümmer, "Does God Really Exist?," 17–34; Herrmann, "Realism, Semantics and Religion," 77–94; and Soskice, "Knowledge and Experience in Science and Religion," 173–84.

6. With regard to the classical debate between realism and nominalism, it is sufficient here to refer to the vast amount of literature on it. An informative general survey is given by Marilyn McCord Adams, "Universals in the Early Fourteenth Century," 411–39.

7. Audi, *Belief, Justification, and Knowledge*, 16–17.

8. Dancy, *An Introduction to Contemporary Epistemology*, 147–55. See also Audi, *Belief, Justification, and Knowledge*, 18, 21, 25, 33.

9. An informative discussion of the role of scientific realism is to be found in Hooker, *A Realistic Theory of Science*, 7–59. An effort to pursue semiotics on a realistic basis is John C. Cahalan's book *Causal Realism*, published in 1985.

10. Chalmers, *What Is This Thing Called Science?*, 161–64.

11. Ibid., 163.

12. Ibid., 164.
13. Hick, "Religious Realism and Non-Realism," 4–5.
14. Ibid., 4.
15. See, e.g., GGT, 10–14; *TCFK*, 61–64, 319–20; *RST*, 39–40, 110–12; and *TS*, 142–43.
16. Gray, "Theology as Science," 66–67. The Scottish philosophy is known precisely through its "commonsense" realism with Thomas Reid (1710–96) as its preeminent figure. Reid served as a professor of moral philosophy at the University of Glasgow, and he attacked his fellow countryman David Hume and his skepticism by denying Hume's thought, claiming that the world is what it is as to its necessities and externalities, although we cannot give a philosophical account of it. Broadie, *The Tradition of Scottish Philosophy*, 105–18. It is also worth noting that Thomas Reid had a strong interest in theology as well, as is apparent in his principal work *An Inquiry into the Human Mind*, esp. 3–4. In a preface to one of his most important books discussing theology-science problematics, Torrance links himself consciously to the Scottish realist tradition, an heir to the work of John Duns Scotus. See *DCO*, x.
17. Miller, "Beyond Dualism," 139–40.
18. Trook, "The Unified Christocentric Field," 3–6. This is not, however, the same as metaphysical realism, as held, for example, by Karl Popper, in which the existence of events and laws governing them, independent of an inquiring human mind, is supposed. Nevertheless, there are certain similarities in the thought of Torrance and Popper. See Popper, *Realism and the Aim of Science*, 79–88.
19. Meardon, "A Study of Space, Time and Incarnation," 49–51. Popper's idea of falsification as a feature of scientific theories is closely linked to what he calls critical realism. He holds that "theories are our own inventions, our own ideas: this has been clearly seen by the epistemological idealists. But some of these theories are so bold that they clash with reality: they are the testable theories of science. And when they clash, then we know that there is a reality: something that can inform us that our ideas are mistaken. And this is why the realist is right." Popper, *Quantum Theory and the Schism in Physics*, 3. See also Lakatos, "Falsification," 91–189.
20. *TS*, 102.
21. Spjuth, *Creation, Contingency and Divine Presence*, 94–101.
22. Ibid., 101.
23. *RET*, 39. Torrance's way of using the terms *post-Newtonian* or *post-Kantian* is another example of his mode of operating with terms that he has not adequately defined.
24. *TCFK*, 31. Here Torrance points to John Locke, whose famous idea was to emphasize sense experience as the sole basis for knowledge. Cf. Locke, *An Essay Concerning Human Understanding*, 33: "Let us then suppose the mind to be, as we say, white paper void of all characters, without any *ideas*. How comes it to be furnished? . . . To this I answer, in one word, from *experience* . . . ; in that all our knowledge is founded, and from that it ultimately derives itself. Our observation, employed either about *external sensible objects, or about the internal operations of our minds perceived and reflected on by ourselves, is that which supplies our understandings with all the materials of thinking*" (italics his).
25. *RET*, 40.
26. *GR*, 9–10.
27. George Hunsinger (*Karl Barth*, 35–42) sees objectivism as one of the basic motifs in Barth's thinking, for instance, in his insistence that "the knowledge of God as confessed by faith is objective in the sense that its basis lies not in human subjectivity but in God" (35).
28. Hunsinger (ibid., 36) explicitly states that Barth was well aware of the circularity of his argument that God has in an exclusive way revealed himself in order to give us knowledge of himself. "He had in no way tried to prove that God has engaged in an act of self-revelation. Rather, in essentially coherentist fashion, he had simply made candid the logic of believing in such an act of divine self-revelation."

29. *TS*, 35.

30. *RET*, 30–31. For Torrance's idea of the function of the idea of compulsion in the practical life of the Church, see "Service in Jesus Christ" (SJC), 1–16.

31. Torrance wishes thereby to restore what he sees as the "original" meaning of subject and object in epistemology. See *TS*, 12.

32. *RST*, 121.

33. *RET*, 54.

34. *GR*, 9; *TS*, 31.

35. "Man acts rationally only under the compulsion of reality." *RET*, 26. For Torrance, this compulsion also means that a correct methodology always must remain open for corrections. See *TS*, viii. This characterization of scientific methodology seems to leave the door open for the possibility of a theory being proved false, or a kind of falsificationism in the Popperian sense, namely, that a theory can never be found false definitely or conclusively but only in a logical-technical or methodological manner. Popper, *Logic of Scientific Discovery*, 27–48; Popper, *Realism and the Aim of Science*, xxii. For Torrance's attitude toward Popper's idea of falsificationism, see *TCFK*, 115–16, 163, 193.

36. Torrance uses the term *scientific hermeneutics*, by which he understands the primary role of the object of investigation. See *DM*, 10–11.

37. *TS*, 120.

38. Ibid., 124. Torrance makes a distinction between two modes of questions: theoretical questioning (*quaestio*) tries to clarify knowledge already existent and to form it into a logically acceptable shape in order to offer a solution to "mental problems" in the Scholastic manner. True scientific enterprise, however, uses mainly active questioning (*interrogatio*) in which structures of reality and their disclosure are sought. *TS*, 34.

39. It has to be admitted, however, that in the philosophy of science a more modest form of realism, compared to that of Torrance's, is known whereby a hypothesis is submitted to a test to establish its validity. "Scientific knowledge, as we have seen, is not arrived at by applying some inductive inference procedure to antecedently collected data, but rather by what is often called 'the method of hypothesis,' i.e. by inventing hypotheses as tentative answers to a problem under study, and then subjecting these to empirical test." Hempel, *Philosophy of Natural Science*, 17. Note that, although the idea of submission is apparent, there is not such a strong notion of reality as entertaining a compulsive force upon the scientist as is to be detected in Torrance's thought.

40. *DM*, 385; *TF*, 26–27; *TRs*, 49.

41. As to the origins of the concept of *homoousion*, Bernard Lonergan has pointed out that before its application in the Nicene debates it simply meant "made of the same stuff," disclosing its materialistic connotation. The Nicene Fathers, however, in their application of the *homoousion* did not accept this interpretation that can be seen as problematic from a naive realist standpoint, holding that reality can be attributed only with those objects that are present to the senses. Critical or dogmatic realism, on the other hand, sees no difficulty here because it regards as real what is known through critical judgment. Lonergan, *The Way to Nicea*, 88–89.

42. GGT, 162.

43. An example of this feature of self-evidence, although with a slightly different emphasis, is Torrance's claim that "No more than you can picture in a picture how a picture pictures what it pictures, without reducing everything to a picture, can you state in statements how statements are related to being, without reducing everything to statements." *TCFK*, 318.

44. Einstein, *Ideas and Opinions*, 266.

45. Jammer, "Einstein and Quantum Physics," 59–76. The field of quantum physics, as to its correct interpretation, is scattered and manifold. The Copenhagen interpretation, with the Dane Niels Bohr as its preeminent figure, has been perhaps the most influential over past decades. Other interpretations, however, exist as well, and a short survey of them is given by Russell,

"Quantum Physics," 348–54. It is worth noting that an apparent development is discernible in Torrance's attitude toward the Copenhagen interpretation of quantum mechanics. In *TS*, one of his first monographs with specifically scientific interests, quantum or nuclear physics, as well as its proponents in the Copenhagen school, are referred to numerous times but never in a critical tone (*TS*, 16, 90–92, 95, 100, 102, 111, 116, 136, 261, 265, 271, 290, 292, 295, 305–6, 318). Quite the contrary is the case in a later work, *DCO*, where Torrance applies consistent criticism to the Copenhagen/Göttingen interpretation, accusing it of lapsing back into determinism (*DCO*, 13–15, 44). This remarkable shift in Torrance's thought is best understood as an indication of his deepened realization of the implications that the Copenhagen interpretation might have for his own system.

46. Heisenberg, *Physics and Philosophy*, 160.

47. The Finnish physicist K. V. Laurikainen has expressed this by saying that it is "incorrect to presuppose that any rational description will ever be able to reach 'reality itself.'" Laurikainen, "Quantum Physics, Philosophy, and the Image of God," 393. He states further, "Objects and their properties (dynamic variables) are in quantum mechanics only symbolic" (395).

48. "What seems to be coming under increasing pressure, at least in the realm of quantum physics (and, I would argue, in cosmology as well), is the further and more specific claim of correspondence, that the structure of theoretical concepts corresponds to some extent with the structure of their references in nature." Russell, "Quantum Physics," 353. The phrase "corresponds to some extent" reveals that Russell does not demand such a strict correspondence from truth-statements as does the traditional form of the correspondence theory.

49. Karl Popper, in whose thinking there are remarkable similarities with Torrance, is also reluctant to accept the idea that true reality must be known in order for the comparison between facts and human statements about them to be made. Popper, *Conjectures and Refutations*, 226.

50. According to Torrance, this is due to a disjunction between form and being in Western thought. TheRe, 172–73.

51. *TS*, 173.

52. Ibid., 167.

53. Reichenbach (*The Theory of Relativity*, 50) is quite close to Torrance in this respect, since he, too, adheres to the indefinability of reality: "The real thing is the thing confronting us; there is no point in trying to define its existence more closely, because what is meant by 'real' can only be experienced."

54. Heisenberg, *Physics and Philosophy*, 166.

55. *DCO*, 13.

56. It is interesting to note that Karl Popper, too, fiercely criticizes the Copenhagen interpretation on a realist basis but that Torrance has not been able to use his argumentation to back his own ideas, due to the fact that Popper's book *Quantum Theory and the Schism in Physics* was published as late as 1982. Like Torrance, he relies heavily on Einstein and his reluctance to accept the Copenhagen interpretation, accusing it of two serious mistakes. Firstly, it has introduced subjectivism into physics, and secondly, it has tried to convince the audience that it offers the final and absolute theory for understanding quantum phenomena, thus presenting the "end-of-the road thesis." Popper, *Quantum Theory and the Schism in Physics*, 1–10. There are many interesting resemblances between the ideas of Torrance and Popper, and one can only imagine how the latter could have influenced the former if the book mentioned had been published several years earlier. Brian H. Meardon ("A Study of Space, Time and Incarnation," 29) also has paid attention to this relationship between Torrance and Popper.

57. John Polkinghorne ("The Quantum World," 335) has stated briefly but illuminatingly that "it is ontology which controls epistemology. (I suppose that is a definition of realism.)" See also *TCFK*, 331.

58. Russell ("Quantum Physics," 353–54) has evaluated the possibilities of the realist approach in quantum physics, and he refers to scholars such as Stanley Jaki, John Polkinghorne, and Chris Isham. He concludes that "the challenge posed by quantum physics to realism continues to be a highly controversial and stimulating area for research."

59. TheRe, 171. See also *RET*, 66.

60. TheRe, 171–72.

61. Ibid., 172; *TF*, 147. Torrance's approach is quite contrary to that of Paul Tillich's, who is in favor of symbols instead of signs. See, e.g., Tillich, *Systematic Theology*, 1:239; and *TS*, 19. Tillich applies this interpretation with regard to the Incarnation by claiming that "the assertion that 'God has become man' is not a paradoxical but a nonsensical statement. It is a combination of words which make sense only if it is not meant to mean what the words say. . . . The unqualified use of the term 'Incarnation' in Christianity creates pagan, or at least superstitious, connotations" (*Systematic Theology*, 2:94–95). See also Barbour, *Myths, Models and Paradigms*, 15–16.

62. *TCFK*, 319–20; TheRe, 171.

63. *TF*, 53. This provides another example of the resemblance in the ideas of Torrance and Karl Popper. The latter uses the term *verisimilitude* to describe the approximate character of human knowledge. "What, then, is the object of our 'rational belief'? It is, I submit, not the truth, but what we may call the truthlikeness (or 'verisimilitude') of the theories of science, so far as they have stood up to severe criticism, including tests. What we believe (rightly or wrongly) is not that Newton's theory or Einstein's theory is true, but that they are good approximations to the truth, though capable of being superseded by better ones." Popper, *Realism and the Aim of Science*, 57.

64. Torrance's realism has been criticized precisely because of its "unprejudiced experience of the object." In Spjuth, *Creation, Contingency and Divine Presence*, 98, Torrance's standpoint is accused of foundationalism and a lack of any tendency toward a holistic epistemology that could be grounded "anywhere besides this socially interwoven network of acts, experiences and convictions." Spjuth has masterfully pointed out one of the problems in Torrance's thought, but he fails to give a positive alternative to Torrance's epistemological solution.

65. Spjuth maintains that "the logic of Torrance's reasoning ends in a realism that never questions the dichotomy between subjectivity and objectivity." According to him, this leads Torrance to consider "the intuitive experience" as the bridge between them. Ibid., 100–101.

66. *TCFK*, 111.

67. Einstein holds that the highest task of a physicist is to find the most general elementary laws. They are not, however, capable of being found through logic but, rather, through intuition supported by feeling which is based on experience. Einstein, *Mein Weltbild*, 143. Einstein's adherence to the idea of intuition with particular respect to religious understanding is also appreciated by Iain Paul, whose thinking has to be considered an example of efforts to unite the physicist's ideas with a special theological tradition and confessional framework, in this case doctrines formulated by Calvin. This makes him claim surprisingly that "we find in Einstein's thoughts on science an intriguing echo of Calvin's trinitarian theology." Paul, *Knowledge of God*, 85. Paul seems to give too great a role to intuition in Einstein's thought. However, in doing this, he is driving Torrance's argument, upon which he is initially dependent, to its ultimate conclusions, resulting in a view that is not far from naïveté: "Einstein's work shows that, if the scientist seeks sympathetic knowledge of the natural order, all that he can do and needs to do is go to the natural phenomena and rely on intuition" (92). See also Paul, *Science and Theology*, 84–93. A more modest approach to Einstein's idea of "freely invented concepts," or intuition and their implications for religious life, has been adopted by the Roman Catholic Stanley L. Jaki, who reflects Einstein's thoughts in a wider context as a factor opposing positivism. Jaki, *Relevance of Physics*, 343–45. A discussion of Einstein's intuitionism is provided by Voorhees and Royce, "Einstein and Epistemology," 25–33.

68. *CTSC*, 65; *TS*, 53, 70.

69. Einstein himself used the phrase "freely *formed* concepts" (italics mine). Einstein, *Ideas and Opinions*, 294.

70. Pais, *"Subtle Is the Lord . . ."*, 13. For a detailed discussion of Einstein's "objective realism," see Morrison, *Science, Theology and the Transcendental Horizon*, 277–349.

71. *TCFK*, 112.

72. Ibid.

73. Polanyi, *Personal Knowledge*, 69–245.

74. Torrance's definition of the term can be found in "Notes on Terms and Concepts" (NTC), 145. Polanyi's importance for Torrance has been evaluated in McGrath, *Thomas F. Torrance*, 228–32.

75. *TCFK*, 112.

76. Ibid., 111.

77. See *RET*, 31.

78. Ibid., 150.

79. *TS*, 120–21.

80. Jonathan Dancy points out that, according to classical foundationalism, "our basic beliefs are beliefs which concern the nature of our own sensory states, our own immediate experience." Dancy, *Epistemology*, 53, 58.

81. The idea of self-evidence with respect to realism and its compelling force does not fall in line with Torrance's frequent use of the so-called Gödel theorems, according to which any logical system is consistent only if it does not contain reasons for its own explanation but derives them from outside. *DCO*, 60; *GGT*, 70; *TCFK*, 98, 136–45. Brian H. Meardon has quite convincingly shown that "for Torrance rationality is closely bound up with the capacity to be objective," leading to another form of circular argumentation: "His case is that rationality exists in the universe because one can give a rational explanation and that rational explanation is valid because the rationality of the universe impresses itself upon us." Meardon, "A Study of Space, Time and Incarnation," 26. Farther down in his work, he claims pointedly that "Torrance fails to show how knowledge of that reality, which discloses itself, is reached," resulting in Torrance's inability to answer the question of what reality is (35).

82. This has resulted in an inadequate overall picture as presented by some scholars, especially with respect to the question of realism and the role of subjective intuition. Although Roland Spjuth's excellent analysis of Torrance's realism (*Creation, Contingency and Divine Presence*, 94–101) accurately discloses some crucial weak points, it fails to enter more deeply into the way in which Torrance uses the concept of intuition and, as a result, cannot offer any positive alternative. The same can be said of Brian H. Meardon's approach, in which a hint of denominational controversy is not totally lacking. Explaining Torrance's view of science-theology relations, he asserts, "Such a view links the whole of the scientific enterprise to Christian theology as practiced in Edinburgh!" Meardon, "A Study of Space, Time and Incarnation," 60. Another example of moderate neglect of Torrance's realist position is James B. Miller's claim, made from the trend-conscious standpoint of postmodernism, that "For Torrance, all inquiry whether scientific or theological is ultimately grounded in a source which is absolutely transcendent of the world, immutable and essentially unrelated to nature and history." Miller, "Beyond Dualism," 141–42. Miller cannot appreciate Torrance's point in which there is a relation between transcendence and immanence, namely intuition, a relation that is based on the application of the concept of *homoousion*, "of the same being."

83. A critical study of the role of intuition in scientific enterprise is made by Mario Bunge, who says, "Intuition is the collection of odds and ends where we place all the intellectual mechanisms which we do not know how to analyze or even name with precision, or which we are not interested in analyzing or naming." Bunge, *Intuition and Science*, 68.

84. See, e.g., Calvin, *Institutio*, I:XIV:22 (133–34).

85. Hing Kau Yeung ("Being and Knowing," 158–59, 197–99) briefly refers to this relation, but he fails to penetrate deeper into the issue.

86. *TS*, 86, 96, 128, 215.

87. Ibid., 127–28.

88. Ibid., 86. For Torrance's interpretation of St. Paul's teaching involving the ideas of predestination, see *Conflict and Agreement in the Church* (CAC), 2:86. A helpful glance at the Reformed doctrine of election is David A. S. Fergusson's article "Predestination: A Scottish Perspective." See also Newbigin, *The Gospel in a Pluralist Society*, 80–88.

89. Calvin, *Institutio*, III:XXI–XXIV (678–728).

90. "Praedestinationem vocamus aeternum Dei decretum, quo apud se constitutum habuit, quid de unoquoque homine fieri vellet. Non enim pari conditione creantur omnes; sed aliis vita aeterna, aliis damnatio aeterna praeordinatur." Ibid., III:XXI, 5 (683). Calvin strongly gives the impression that predestination took place in the act of creation, not after the Fall.

91. "Quanquam satis iam liquet Deum occulto consilio libere quos vult eligere, aliis reiectis, nondum tamen nisi dimidia ex parte exposita est gratuita eius electio, donec ad singulas personas ventum fuerit, quibus Deus non modo salutem offert, sed ita signat, ut suspensa vel dubia non sit effectus certitudo." Ibid., III:XXI:7 (685).

92. Ibid., III:XXIV:5 (716).

93. Although Calvin states explicitly that he holds to the view that God's election and reprobation took place in eternity, i.e., before the creation and the fall of the human race, within the Reformed tradition there emerged a strict conflict between supra- and infralapsarianism. While the former party maintained that God's eternal decree really took place before the creation and the Fall, the latter party claimed that it was not so until the Fall. Berkouwer, *Divine Election*, 258–59. Berkouwer himself avoids taking a fixed stand in the controversy but tries to give the impression that both aspects are present in the Reformed tradition in a positive way (273–77). It seems obvious, however, that the supralapsarian view is more generally, and also officially, accepted by many Reformed theologians. See, for instance, Heppe, *Reformed Dogmatics*, 155–56: "Predestination is eternal, i.e. it is fixed from within eternity and for eternity, it is unsearchable and unchangeable."

94. According to Alister McGrath, the doctrine of predestination cannot be seen as the center of Calvin's theology. His interpretation of the reformer emphasizes the pastoral character of Calvin's idea of predestination: the fact that Calvin placed his discussion of predestination in the context of God's grace gives McGrath reason to claim that the reformer consciously avoided discussing the issue apart from grace as a consequence of divine election or reprobation but regards God's mercy as an essential presupposition of predestination. McGrath makes the conclusion that Calvin's teaching of predestination should be seen as an interpretation, formulated in the light of the Bible's message, of human experience rather than as a conclusion reached on the basis of preconceived views concerning God's omnipotence. McGrath, *A Life of John Calvin*, 167–69. It can be argued, however, that the status of the doctrine of election occupies a more central place in Calvin's system than McGrath suggests. Although Calvin placed his discussion of election after that of God's grace and our justification, he nevertheless wrote three long chapters in explicit defense of his views of eternal election, making efforts to refute opposing views. This indicates clearly that more is at stake than a mere interpretation of a pastoral nature.

95. This rigorous mode of Calvinist predestination has its plainest expression in the so-called *Westminster Confession of Faith*, originally published in 1647 to unite the Presbyterian churches in England, where it is stated that "III. By the decree of God, for the manifestation of his glory, some men and angels are predestinated unto everlasting life, and others foreordained to everlasting death. IV. These angels and men, thus predestinated and foreordained, are par-

ticularly and unchangeably designed; and their number is so certain and definite, that it cannot be either increased or diminished." *Confession*, III:III–IV (8).

96. The German sociologist Max Weber's well-known argument emphasizes the Calvinist doctrine of predestination as the basis for the Puritan ethic of work. Weber, *The Protestant Ethic*, 98–112. An interesting detail is that Weber explicitly mentions the *Westminster Confession* as the best exponent of the essence of Calvinism, a view that has to be seen as at least questionable (99–101).

97. *SF*, lxxviii.

98. Ibid.

99. Ibid., lxxvii.

100. *CDM*, 20. See also *TS*, 128 (n. 2).

101. CTSC, 127.

102. Ibid., 128.

103. Barth, *Church Dogmatics*, 4:1:50–51, 516.

104. Ibid., 2:2:118.

105. Ibid., 2:2:162.

106. Ibid., 2:2:295, 417, 422, 476.

107. This feature in Barth's thought has been emphasized by G. C. Berkouwer (*The Triumph of Grace*, 262–96), who interprets Barth's position by pointing out that "as sin is ontologically impossible because it cannot achieve what it aims to achieve—man's separation from the grace of God—so unbelief is an impossible reality because it cannot touch what in God's decision has become irrevocable truth, namely, that God's grace is for man, for all men" (266).

108. Barth, *Church Dogmatics*, 2:2:20–21.

109. Ibid., 2:2:22.

110. *DCO*, 29.

111. Ibid., 59.

112. Heppe, *Reformed Dogmatics*, 152.

113. Especially illuminating in this respect is Torrance's interpretation of Athanasius in *TF*, 103.

114. Barth himself suggests that not only men and angels but also the rest of creation is elected by God. He states that the righteousness of God "utterly crushes us. In it God asserts and vindicates His own worth over against the creature. Yet in the election of the creature even this righteousness reveals itself as the grace and loving-kindness and favour of God directed towards it." Barth, *Church Dogmatics*, 2:2:33. He fails, however, to develop this line of thought further, thus leaving the task to Torrance.

115. GGT, 9.

116. *TS*, 214.

117. Ibid., 215.

118. Torrance speaks of the vicarious humanity of Christ, emphasizing Christ's role as our human response to God's calling. This means that what we call our regeneration or our conversion is actually our participation in the conversion or regeneration of our humanity carried out by Jesus. MC, 86.

119. *TS*, 71; *RST*, 111.

120. *RST*, 137.

121. James B. Miller ("Beyond Dualism," 139) calls Torrance's position revelational positivism, stressing thereby his willingness to defend the a posteriori character of theology and other scientific enterprises. (Compare this to *TS*, 33.) However, due to the complicated and encumbered nature of the term *positivism*, it does not seem very appropriate to use it to define Torrance's views. One should remember, as well, how Torrance wants to maintain a considerable distance from positivism as such and considers it the opposite pole of realism. CTSC, 125–26; GGT, 35; *RET*, 10, 12; *TRs*, 67.

122. The idea of the controversy between Christian realism and Greek "mythological" thinking was originally exposed in 1904 by the Frenchman Lucien Laberthonnière (1860–1932) in his book *Le Réalisme chrétien et l'idéalisme grec*. Macquarrie, "Laberthonnière," 372.

123. Athanasius, "Contra Arianos," I:1–4 (11–20).

124. *TF*, 119; *RET*, 27.

125. McInerny, "Dualism," 1075.

126. Fontaine, *The Light and the Dark*, 1:ix. Fontaine maintains that the history of the concept can be dated back to the Greek author Plutarch (c. 45/50–125 AD.), who concentrated in his studies on Egyptian myths. For this and other features in the history of the term, see Fontaine, *The Light and the Dark*, 6:xiii–xli.

127. Bianchi, *Selected Essays*, 50.

128. Simons, "Dualism," 370–71. In Hall, "Monism and Pluralism," 363, monism is defined as "a name for a group of views in metaphysics that stress the oneness or unity of reality in some sense," a view that can be traced back to pre-Socratic times in Greek philosophy. The actual emergence of the term is accredited to Christian Wolff (1679–1754), who used it to analyze idealism and materialism and their reductionistic tendencies to interpret the universe from a single standpoint.

129. Simons, "Dualism," 370–71.

130. Lanczkowski, "Dualismus," 199–202.

131. Manes or Mani (216–276/277) was born in Persia, where he came under Zoroastrian influence, though he finally opposed it so that he was forced into exile in India. His teaching quickly gained remarkable success in the East. His thought has been described as a radical offshoot of Eastern gnosticism. It was based on belief in the eternal struggle between light and darkness. Because Satan had stolen particles of light from the World of Light and imprisoned them in the human brain, they had to be released from Satanic captivity. In that task Jesus, Buddha, the Prophets, and Mani were persons sent to accomplish the mission in favor of the light. Mani's speculation is an example of one of the most comprehensive and most organized gnostic systems. St. Augustine, before his conversion to Christianity, was deeply involved in Manichaeism. Kelly, *Early Christian Doctrines*, 13–14. Augustine himself writes of this stage of his life in the fifth book of his *Confessions*.

132. Complementarity in physics is represented as an effort to consider light as consisting of two differing modes of existence, that of wave and that of particle. John Polkinghorne (*Science and Christian Belief*, 133) has pointed out that this has been used analogously in Christology to express the indwelling of both human and divine elements in the Person of Christ. See also MacKinnon, "Complementarity," 255–70; and Loder and Neidhardt, "Barth, Bohr, and Dialectic," 271–89.

133. Lanczkowski, "Dualismus," 202.

134. The nature of dualism is described along similar lines by Roland Hall ("Monism and Pluralism," 364), who maintains that dualism is the result of failed monism. According to Hall, the mere existence of dualism implies efforts to find satisfying solutions to bridge the two poles. He seems very suspicious of the possibility of a complementary dualism as described by Lanczkowski, and he states explicitly that dualism as a consciously chosen starting point for philosophical reflection is simply an impossibility (364).

135. Fontaine, *The Light and the Dark*, 1:x.

136. McInerny, "Dualism," 1073.

137. Ibid., 1075.

138. Mensching, "Dualismus," 272.

139. G. Gloege directs attention precisely to this detail in his own definition of the concept of dualism. He speaks of tension (*Spannung*) as a distinctive character of polarity. Gloege, "Dualismus," 274.

140. Mensching, "Dualismus," 272–74.

141. John Hick describes how monistic and dualistic modes of thought are applied in efforts to explain the problem of theodicy within the Christian faith. He states that Christian theology has always moved between monistic and dualistic attempts at explanation, while neither of them is really compatible with the monotheistic belief in a good God. Hick, *Evil and the God of Love*, 15.

142. Both Mensching and Simons quote at this point 4 Ezra 7:50: "The most High has not made one aeon but two." Mensching, "Dualismus," 273; Simons, "Dualism," 373.

143. Simons, "Dualism," 374.

144. McInerny claims that "in modern times, dualism is espoused most frequently in dealing with the problem of knowledge," thus indicating the epistemological implications of dualism. McInerny, "Dualism," 1073.

145. Gloege, "Dualismus," 274.

146. Simons, "Dualism," 374.

147. Gloege, "Dualismus," 274–75.

148. Simons, "Dualism," 374.

149. Simons expresses this view explicitly when he defines an agenda for modern theology which, in his opinion, should search for solution to "the dualism between Church and world, word and sacrament, office and charism and so on." Simons, "Dualism," 374. This passage is also clear evidence of the difficulty of defining the concept of dualism and the ambiguous way in which it can be applied. When at the beginning Simons defined dualism as a view reducing reality "to two equally primordial and opposed principles," these models of dualisms mentioned above (Church/world, etc.) are extremely difficult to accept with such a definition.

150. McInerny, "Dualism," 1075.

151. This detail becomes apparent as soon as one examines Torrance's book *Space, Time and Incarnation*, one of his first treatises concerning dualism and the relations between theology and the natural sciences. Neither a satisfactory account of dualism nor an effort to define the term is provided in works on Torrance. Alan G. Marley admits that "the rejection of dualism is fundamental to Torrance's view of things," but, nevertheless, no account is given of dualism itself. Marley, *T. F. Torrance in a Nutshell*, ii. Accordingly, such definitions are lacking in the treatises on Torrance by such writers as Stamps ("The Sacrament of the Word Made Flesh"), Meardon ("A Study of Space, Time and Incarnation"), Miller ("Beyond Dualism"), and Richardson ("Trinitarian Reality"). Wolfgang Achtner widely uses Torrance's concept of dualism throughout his work, even criticizing Torrance's usage of it as "a less fortunate concept," but he fails to provide any further analysis of Torrance's view of the term. Achtner, *Physik, Mystik und Christentum*, 43–107, 222.

152. An example of the misleading results of an application of the ambiguous concept of dualism is provided by Richard A. Muller, who in his heavy criticism of Torrance gives no thought to his own way of using the term but starts from a very general definition. This drives Muller to claim that Torrance is actually a dualist because he draws distinctions between God and the creation. Muller, "The Barth Legacy," 695. The same claim can also be found in Achtner, *Physik, Mystik und Christentum*, 75–79. Torrance's use of the concept of dualism, however, is much more complicated and diverse, as will be shown.

153. NTC, 136.

154. See *TF*, 47. See also CFM, 35. Carver T. Yu (*Being and Relation*, xiv) presents a Chinese view of dualism in Western culture, seeing the separation of matter and spirit as its reason.

155. *TF*, 47. It can be claimed that one of the disputable features in Torrance's thought lies in his opinion that Aristotle was basically a dualist, although he is generally seen as a keen opponent of Plato and his dualistic cosmology. A vivid example of the discussion within Aristotelian scholarship concerning the issue of dualism is offered by Howard Robinson ("Aristote-

lian Dualism," 123–44) and Martha C. Nussbaum ("Aristotelian Dualism: Reply," 197–207), who disagree upon the question of whether there is any Cartesian-like mind-body distinction in Aristotle.

156. Torrance, *Belief in Science and in Christian Life.*

157. *TF*, 47.

158. *TRc*, 224.

159. It is worth noting that Torrance is careful not to emphasize the controversy between Athanasius and the person Arius. He prefers to use the term "the Arians" or "the Arian theology." In this way he is able to avoid strict claims concerning the exact teaching of Arius himself, admitting that Arius's theology is not unambiguous. Williams (*Arius*, 82) has stated that even "'Arianism' as a coherent system, founded by a single great figure and sustained by his disciples, is a fantasy—more exactly, a fantasy based on the polemic of Nicene writers, above all Athanasius." Lonergan notices the same feature, namely that the Arians as a group did not share the position of Arius himself, the basic tenet of which was that the Son is a creature. The name "Arians" was ascribed by the Nicene Fathers to all their opponents, who basically shared only one common factor, their opposition to the theological formulations of the Council of Nicea. Lonergan, *The Way to Nicea*, 68.

160. *TRc*, 224–25.

161. *TF*, 47.

162. Williams, *Arius*, 230.

163. Ibid., 231. In fact, Torrance describes Nicene theology as having exactly the same concern. See *TF*, 17.

164. Lonergan, *The Way to Nicea*, 71.

165. John D. Zizioulas sees the problem of Greek influence upon the development of Christian dogma quite differently. Unlike Torrance, he does not regard Greek thinking as basically dualistic but rather as monistic, and it is precisely this, according to him, that made the task of the Nicene Fathers so difficult. He argues further that the sharp separation of God from the world was not due to Greek thinking in general but to gnosticism, "the other great danger of this period." Zizioulas, *Being and Communion*, 16.

166. As Rowan Williams has pointed out, Athanasius's interpretation of Arius is reductive and one-sided. Williams, *Arius*, 82.

167. E. P. Meijering has shown that although Athanasius explicitly and heavily criticizes the Platonists, philosophers who consciously continued the tradition of Plato, he implicitly uses in his argumentation conceptual tools originally derived from his Platonic education. Meijering, *Orthodoxy and Platonism in Athanasius*, 114–31. Meijering states also that the Church Fathers became Platonized in their theology in a neutral or formal way. This means that they applied the language, the thoughts, and the arguments of the Platonists. With this idea Meijering does not want to offer any definite answer to the question of whether the actual substance of the Church Fathers' theology was affected as well. Meijering, *God Being History*, 135. It is, in any case, true that the Greek Fathers, Athanasius included, fought against that mode of radical dualism which was present in Gnosticism. Fontaine, *The Light and the Dark*, 8: 90–91, 284–85, 296–300.

168. *TF*, 119–20, 122–25.

169. MC, 67.

170. Ibid., 69. D. M. Baillie, who was professor of systematic theology at the University of St. Andrews in the 1940s and who had an impact upon Torrance's thinking, in 1947 published a book entitled *God Was in Christ*. He argues that in the case of the Incarnation we are dealing with a mystery and, to be precise, with a deep paradox. Baillie, *God Was in Christ*, 106–32. Torrance rejects this kind of understanding, implying that it is just the result of irreconciled logical thought that has not found a solution. (For Torrance's relation to the Baillie brothers, John and Donald B., see McGrath, *Thomas F. Torrance*, 34–37.) Kang Phee Seng puts this problem

as follows: "For the dualist thinkers, the dilemma is—How can the eternal God who is impassible and changeless be thought of as actually entering the spatio-temporal history of this changing and decaying world and alas, even within our creaturely and contingent order?" Kang, "The Epistemological Significance of ὁμοούσιον," 341.

171. On how Torrance uses the phrase "logical bridge" between concept and experience when criticizing logico-deductive testing of scientific theories, see *RST*, 49–50.

172. *DCO*, 33; *TF*, 98–102.

173. *DCO*, 4.

174. Torrance also calls this defect an ultra-transcendentalist view of God. MC, xi.

175. *CDG*, 79–80; GGT, 160. "The *homoousion*, however, is to be taken along with a cognate conception about the indissoluble union of God and man in the one Person of Christ, to which the Church later gave theological formulation as the hypostatic union (ἕνωσις ὑποστατική)." *CDG*, 94.

176. *TF*, 132–45, 154–90.

177. GGT, 161; *RST*, 184–87; *TF*, 33, 186.

178. GGT, 161; TheRe, 182–85, 189.

179. Torrance very rarely refers to the Chalcedonian settlement of the two co-equal natures in Christ, the event in which the dilemma was reflected more thoroughly than in the Councils of Nicea or Constantinople. In GGT, 61, 126–27, he maintains that Byzantine and Augustinian dualism had an unfortunate effect on the Chalcedonian formula, leading to the monophysite controversy, which, in Torrance's opinion, could not have arisen at all if dualistic frames of thought had been set aside. On the Chalcedonian formula, see Kelly, *Early Christian Doctrines*, 338–43.

180. Here Torrance makes use of a distinction made originally by Friedrich Schleiermacher. See Schleiermacher, *The Christian Faith*, 1:97–101.

181. *TF*, 112–13. On Ebionitism and Docetism, see Kelly, *Early Christian Doctrines*, 138–42.

182. *TF*, 113.

183. It is worth noting that Paul Tillich also presupposes a balanced view of the two natures in the person of Christ. However, unlike Torrance, he emphasizes the tension between the poles and does not accredit the concept of *homoousion* with such a problem-free status as Torrance does. Tillich, *Systematic Theology*, 2:142–45.

184. MC, 47.

185. See the General Foreword to the series Theology and Science at the Frontiers of Knowledge, for instance, in *RST*, ix.

186. Robert Stamps, who has studied Torrance's eucharistic theology, has certainly not paid sufficient attention to this problem when he says that, despite his fierce rejection of dualism, Torrance is obliged to make certain necessary applications of it in his eucharistic theology. Stamps goes further and claims that they are understandable only as a "quasi-Hellenistic contrast of divine/human." Stamps, "The Sacrament of the Word Made Flesh," 81.

187. See, e.g., GGT, 19; *TF*, 47; and *TRc*, 221. It is worth noting that Karl Barth uses the concept of dualism, as well, to describe something negative or objectionable when he comments on the so-called Federal theology, and especially Johannes Coccejus's impact on it. Accordingly, Barth accuses the seventh article of the *Westminster Confession* of giving confessional status to this dualism, thereby making it "established." Barth, *Church Dogmatics*, 4:1:59, 65. On Torrance's view of Barth's relation to dualism, see GGT, 87–95. See also Hunsinger, *Karl Barth*, 177.

188. *GR*, 29–31. See the section "Theology and Science" in chapter 2.

189. *TRc*, 267–68.

190. Ibid., 268.

191. *GR*, 30–31; *TRc*, 269–70.

192. *TRc*, 268.

193. *GR*, 31.

194. On the one hand, here Torrance echoes Barth's rejection of independent worldviews as allies of Christian faith: "Even where (in certain types of thought among certain Christians and movements) we think we detect an absolute union of faith with this or that world-view, we are not really dealing with faith at all, but with a partial deviation from faith," Barth, *Church Dogmatics*, 3:2:9. On the other hand, Torrance does not reject talk of cosmologies or cosmological outlooks as does Barth, who claims that cosmology "can arise only in the sterile corner where the Word of God with its special revelation has not yet found, or has lost again, hearing and obedience on the part of man." Barth, *Church Dogmatics*, 3:2:11.

195. As a source-critical remark, it should be noted that Torrance's analysis of Newton's theological thought is based almost exclusively on certain parts of the *Mathematical Principles of Natural Philosophy* (*Principia Mathematica Philosophiae Naturalis*) and on his other remarkable work, *Opticks*. Frank E. Manuel points out that far-reaching conclusions about Newton's theological thought reached on the basis of these few published texts (and the Clarke-Leibniz correspondence added to them), which have been "chewed and re-chewed ad nauseam," can give only a rather stereotyped picture of his religious thinking. Manuel, *The Religion of Isaac Newton*, 8–11.

196. *GGT*, 23; *TCFK*, 12–15.

197. *GGT*, 67–68.

198. In theology, even nowadays, there are approaches, e.g., fundamentalism, which are at least analogically connected with Newton and his thinking. Popkin, *Essays on Isaac Newton's Theology*, 176.

199. A good example of Newton's spiritual intentions are the first lines of one of his letters where he gives reasons why he wrote his famous book, *Mathematical Principles of Natural Philosophy*: "When I wrote my treatise about our system, I had an eye upon such principles as might work with considering men for the belief of a Deity; and nothing can rejoice me more than to find it useful for that purpose." Newton, "God and Gravity," 46. See Scheurer and Debrock, *Newton's Scientific and Philosophical Legacy*, v. "Newton's aim goes beyond explaining the phenomena of nature, important though this is. Ultimately, he wished to render the Creator and the whole of creation intelligible to human understanding." J. E. McGuire, *Tradition and Innovation*, 39. A more modest approach is adopted by James Byrne (*Glory, Jest and Riddle*, 155), who maintains that "Newton was too methodical a scientist to allow his theological concerns to dominate his empirical investigations." In spite of diverse opinions on the relations between scientific and religious intentions in Newton's thought, scholars agree upon the central role of his religious conviction. A concise description of Newton's religious development is to be found in Christianson, *In the Presence of the Creator*, 246–65. See also Dobbs and Jacob, *Newton and the Culture of Newtonianism*, 32–34. On pages 95–101 is given an account of a later Newtonian interpretation of Christian faith.

200. Quite illuminating is the passage in which Torrance describes the significance of the Newtonian outlook of the universe for contemporary theology which has injected deistic tendencies into Western thought. *GGT*, 71.

201. *GGT*, 68; see also *TCFK*, 20, where Torrance characterizes this distinction as "fateful" and continues by saying that it drove science and common sense far away from each other.

202. *GGT*, 68.

203. *TCFK*, 20.

204. Ibid., 24.

205. Ibid., 16. Byrne (*Glory, Jest and Riddle*, 151–52) sees Newton's approach as an empiricist reply to Cartesian rationalism, a reply that avoided rationalistic speculations, *hypotheses*. Newton, *The Mathematical Principles*, 446–47. Torrance, interestingly enough, has noted Newton's

ambiguity concerning the use of hypotheses: "It is certainly clear that he insisted on the rigid exclusion of hypotheses from his 'system,' for the feigning of hypotheses, as he expressed it, is not relevant to mechanical explanation." Nevertheless, on the other hand, Newton uses hypotheses, e.g., in his *Principia,* and therefore Torrance concludes that the scientist deals with a hypothesis insofar as it is understood *scientifically,* and not as "an imagined assumption or a postulate advanced without empirical grounds but a *compound question* directed through theorems to the phenomena being investigated." *TCFK,* 17.

206. Ibid., 24.

207. Newton, *The Mathematical Principles,* 445.

208. Koyré, *From the Closed World,* 225–28. See also Stanley L. Jaki's modest interpretation (in *Science and Creation,* 239), where he simply states that Newton entertained the idea that "infinite empty space is anchored in the Creator's infinity." McGuire (*Tradition and Innovation,* 13) emphasizes that "Newton is not embracing pantheism, since he wishes to assert the contrary truth that God is not identical with space." McGuire's entire discussion of the complex relation of the idea of space and time to God's presence (1–39) is illuminating in underscoring the problems that have led to various interpretations of Newton's true intentions and thoughts. See also White, *Isaac Newton,* 154–55.

209. *TCFK,* 25.

210. GGT, 68; *TCFK,* 24.

211. GGT, 68; *TCFK,* 26, 270.

212. Newton, *Opticks,* 370.

213. A concise description of the fate of Newtonian absolute time in later scientific developments is given in Čapek, "What Survives from Absolute Time," 309–19.

214. Manuel, *The Religion of Isaac Newton,* 77.

215. Craig, *Mind of God,* 31.

216. Manuel (*The Religion of Isaac Newton,* 77–78) asks: "Did Newton mean that God intimately discerns and clearly sees things in infinite space in His sensorium, or 'as it were' in His sensorium? Did he later add the word *tanquam,* the 'as it were' on page 315 of some copies of the 1706 *Optics* with the intention of covering himself?" Manuel does not give any definite answer but he points out that, after all, "Newton as *homo religiosus* could not have cared less about such trivia."

217. See, e.g., Richard H. Popkin's Introduction to *Essays on the Context, Nature, and Influence of Isaac Newton's Theology,* written by him and James E. Force, where this more precise approach, at least from the standpoint of source criticism, and its consequences, have been explained. This suggests that Newton's religious and theological thought bears more weight than Torrance has understood, although, let it be admitted, in many ways he is combating the ideas and approaches (fundamentalism, for instance) that are mentioned by Popkin as propagating Newton's influence. See *RET,* 16–17.

218. GGT, 68. A provocative interpretation of the reasons that led Newton to abandon Christ's true divinity is presented in White, *Isaac Newton,* 153–54.

219. *RET,* 14–15; GGT, 76–78; MC, 1–2.

220. *STI,* 39. The problem here is that Torrance does not give any notion where to find such ideas in Newton's writings. The General Scholium at the end of the *Principia,* which, alongside the first Scholium, is the main source for Torrance's interpretation, does not give a hint of any Christological questions, and therefore it is not easy to distinguish Newton's real ideas from Torrance's interpretation of them.

221. Ibid., 38.

222. Ibid., 60.

223. Newton's knowledge of Early Church theology is revealed by the fact that dozens of copies of patristic works were included in his library. Christianson, *In the Presence of the Creator,* 247.

224. *STI*, 39-40.

225. Byrne, *Glory, Jest and Riddle*, 154–55; White, *Isaac Newton*, 149–54. Details of how strongly Newton favored Arius are, however, under dispute. While James E. Force (*Essays on Isaac Newton's Theology*, 78–80) holds that Newton favored Arian doctrine and, in fact, sees Newton as a representative of Arianism, Frank E. Manuel (*The Religion of Isaac Newton*, 58) says that "while Newton's chief villain in the history of the Church was Athanasius rather than Arius, he censured both for having introduced metaphysical subtleties into their disputes and corrupted the plain language of Scripture." Manuel's claim contains too much mitigation because Newton paid special critical attention to Athanasius's view of the *homoousion*, claiming it to be a result of conscious efforts of Church authorities to ensure their power. Christianson, *In the Presence of the Creator*, 252–53. Newton even resorted to unconvincing arguments ad hominem when he tried to cast a shadow upon Athanasius by accusing him of immoral actions, even a murder, toward his opponents. Gjertsen, *The Newton Handbook*, 422–23.

226. *STI*, 2.

227. James E. Force argues that in order to understand Newton's theology, natural philosophy, or political views, the great underlying key theme in his thinking has to be remembered, and that is the regulative principle of the dominion of the Lord, in which strong emphasis is laid on the monotheistic oneness of God and his absolute Lordship over all his creatures. Force, *Essays on Isaac Newton's Theology*, 95–96. Another view is that for Newton there was a close interaction between God and the world, namely that of divine rule and human obedience. Newton himself writes in one of his manuscripts that the word *God* relates primarily to his dominion to teach us obedience. Manuel, *The Religion of Isaac Newton*, 21.

228. Despite its obvious defects, Torrance's theological approach is able to dig deeper into the problem of Newton's Arianism than, e.g., that of James E. Force (*Essays on Isaac Newton's Theology*, 122–28) who simply concludes his discussion of Newton's Arianism with his rejection of the idea of *homoousion* and his adherence to anti-Trinitarianism.

229. *TF*, 115.

230. *TF*, 48, 85–86. Another good example is in *DM*, 188, where Torrance says that the radical dualism between κόσμος νοητός and κόσμος αἰσθητός lies at the core of Arius's theology. Torrance thus suggests that this Platonic division makes Arius's thought primarily a representative of Greek thinking.

231. *TRc*, 268-69.

232. *TRc*, 268.

233. Varying views concerning the alleged dualism in Augustine's thought have been presented, e.g., by Crawford Knox (*Changing Christian Paradigms*, 66) and Lesslie Newbigin (*The Other Side of 1984*, 23–38).

234. Richard H. Popkin has tried to trace the origins of Newton's theology to Judaism and especially to the mediaeval Jewish philosopher Moses Maimonides. Popkin, *Essays on Isaac Newton's Theology*, 1–6. J. E. McGuire's and M. Tamny's analysis of one of the early writings of Newton, *Quaestiones quaedam philosophicae* (1664–65), containing tentative ideas resulting in his mature view of the structure of the universe, reveals that the most important intellectual factors influencing Newton's thought were Epicureanism, Cartesianism, and Henry More's Platonic philosophy of nature. McGuire & Tamny, *Certain Philosophical Questions*, 113–27.

235. *DCO*, 10.

236. Byrne, *Glory, Jest and Riddle*, 100–101. A good glimpse at the manner in which deism was defined in Newton's own time is provided by Samuel Clarke, a student of Newton, who divided deists into four categories. The first kind of deist consists of those who hold that God is eternal, infinite, and intelligent but not at all interested in the government of the world. In the second category belong those who believe in God's providence but restrict it to natural things, which means exclusion of any moral interest. The third kind of deists are those "who believe in

a deity of total dominion whose all-powerful providential government of the world extends into both the natural and moral realms. . . . Deists of this sort go on to deny any future state for mankind, however." Deists of the fourth group believe in God as Creator, the God of Providence and all Christian dogmas with one significant deviation, namely, that they do not believe in any divine revelation. Force, *Essays on Isaac Newton's Theology*, 44.

237. GGT, 68.

238. *DCO*, 10.

239. "Newton . . . comes to view Jesus Christ as neither co-substantial nor co-eternal with God. This Arian position is definitely unorthodox, even heretical, but it is clearly not deistic. Even in his Arianism Newton takes the Bible seriously." Force, *Essays on Isaac Newton's Theology*, 60. See also Byrne, *Glory, Jest and Riddle*, 154–55.

240. *DCO*, 10.

241. *TF*, 105–9.

242. *TF*, 106.

243. "We know him only by his most wise and excellent contrivances of things, and final causes; we admire him for his perfections; but we reverence and adore him on account of his dominion: for we adore him as his servants; and a god without dominion, providence and final cause, is nothing else but Fate and Nature." Newton, *The Mathematical Principles*, 446.

244. Manuel, *The Religion of Isaac Newton*, 17; White, *Isaac Newton*, 154.

245. Particularly revealing is Torrance's argumentation in *STI*, 38–40.

246. GGT, 69.

247. GGT, 69.

248. Byrne, *Glory, Jest and Riddle*, 156–57.

249. GGT, 70.

250. *TCFK*, 28.

251. See *DCO*, 19.

252. *DCO*, 72.

253. Roland Spjuth (*Creation, Contingency and Divine Presence*, 175) claims that "it seems that Torrance surprisingly continues the tradition from Newton that speaks of God's activity at the borders or edges and limits, of empirical reality where divine act can serve as a principle of explanation, for example as an explanation for the ordered outcome of 'random movements' within quark."

Chapter 5

1. *RST*, 43–44.
2. *TRs*, 56–57.
3. GGT, 37.
4. GGT, 36
5. *RET*, 11–12.
6. Hing Kau Yeung ("Being and Knowing," 133) is correct in holding that Torrance "does not seek to build theology upon modern science," but he misses the point when he states that Torrance uses modern science only for "analogical purposes." There is certainly an apparent analogy, but something more is involved: there is also a connection between theology and the natural sciences through the concept of *homoousion*. The reminding force of modern science, then, is not based on arbitrary parallelism but essentially on the common ground in God's Incarnation.
7. *RST*, ix.
8. GGT, 16–18, 162; *TCFK*, 248.
9. Torrance, "Introduction," in *A Dynamical Theory of the Electromagnetic Field* (*DTE*), ix. For Maxwell's own definition of his theory, see page 34 in the same volume.

10. *RST*, 80, 148. Chalmers, *What Is This Thing Called Science?*, 56; Hempel, *Philosophy of Natural Science*, 76–77; Peat, *Superstrings*, 72; Polkinghorne, *Reason and Reality*, 25. Along very similar lines, Maxwell is also appreciated by Anthony M. Alioto (*A History of Western Science*, 315), who makes the bold claim that Maxwell "was possibly the greatest physicist of the nineteenth century (if not the equal of Newton)" (307). See also Kuhn, *The Structure of Scientific Revolutions*, 6–7.

11. *TCFK*, 232. The concepts of field, space, and time applied in a theological context as expressions of God's dynamic nature have been used by Wolfhart Pannenberg (*Systematic Theology*, 2:79–102), not from a Christological but a pneumatological standpoint.

12. *TCFK*, 235.

13. Maxwell, *A Dynamical Theory of the Electromagnetic Field*, 34. The idea of ether occupied a firm position in contemporary scientific thinking which could not but affect Maxwell's own thoughts. Nevertheless, the fact that his equations do not presuppose any ethereal medium meant a determinate mutation in evaluating its role, a feature that has been commented on as follows: "Maxwell's own theory placed the field concept in a central position. It did not abolish the aether, but it did greatly simplify it." Pais, *"Subtle Is the Lord . . ."*, 119.

14. *TCFK*, 236.

15. Maxwell, *A Dynamical Theory of the Electromagnetic Field*, 34 (italics mine).

16. Torrance admits that Maxwell was not able to distance himself totally from his earlier mechanical thought (*TCFK*, 234). It has to be noted, however, that in the present argument the point is not Maxwell's mechanics but his conception of space.

17. See Čapek, The Philosophical Impact of Contemporary Physics, 68–69, where he quotes Maxwell's "Matter and Motion," ch. 1, sec. 18.

18. *TCFK*, 234.

19. *TRs*, 73.

20. Einstein, *Ideas and Opinions*, 284–85.

21. Douglas Alan Trook has drawn far-reaching conclusions concerning Torrance's use of the concept of field, discernible even in the title of his dissertation: "The Unified Christocentric Field." Trook sees the concept of field as the primary and most important tool with which to understand Torrance's theology. He refers to a physical magnetic field centered around a magnet with its "north" and "south" poles and lines of forces originating from it, and claims that, according to Torrance, in the "unified Christocentric field" the circumstances are similar: Christ is like a magnet, his divinity is "north" and his humanity its "south" pole and from this center his redemptive power extends to all times and all places like currents of force in an ordinary magnetic field. Trook, "The Unified Christocentric Field," 199. Trook gives the impression that Torrance's thinking is based on an imprecise mixture of theological and scientific concepts, but that is certainly not the case. Torrance's intentions are purely theological, in the doctrine of the Incarnation with its concept of *homoousion*, and if an overall pattern for understanding his thinking is sought, its construction should be based on theology and not on the natural sciences or even a compound of them.

22. *DCO*, 91–93.

23. *TCFK*, 224. Here Torrance refers to Maxwell's illustration where the universe can be compared to a book rather than to a magazine, thus indicating the idea that nature is an intelligible whole where its distinct parts throw light on the understanding of each of them. Tolstoy (*James Clerk Maxwell*, 76–77) interprets this same passage quite contrarily by saying that, for Maxwell, "the intelligibility of the universe is questionable."

24. *RST*, 75.

25. Popper, *Quantum Theory and the Schism in Physics*, 164.

26. Northrop, "Causation, Determinism, and the 'Good'," 206–7.

27. *DCO*, 99–100.

28. *DCO*, 100.

29. *RET*, 43.

30. Of internal relations in philosophical discussion, see Sprigge, "Relation—III: Internal Relations," 781-82.

31. For instance, both Milič Čapek (*The Philosophical Impact of Contemporary Physics*, 297–98) and John Polkinghorne (*Reason and Revelation*, 85) see Einstein as a direct heir of the Newtonian scientific paradigm. Polkinghorne states that "Einstein was the last of the ancients" ("The Quantum World," 333). There are fairly weighty points reminding one of the ultimate reason why Einstein preferred an interpretation that came to be viewed as deterministic. He was not ultimately satisfied with the quantum theory but regarded it as essentially incomplete, which made him opt for a theory of greater explanatory power. See Fine, *The Shaky Game*, 5; and Einstein, *Ideas and Opinions*, 269–70. Torrance, in turn, refers to Wolfgang Pauli, another eminent physicist of the twentieth century, mentioning his conviction that to accuse Einstein of determinism is a "fatal mistake." *DCO*, 13; *GGT*, 113.

32. *GGT*, 112–13; *DCO*, 12–14, 100–103. A lively description of the debate between Einstein and Bohr is given in Pais, *"Subtle Is the Lord . . ."*, 410–12, 440–57. See also Jammer, *The Philosophy of Quantum Mechanics*, 108–58. One of the most authoritative defenders of a deterministic interpretation of the quantum theory is David Bohm. See his book *Wholeness and the Implicate Order*, 65–110.

33. *DCO*, 110-11

34. Pais, *"Subtle Is the Lord . . ."*, 411–12.

35. In one of his letters Einstein maintains, "For us causal connections only exist as features of the theoretical constructs." Fine, *The Shaky Game*, 87–88.

36. See Bridgman, "Determinism in Modern Science," 57–75.

37. See, e.g., John Polkinghorne, "The Quantum World," *Reason and Revelation*, and *Science and Christian Belief*; Werner Heisenberg, *Physics and Philosophy*; Robert J. Russell, "Quantum Physics in Philosophical and Theological Perspective"; C. J. Isham, "Creation of the Universe as a Quantum Process"; and K. V. Laurikainen, "Quantum Physics, Philosophy and the Image of God" and *Message of Atoms*.

38. Polkinghorne (*Science and Providence*, 27–28) points out that, for some, the idea of probability in quantum theory has provided means for making more room for God's providence as well as the freedom of humankind. He contends, however, that this approach has not been found very appealing. For the implications of probability in modern science, see Gigerenzer, *The Empire of Chance*, 271–92.

39. "If one accepts probability as ultimate, he has, by the very definition of probability, given up the possibility of explanation." Bridgman, "Determinism in Modern Science," 71.

40. Another interesting line of thought in Torrance's theology is the link he draws between Einstein's understanding of light as the great constant in the universe and the patristic theology of light according to which God is the uncreated Light illuminating reality so that created beings can be perceived. *GGT*, 128–29. A lengthier discussion is available in *CTSC*, 73–104. For Einstein's ambition to find a unification theory without any constants, see Barrow, *Theories of Everything*, 86–92.

41. This proves the correctness of James B. Miller's view that Torrance is committed to a theological position in spite of his affection for Einstein's scientific system. Miller, "Beyond Dualism," 141.

42. *CTSC*, 12.

43. *TRc*, 270.

44. *TCFK*, 37.

45. *GGT*, 25-26; *TCFK*, 36-37.

46. *TCFK*, 38.

47. "Now by means of mere relations, a thing cannot be known in itself; and it may therefore be fairly concluded, that, as through the external sense nothing but mere representations of relations are given to us, the said external sense in its representation can contain only the relation of the object to the subject, but not the essential nature of the object as a thing in itself." Kant, *Critique of Pure Reason*, A49/B66 (65).

48. GGT, 25–26.

49. Kant, *Critique of Pure Reason*, A36–40/B53–57 (58–60).

50. Earman, *World Enough and Space-Time*, 137–53; Sklar, *Philosophy and Spacetime Physics*, 234–48.

51. The complex situation as to Kant's philosophy of space and time is well described in Earman, *World Enough and Space-Time*, 137.

52. Kant, *Critique of Pure Reason*, A38–40/B55–57 (59–60).

53. Ibid., A23/B37 (50).

54. Ibid., A29–31/B45–46 (54–55).

55. Ibid., A47/B64 (64).

56. Copleston, *A History of Philosophy*, 6:I:30–36.

57. Ibid., 6:I:36.

58. *TCFK*, 271–72. Torrance refers to the so-called Copernican revolution in Kant's thought, a phrase underlining the novel insight giving primacy to the human mind. Frederick Copleston (*A History of Philosophy*, 6:I:20–21) reminds us that "Kant's 'Copernican revolution' does not imply the view that reality can be reduced to the human mind and its ideas. He is not suggesting that the human mind creates things, as far as their existence is concerned, by thinking them. What he is suggesting is that we cannot know things, that they cannot be objects of knowledge for us, except in so far as they are subjected to certain a priori conditions of knowledge on the part of the subject."

59. It should be remembered that Torrance is aware of his inability, as well as of its being unnecessary, to reverse the wheel of history, for example, when he admits that Kant or Newton cannot be dismissed. *TCFK*, 45.

60. Ibid., 265.

61. Ibid., 62, 64.

62. Ibid., 265. On the relationship between Einstein's science and Kant's epistemology, see Morrison, "Einstein on Kant," 47–57. See also Reichenbach, *The Theory of Relativity.*

63. *TCFK*, 73.

64. Ibid., 74.

65. "The belief in an external world independent of the perceiving subject is the basis of all natural science. . . . It follows from this that our notions of physical reality can never be final. We must always be ready to change these notions—that is to say, axiomatic basis of physics." Einstein, *Ideas and Opinions*, 266. See also Fleming, "Einstein and the Limits of Reason," 81–83; and Morrison, "Einstein on Kant," 53–54.

66. *TCFK*, 68–71.

67. TheRe, 183.

68. *TCFK*, 84.

69. Ibid., 85; TheRe, 183.

70. *DCO*, 100.

71. *TCFK*, 67.

72. *DCO*, 13. Werner Heisenberg (*Physics and Philosophy*, 57) formulates this understanding succinctly when he states that using measuring apparatus made by humans "emphasizes a subjective element in the description of atomic events, since the measuring device has been constructed by the observer, and we have to remember that what we observe is not nature in itself but nature exposed to our method of questioning."

73. DCO, 101.

74. TheRe, 191.

75. *RST*, 53.

76. Einstein, *Ideas and Opinions*, 293–95. Torrance describes functioning on these levels as the characteristic feature of any science. *TCFK*, 322.

77. GGT, 169–70.

78. *GR*, 24. In Torrance's imagination, entering a new level sometimes means "descending" but most often "ascending." See also GGT, 168–71; *RST*, 147; *STR*, 188-91; *TS*, 259; and Einstein, *Ideas and Opinions*, 293–95. Torrance's application differs from Einstein's original version in that Torrance mentions only these three levels while Einstein held that their number can be infinite.

79. Referring to different "levels" in connection with epistemology suggested by contemporary science is not unique. For example, Milič Čapek, when describing the philosophical implications of modern physics, summarizes them with regard to ontological questions and points out that "the objective substrate of physical phenomena cannot be described in imaginative terms; all sensory qualities are basically on the same phenomenal level, which is a result of interaction of our conscious organism and the transphenomenal physical processes." Čapek, *The Philosophical Impact of Contemporary Physics*, 398. According to Čapek, physical processes have an ontological slant—they are, as he claims, "transphenomenal." The human senses, however, operate only on a phenomenal level although they could not do that without interaction with the transcendent level. It seems evident that whenever ontological questions are brought into close relation with epistemology, a certain stratification of the process of conception is required to explain the bearing of reality upon the human mind.

80. *STR*, 188.

81. GGT, 171.

82. *STR*, 188–89.

83. The relation between different layers in Torrance's thought has been illuminatingly studied from the standpoint of analogical correlation by Roland Spjuth (*Creation, Contingency, and Divine Presence*, 54–57). In the present discussion on the content of the idea of reminder, the concept of analogy is not regarded as fruitful enough, and therefore the idea of openness is preferred. Morrison ("The Self-Given Knowledge of God," 107) pays attention to this stratified structure, but a certain unclarity makes his exposition disjointed. He claims that the truth on one level is dependent on a higher level, but this is hardly the point Torrance wants to make when he holds that on each level the truth value comes, as it were, from "below," from empirical reality. The aim of higher levels is only to help achieve a larger conceptual and theoretical unity and coherence. *RET*, 35-36; *TCFK*, 305, 325. See also Richardson, "Trinitarian Reality," 86.

84. *TS*, 15.

85. GGT, 170.

86. Gunton, *Trinitarian Theology*, 1–14; Schwöbel, "The Renaissance of Trinitarian Theology," 1; Schwöbel, "Christology and Trinitarian Thought," 113–19. See also D. M. MacKinnon's article "The Relation of the Doctrines of the Incarnation and the Trinity" in Torrance's *Festschrift*.

87. See, e.g., Baillie, *God Was in Christ*; D'Costa, *The Meeting of Religions and the Trinity*; Pannenberg, *Jesus–God and Man*; and Weinandy, *The Father's Spirit*. The discussion in Küng, *On Being a Christian*, 119–65, 411–57, is also illuminating.

88. CDG, 17.

89. MC, 1.

90. CDG, 1.

91. See, e.g., *RET*, 36. An illuminating and authoritative introduction to the conception of theology as science and its history is given in Gray, "Theology as Science," 17-61.

92. *TS*, 33.

93. FB, 10. See also *TF*, 19–20.

94. FB, 12. John Polkinghorne (*Reason and Revelation*, 49) makes a similar statement: "An act of faith is necessary for the scientist: a commitment to the metaphysical belief that the world is intelligible and open to our rational exploration." See also Richard Swinburne's discussion of rational belief in Swinburne, *The Existence of God*, 33–103.

95. This proves justified the claim made by Brian J. A. Gray ("Theology as Science," 382) that "Torrance has an extremely intellectualist understanding of faith. The dominant character of faith, as he presents it, is its rationality." Gray continues by criticizing Torrance for neglecting mystical elements in Christian faith and experience (382–87). Torrance's emphasis on the rational feature of faith is not, however, anything alien to Reformed theology in general, for already in Calvin the role of reason is significant. See, e.g., Calvin, *Institutio*, I:VII:5, and III:VII:2 (60 and 506). It is useful to note that Thomas Aquinas did not see faith as a thoroughly rational *habitus mentis* or as a submode of rationality, but, rather, he holds reason to be its servant, as is indicated in Thomas's version of the famous thought: "Cum enim gratia non tollat naturam, sed perficiat, oportet quod naturalis ratio subserviat fidei." Aquinas, *Summa Theologica*, I:1:8 ad 2 (11).

96. *TS*, 33.

97. Ibid., 51.

98. Ibid., 47.

99. The matter is viewed quite differently by Karl Rahner, who holds that only the natural sciences, while researching concrete single phenomena and their relations, are based on a posteriori experience. On the other hand, theology asks characteristically a priori questions concerning reality and its basis. Rahner, "Weltall–Erde–Mensch," 37–42.

100. *TS*, 174.

101. Pannenberg, *Jesus–God and Man*, 28.

102. *RET*, 39.

103. Ludwig Feuerbach, one of the first thinkers to reflect on the psychology of religion, made precisely this transposition, so fiercely objected to by Torrance. "I, on the other hand, do not inquire what the real, natural Christ was or may have been in distinction from what he has been made or has become in Supernaturalism; on the contrary, I accept the Christ of religion, but I show that this superhuman being is nothing else than a product and reflex of the supernatural human mind." Feuerbach, *The Essence of Christianity*, xli.

104. Immanuel Kant held that a concept remains empty of any content and positive reference to its object if it does not belong to experience. But, unlike Torrance, he made the distinction between empirical concepts, having to do with a posteriori thought and sense experience, and pure concepts, which are considered the form of human thinking in the a priori sense. Kant, then, does not dismiss *a priori* thinking but guarantees a role for it in human knowledge. Kant, *Critique of Pure Reason*, A220/B267 (191). Torrance, however, cannot find any place for it–in a truly scientific enterprise there simply is no a priori thinking because everything must depend upon the dictates of objective reality.

105. Bryan J. A. Gray ("Theology as Science," 387) has pointed out that "as it stands, Torrance's efforts to pin theology down to one scientifically apprehendable area have rather unfortunate results for non-Christian religions." A similar observation is also made in Yeung, "Being and Knowing," 244.

106. *TS*, 285.

107. Ibid., 285–86.

108. Ibid., 184.

109. Ibid., 294–312. Here the differing objectivity and subjectivity of theology is described as pertaining to the uniqueness of the Object of theology, namely God, who demands at least as rigorous attentiveness from subjects as from objects of the created world, objects of the natural sciences.

110. *TS*, 33.

111. *TF*, 3.

112. *CDG*, x. Another example of a rational approach in explaining the revelation is Swinburne's book *Revelation*.

113. Nancey Murphy ("What Has Theology to Learn from Scientific Method?," 101–26) discusses the problems inherent in the view which conceives religious experience as capable of providing appropriate knowledge of God. She discerns two major difficulties, namely the problems of circularity and subjectivity, to which she offers her own solution. Although there is considerable agreement between Murphy and Torrance, the latter could hardly accept the view that theology was less objective or less reliable than, for example, the natural sciences. Richard Swinburne, too, has made noteworthy attempts to reason for the reliability of religious experiences. He claims that how things "seem to be is good grounds for a belief about how things are" (*The Existence of God*, 254). On this basis he continues that religious experience as such tells about the existence of God so that "the onus of proof is on the atheist" (265).

114. George Hunsinger has detected a motif of particularism in Barth's thought, whereby God's truth is always particular in the sense that it does not require any human efforts to verify it. Hunsinger, *Karl Barth*, 70. See also Wolfhart Pannenberg, *Theology and the Philosophy of Science*, 265–76. Eberhard Jüngel (*Karl Barth*, 11) reminds us, with a tone of appreciation, that claims about truth in Barth's writings "are always one-sided, but never one-dimensional."

115. See, e.g., *TS*, 55. Karl Barth has stated: "Theology follows the talk of the Church to the extent that in its question as to the correctness of its utterance it does not measure it by an alien standard but by its own source and object." Barth, *Church Dogmatics*, 1:1:4. The Swede Hjalmar Lindroth (*Kyrklig Dogmatik*, 1:10–11) criticizes Barth's view of the scientific nature of theology as depending upon its object and prefers to replace it with the idea of *objective argumentation*. He also points out that while the concept *kirchlich* in the name of Barth's main work refers to theology or preferably dogmatics as a function of the Church "in the room of the Church," his own understanding refers to the word *kyrklig* in the sense that dogmatics stands "in the time of the Church."

116. E.g., *DCO*, 52, 54; *TCFK*, 136–45. Of the Gödel theorems, see Chaitin, "Number and Randomness," 16–20; Polkinghorne, *Science and Theology*, 59–60, 69–70; and Schins, "Mathematics," 50–54.

117. *RST*, 123–24.

118. *TS*, 135.

119. *RST*, 103.

120. Ibid.

121. Ibid., 105.

122. On the Christian community as the starting point of doctrinal reflection, see Schwöbel, "Christology and Trinitarian Thought," 139, 145–46.

123. *RST*, 122. However rational Torrance's approach can be claimed to be, the mystical element is not totally abandoned. See, e.g., *RST*, 123.

124. *TS*, 41. See also *TS*, 45, 97–98, 129.

125. Ibid., 98.

126. *RST*, 107.

127. *GR*, 166–67. See also Thiselton, *The Two Horizons*, 91–92.

128. *GGT*, 154.

129. Hunsinger, *Karl Barth*, 7.

130. Jenson, "Karl Barth," 30–35; Küng, "Karl Barth," 24.

131. Torrance finds a slant of subordinationism in Barth's conception of the Holy Trinity which he regards "as a hang-over from Latin theology but also from St Basil's doctrine of the Trinity." Torrance, *Karl Barth, Biblical and Evangelical Theologian* (*KB*), 131.

132. *TCFK*, vii. See also Torrance, *Karl Barth: An Introduction to His Early Theology* (*KBI*), 105–97.

133. *TCFK*, 280.

134. Efforts to draw any kind of direct links between Athanasius and Barth are criticized by Charles Kannengiesser (*Arius and Athanasius*, 7:112), who holds that "I would not like to anticipate in Athanasius a Karl Barth or a Pannenberg, but Athanasian Christocentrism remains an astonishing innovation in the context of the ancient theological tradition of Alexandria." See also Meijering, *Orthodoxy and Platonism in Athanasius*, 150–60.

135. *KB*, 160–62. This comparison is strongly criticized by Richard A. Muller ("The Barth Legacy," 703–4.) See the reply of John D. Morrison ("The Self-Given Knowledge of God," 345–46) to Muller on behalf of Torrance. Muller's at times harsh critique seems to be based, to a large extent, on his efforts to view theology as a readily conceivable system of doctrinal expressions.

136. GGT, 91–93.

137. The rise of non-Euclidean geometries started as early as the eighteenth century as a result of the mathematization of geometry beyond the scope of Euclid's axioms. Eventually it attempted to construct a new understanding of the universe in its manifold dimensionality, which Euclidean geometry, operating with pure lines and circles only on a plane and not in three-dimensional space and its curvature, was not able to provide. Among the most influential non-Euclidean geometries are the Lobatchevskian and Riemannian geometries. Sklar, *Space, Time, and Spacetime*, 16–25.

138. A concise and critical exposition of the relationship between Barth's and Torrance's ways of conceiving the doctrine of Trinity as the basis of theological inquiry is given by Kurt A. Richardson ("Trinitarian Reality," 113–26). According to him, "Torrance believes that the homoousion with the Father and the Holy Spirit in [*sic*] a perfect unity of being and action in Barth's CD as its unifying dogmatic principle" (117).

139. GGT, 148.

140. This view is also shared by other theologians who are not necessarily and explicitly Barthians, a fact which means that what Torrance understands by strictly scientific theology in the doctrine of Trinity can be wholly appreciated within other branches of Christianity, too. Cf. a definition by the Lutheran Robert W. Jenson ("What Is the Point of Trinitarian Theology?," 31): "'Theology' is not initially a second-level discourse. . . . initially it is rather the first-level act of calling God by the triune name, and of making prayers and sacrifices that follow the triune logic and use the triune rhetoric. And in this mode, trinitarian theology does not have a point, it is the point."

141. In Torrance scholarship, it has not been uncommon to apply to his theology both of the epithets *Christological* and *Trinitarian*. See, e.g., Kruger, "The Doctrine of the Knowledge of God," 387–88; Morrison, "The Self-Given Knowledge of God"; and Palma, "Thomas F. Torrance's Reformed Theology,"13–15. Palma first mentions the Trinitarian aspect, but in this work it is argued that Torrance himself considers the Christological side primary.

142. Gray claims (in "Theology as Science," 392) that "Torrance's theological method reflects at times the tension evident in Barth's writings between a desire for a Trinitarian approach to knowledge of God and its actual working out in purely Christological terms." Gray's account, however, is not fully adequate when Torrance's later writings are analyzed.

143. CDG, 21.

144. In an earlier text Torrance is quite determined to suggest a more subordinate status for pneumatology when he holds that "the doctrine of Spirit is really Christology . . . applied to the Church as the Body of Christ." *Royal Priesthood* (*RP*), 25. The words reflect the mood in the first edition of the book, published in 1955.

145. CDG, 18.

146. Ibid., 30.

147. *GR*, 173–74; *TS*, 524; Kruger, "The Doctrine of the Knowledge of God," 383–85. This is not, of course, to neglect the soteriological role of the Holy Spirit as seen by Torrance, since the divine Person is conceived as granting us not only the communion of knowledge but also that of salvific participation in God's own Trinitarian life through θέωσις. *CDG*, 96–97.

148. *CDG*, 100–101.

149. Ibid., 62–63.

150. *GR*, 167. See Thiselton, *The Two Horizons*, 91.

151. *CDG*, 96–97, 100–101, 107, 125–27, 147, 158, 186–88, 197; *TF*, 199, 202, 215, 216, 236; *TP*, 10, 14, 25, 40, 121.

152. *CDG*, 88–111.

153. *GGT*, 166–68.

154. McGrath, *Christian Theology*, 255.

155. *CDG*, 89.

156. Ibid., 156–57.

157. This is evident, e.g., in Torrance's claim that the relations in the Godhead make it deeply personal, suggesting that in a purely human dimension, too, to be a person means to be in relation to another person. *CDG*, 102; *RST*, 110, 171–80. The relational thinking of the Church Fathers, who had to reflect thoroughly on the Incarnation, was the primary reason for the birth of the modern concept of person. *CTSC*, 50. Torrance refers to Christ, saying that He alone is the Person who gives us our personality. *TS*, 147. On the relation between concept of person and the doctrine of the Trinity, see, e.g., Gunton, "Relation and Relativity," 98–100; Gunton, *The Promise of Trinitarian Theology*, 83–99; Hallamaa, "The Concept of Person," 140–59; and Zizioulas, *Being and Communion*, 27–65. For Torrance's further application of the concept of person with special reference to ethics, see his booklet *The Soul and the Person of the Unborn Child* (SPUC), 14–19.

158. Torrance does not see any problem in his insistence upon the primacy of a relational understanding of the Trinity. It has, however, been pointed out that, in medieval western theology, the doctrine of the Persons of the Trinity was, to a large extent, forgotten. Thereby the Persons lost their distinctive reality and became reduced to abstract relations. Knuuttila and Saarinen, "Innertrinitarische Theologie," 243. In the present ecumenical discussion, varying opinions have been presented with regard to the question of whether it was the substantival or the relational interpretation of the Trinity that should be considered original. See Martikainen, "Kommunion der Liebe," 287–314.

159. *RET*, 42.

160. MC, 1–2.

161. *DM*, 10.

162. *RET*, 16.

163. Ibid., 14–17.

164. Torrance, "Hermeneutics according to F. D. E. Schleiermacher" (HAS), 266.

165. See *TS*, 34–43, in which one of the most representative accounts of Torrance's conception of objectivity is given.

166. *TS*, 49. A rational exposition of the Christian Creed is presented by Richard Swinburne (*The Existence of God*, 143–72).

167. One of Torrance's goals is to use theology in a persuasive manner, an approach that certainly makes him, even implicitly, an apologist, contrary to his explicit indifference toward apologetics in general. "We communicate with other minds only when we can get them to submit their thought to the same rationality in things that we have experienced. Thus communication from the very start involves a persuasive element." *GR*, 196. See also *GR*, 206.

168. Gray, "Theology as Science," 377–82.

169. David C. Lindberg (*The Beginnings of Western Science*, 1–2) presents several different characteristics that have commonly been accredited to science, although no general consensus among scientists or philosophers of science has been achieved. These features include, for instance, a certain pattern of behavior by which the environment is controlled (technology), a certain body of theoretical knowledge, the search for universal quasilegal rules, or an effort to apply rigor, precision, and objectivity to thinking on the basis of scientific observation.

170. *GR*, 112. See also CAC, 2:27–30.

171. The World Council of Churches has made a serious attempt to relate science and theology or Christian faith to each other. In 1979 a conference was held in Massachusetts resulting in a two-volume report entitled *Faith and Science in an Unjust World*. As indicated in the title, primary attention was given to the ethical aspects of this interrelation.

172. *GR*, 121.

173. Ibid., 123–31.

174. See also *TCFK*, 326–28, where Torrance explains the Catholic doctrine of transsubstantiation as having originated from an alien cultural framework.

175. *TRc*, 29.

176. *GR*, 125.

177. Ibid.

178. Ibid., 126. See also Torrance, "The Arnoldshain Theses on Holy Communion" (AT), 4–21.

179. *GR*, 126; *STI*, 31.

180. See Calvin, *Institutio*, IV:27:16–20 (1015–19).

181. Heick, *A History of Christian Thought*, 1:435–36. It should be remembered that Heick's interpretation is directed by his conviction that Calvin was nearer to Zwingli than to Luther, contrary to the way German Union theologians would like to interpret the dilemma.

182. Metzke, *Coincidentia oppositorum*, 191–92. Compare this to *TRc*, 124–29. Torrance's most extensive discussion of Luther's theology, analyzing the reformer's eschatology, is in *Kingdom and Church* (KC), 7–72.

183. Luther, "Vom Abendmahl Christi," 327–30; Hilgenfeld, *Mittelalterlich-traditionelle Elemente*, 183–209; Schwab, *Entwicklung und Gestalt*, 262–66. For Luther's conception of doctrinal formulae, see Työrinoja, "Nova vocabula," 221–36.

184. *TRc*, 7.

185. Ibid., 11–12.

186. Divisions are something sinful against which the churches have to fight, compelled by the authority of Christ himself. "[O]ur disagreements must come under the judgements of Christ, for the very oneness of Christ condemns division; but if the doctrine of Christ includes His Cross, as indeed it does, then in the very heart of Christology we are provided not only with judgement upon our divisions but with the mighty act of God which assumed upon Himself the sin that separates and divides and contradicts in order to overcome it and set it aside in reconciliation and recreation." CAC, 1:13.

187. Torrance, "Memoranda" in *Theological Dialogue* (MEM), 3. Torrance has reflected on the role of the absolute notion of space, applying its criticism to the doctrinal consultations between the Reformed and Orthodox Churches. See MEM, 12–13.

188. Ibid., 7.

189. See Popper, *Realism and the Aim of Science*, 57.

190. *RET*, 18. Torrance sees this feature as a direct implication of God's unconditional grace.

191. Something similar has been proposed, for example, by George A. Lindbeck, who has made efforts to construct what he calls a "cultural-linguistic" approach to religious doctrine, or a "regulative" or "rule" theory. His ambitious aim is to contribute to "reconciliation without

capitulation" by removing "propositional" and "experiental expressive" modes of religious doctrines. Lindbeck, *The Nature of Doctrine*, 15–25. Torrance's discussion differs, however, in many respects from that of Lindbeck's, not least in the fact that in crucial doctrinal formulations, for example, in the Nicaenum, he discerns positive truth claims, for instance, when he states: "Doctrines regulate truth claims by excluding some and permitting others, but the logic of their communally authoritative use hinders or prevents them from specifying positively what is to be affirmed." Lindbeck, *The Nature of Doctrine*, 19. The relation between Torrance's and Lindbeck's approaches has been studied by Colyer in his work "The Nature of Doctrine in the Theology of T. F. Torrance." John Polkinghorne (*Science and Christian Belief*, 49) has maintained that postmodernists "tend to prescind from questions of truth, as we can see from the coyness exhibited by Lindbeck about the truth claims of theology." Similar criticism has also been expressed in Yu, "The Principle of Relativity," 180–82.

192. For a detailed analysis of the method applied in the Leuenberg process, see Mannermaa, *Von Preussen nach Leuenberg*, especially pages 155–75. In the American context, too, the problem in the mode of rapprochement between the two churches is similar. See Bridston and Nafzger, "Lutheran-Reformed Dialogue," 33–45.

193. *TRc*, 36.

194. MEM, 7.

195. *CDG*, 185–86; *TF*, 241–47. See also Chapman, "Book Review: The Trinitarian Faith," 501. It was Augustine in particular who "more equivocally than any other of the Western fathers before him, taught the doctrine of the double procession of the Spirit from the Father and the Son (*filioque*)." Kelly, *Early Christian Doctrines*, 275. Kelly also points out that the seeds of the problem were implanted by the Greek Father Gregory of Nyssa (262).

196. *TF*, 241.

197. Ibid., 242.

198. *CDG*, 186.

199. *TP*, 126.

200. Ibid., 20.

201. "Agreed Statement" in Torrance, *Theological Dialogue*, 2:219–26. Also to be found in *TP*, 115–22. Compare these to Torrance's earlier stance, to be found in "Comment" in the *New Valamo Consultation* (COM), 41–42.

202. "Agreed Statement" in Torrance, *Theological Dialogue*, 2:223; *TP*, 119.

203. Torrance, *Theological Dialogue*, 2:231.

Bibliography

Works by Torrance

AT "The Arnoldshain Theses on Holy Communion." *Scottish Journal of Theology* 15 (1962): 4–21.

CAC *Conflict and Agreement in the Church.* 2 vols. London: Lutterworth Press, 1959–60.

CDG *The Christian Doctrine of God, One Being Three Persons.* Edinburgh: T & T Clark, 1996.

CDM *Calvin's Doctrine of Man.* London: Lutterworth Press, 1949.

CFM *The Christian Frame of Mind: Reason, Order, and Openness in Theology and Natural Science.* Introduction by W. Jim Neidhardt. Colorado Springs, Colo.: Helmers & Howard, 1989.

CLU "The Christ Who Loves Us." In *A Passion for Christ: The Vision That Ignites Ministry*, ed. Gerrit Dawson and Jock Stein, 8–20. Edinburgh: Handsel Press, 1999.

COM "Comment." In *The New Valamo Consultation: The Ecumenical Nature of the Orthodox Witness*, 41–42. Geneva: World Council of Churches, 1977.

CT "Introduction." In *The Calvin Treasury: Selections from the Institutes of the Christian Religion*, ed. William F. Keesecker, v–xiii. London: SCM Press, 1963.

CTSC *Christian Theology and Scientific Culture.* Belfast: Christian Journals Limited, 1980.

DCO *Divine and Contingent Order.* Oxford: Oxford University Press, 1981.

DM *Divine Meaning: Studies in Patristic Hermeneutics.* Edinburgh: T & T Clark, 1995.

DTE "Introduction." In *A Dynamical Theory of the Electromagnetic Field*, ed. Thomas F. Torrance, 1–27. Edinburgh: Scottish Academic Press, 1982.

EG "Einstein and God." *Reflections* 1 (1998): 2–25.

FB "The Framework of Belief." In *Belief in Science and in Christian Life: The Relevance of Michael Polanyi's Thought for Christian Faith and Life*, ed. Thomas F. Torrance, 1–27. Edinburgh: Handsel Press, 1980.

GGT *The Ground and Grammar of Theology.* Charlottesville: University Press of Virginia, 1980.

GR *God and Rationality.* London: Oxford University Press, 1971.

HAS "Hermeneutics according to F. D. E. Schleiermacher." *Scottish Journal of Theology* 21 (1968): 257–67.

HJC *The Hermeneutics of John Calvin.* Monograph Supplements to the Scottish Journal of Theology. Edinburgh: Scottish Academic Press, 1988.

INC "Introduction." In *The Incarnation: Ecumenical Studies in the Nicene-Constantinopolitan Creed A.D. 381*, ed. Thomas F. Torrance, xi–xxii. Edinburgh: Handsel Press, 1981.

KB *Karl Barth, Biblical and Evangelical Theologian.* Edinburgh: T & T Clark, 1990.

KBI *Karl Barth: An Introduction to His Early Theology, 1910–1931*. London: SCM Press, 1962.

KC *Kingdom and Church: A Study in the Theology of the Reformation*. Edinburgh: Oliver and Boyd, 1956.

MC *The Mediation of Christ*. Revised ed. Edinburgh: T & T Clark, 1992.

MEM "Memoranda on Orthodox/Reformed Relations." In *Theological Dialogue between Orthodox and Reformed Churches*, vol. 1, ed. Thomas F. Torrance, 3–18. Edinburgh: Scottish Academic Press, 1985.

NTC "Notes on Terms and Concepts." In *Belief in Science and in Christian Life: The Relevance of Michael Polanyi's Thought for Christian Faith and Life*, ed. Thomas F. Torrance, 133–47. Edinburgh: Handsel Press, 1980.

PCT *Preaching Christ Today: The Gospel and Scientific Thinking*. Grand Rapids, Mich.: Eerdmans, 1994.

PJC "Preaching Jesus Christ." In *A Passion for Christ: The Vision That Ignites Ministry*, ed. Gerrit Dawson and Jock Stein, 23–32. Edinburgh: Handsel Press, 1999.

RCF "Reaffirming the Cornerstone of Faith." *The Scotsman* (3 Jan. 1994): 10.

RET *Reality and Evangelical Theology*. Philadelphia: Westminster Press, 1982.

RP *Royal Priesthood: A Theology of Ordained Ministry*. 2d ed. Edinburgh: T & T Clark, 1993.

RST *Reality and Scientific Theology*. Theology and Science at the Frontiers of Knowledge, no. 1. Edinburgh: Scottish Academic Press, 1985.

SAG "Science and Access to God." *Jian Dao* 10 (1998): 43–58.

SC "Scientists Commemorating the Third Centenary of Sir Isaac Newton's *Philosophiae Naturalis Principia Mathematica*." In *John Paul II on Science and Religion: Reflections on the New View from Rome*, ed. Robert J. Russell, William R. Stoeger, S.J., and George V. Coyne, 105–12. Vatican: Vatican Observatory, 1990.

SF "Introduction." In *The School of Faith: The Catechisms of the Reformed Church*, ed. Thomas F. Torrance, xi–cxxvi. London: James Clarke, 1959.

SJC "Service in Jesus Christ." In *Service in Christ: Essays Presented to Karl Barth on His 80th Birthday*, ed. James I. McCord and T. H. Parker, 1–16. London: Epworth Press, 1966.

SPUC *The Soul and the Person of the Unborn Child*. Edinburgh: Handsel Press, 1999.

ST *Scottish Theology: From John Knox to John McLeod Campbell*. Edinburgh: T & T Clark, 1996.

STI *Space, Time and Incarnation*. London: Oxford University Press, 1969.

STR *Space, Time and Resurrection*. Edinburgh: Handsel Press, 1976.

TCFK *Transformation and Convergence in the Frame of Knowledge: Explorations in the Interrelations of Scientific and Theological Enterprise*. Belfast: Christian Journals Limited, 1984.

TF *The Trinitarian Faith: The Evangelical Theology of the Ancient Catholic Church*. Edinburgh: T & T Clark, 1988.

TheRe "Theological Realism." In *The Philosophical Frontiers of Christian Theology: Essays Presented to D. M. McKinnon*, eds. Brian Hebblethwaite and Stewart Sutherland, 169–96. Cambridge: Cambridge University Press, 1982.

TP *Trinitarian Perspectives: Toward Doctrinal Agreement*. Edinburgh: T & T Clark, 1994.

TRc *Theology in Reconciliation: Essays towards Evangelical and Catholic Unity in East and West*. London: Geoffrey Chapman, 1975.

TRs *Theology in Reconstruction*. London: SCM Press, 1965.

TS *Theological Science*. Oxford: Oxford University Press, 1969.

TSB "Transfinite Significance of Beauty in Science and Theology." In *L'Art, La Science et la Métaphysique: Études offertes à André Mercier à l'occasion de son quatre-vingtième anniversaire,* ed. Luz García Alonso, Evanghelos Moutsopoulos, and Gerhard Seel, 393-418. Bern: Peter Lang, 1993.

TSHR "Time in Scientific and Historical Research." In *Gottes Zukunft–Zukunft der Welt: Festschrift für Jürgen Moltmann*, ed. Hermann Deuser, 292-97. München: Kaiser, 1986.

TVB "The Truth of the Virgin Birth." *The Herald* (14 Jan. 1994): 11.

Works by Others

Achtner, Wolfgang. *Physik, Mystik und Christentum: Eine Darstellung und Diskussion der natürlichen Theologie bei T. F. Torrance*. European University Studies, Series 23 Theology, vol. 438. Frankfurt: Peter Lang, 1991.

Adams, Marilyn McCord. "Universals in the Early Fourteenth Century." In *The Cambridge History of Later Medieval Philosophy: From the Rediscovery of Aristotle to the Disintegration of Scholasticism, 1100–1600*, ed. Norman Kretzmann, Anthony Kenny, and Jan Pinborg, 411-39. Cambridge: Cambridge University Press, 1982.

Aertsen, Jan. *Nature and Creature: Thomas Aquinas's Way of Thought*. Studien und Texte zur Geistesgeschichte des Mittelalters, vol. 21. Leiden: E. J. Brill, 1988.

Alioto, Anthony M. *A History of Western Science*. Englewood Cliffs, N.J.: Prentice-Hall, 1987.

Allen, E. L. *The Sovereignty of God and the Word of God: A Guide to the Thought of Karl Barth*. London: Hodder & Stoughton, 1967.

Aquinas, Thomas. *Summa Theologica. Prima pars*. 1951. Reprint. Madrid: Biblioteca de Autores Cristianos, 1978.

Ariew, Roger. "Preface." In *Medieval Cosmology: Theories of Infinity, Place, Time, Void, and the Plurality of Worlds*, by Pierre Duhem, ed. Roger Ariew, xix-xxxi. Chicago: University of Chicago Press, 1987.

Ashworth, William B., Jr. "Catholicism and Early Modern Science." In *God and Nature: Historical Essays on the Encounter between Christianity and Science*, ed. David C. Lindberg and Ronald L. Numbers, 136-66. Berkeley: University of California Press, 1986.

Athanasius. "Contra Arianos." In *Patrologiae cursus completus. Series Graeca*, vol. 26, ed. J.-P. Migne, 11-527. Paris: Garnier, 1887.

——. *De Gentes* and *De Incarnatione*, ed. Robert W. Thomson. Oxford: Clarendon Press, 1971.

——. "De Decretis." In *Patrologiae cursus completus. Series Graeca*, vol. 25, ed. J.-P. Migne, 415-75. Paris: Garnier, 1884.

Audi, Robert. *Belief, Justification, and Knowledge: An Introduction to Epistemology*. Wadsworth Basic Issues in Philosophy Series. Belmont, Calif.: Wadsworth Publishing Company, 1988.

Augustine. *The Confessions*. New York: New American Library, 1963.

Baillie, Donald M. *God Was in Christ: An Essay on Incarnation and Atonement*. London: Faber and Faber, 1948.

Balthasar, Hans Urs von. *Karl Barth: Darstellung und Deutung seiner Theologie*. Köln: Verlag Jacob Hegner, 1962.

Barbour, Ian G. *Myths, Models and Paradigms: The Nature of Scientific and Religious Language*. London: SCM Press, 1974.

——. *Religion in an Age of Science*. Gifford Lectures, 1989-91, vol. 1. San Francisco: Harper, 1990.

——. "Ways of Relating Science and Theology." In *Physics, Philosophy, and Theology: A Common Quest for Understanding*, ed. Robert J. Russell, William R. Stoeger, S.J., and George V. Coyne, S.J., 21-48. Vatican: Vatican Observatory, 1988.

Barnes, T. D. "The Career of Athanasius." In *Studia Patristica*, vol. 21, ed. Elizabeth A. Livingstone, 390-401. Leuven: Peeters Press, 1989.

Barrow, John D. *Theories of Everything: The Quest for Ultimate Explanation*. Oxford: Clarendon Press, 1990.

Barth, Karl. *Church Dogmatics*, vol. 1, part 1, ed. G. W. Bromiley and T. F. Torrance. 2d ed. 1975. Reprint. Edinburgh: T & T Clark, 1995.

——. *Church Dogmatics*, vol. 2, part 2, ed. G. W. Bromiley and T. F. Torrance. 1957. Reprint. Edinburgh: T & T Clark, 1994.

——. *Church Dogmatics*, vol. 3, part 2, ed. G. W. Bromiley and T. F. Torrance. 1960. Reprint. Edinburgh: T & T Clark, 1994.

——. *Church Dogmatics*, vol. 4, part 1, ed. G. W. Bromiley and T. F. Torrance. Edinburgh: T & T Clark, 1956.

——. *Evangelical Theology: An Introduction*. 1963. Reprint. London: Collins, 1965.

——. *The Humanity of God*. London: Collins, 1961.

Berkouwer, G. C. *Divine Election*. Studies in Dogmatics Series. Grand Rapids, Mich.: Eerdmans, 1960.

——. *The Triumph of Grace in the Theology of Karl Barth*. London: Paternoster Press, 1956.

Bianchi, Ugo. *Selected Essays on Gnosticism, Dualism and Mysteriosophy*. Studies in the History of Religions Series, no. 38. Leiden: E. J. Brill, 1978.

Bienert, Wolfgang A. "Zur Logos-Christologie des Athanasius von Alexandrien in *Contra Gentes* und *De Incarnatione*." In *Studia Patristica*, vol. 21, ed. Elizabeth A. Livingstone, 402–19. Leuven: Peeters Press, 1989.

Bohm, David. *Wholeness and the Implicate Order*. 1980. Reprint. London: ARK Paperbacks, 1988.

Bowker, John. "Did God Create This Universe?" In *The Sciences and Theology in the Twentieth Century*, ed. A. R. Peacocke, 98–126. Stocksfield, Northumberland: Oriel Press, 1981.

Bridgman, Percy W. "Determinism in Modern Science." In *Determinism and Freedom in the Age of Modern Science*, ed. Sidney Hook, 57–75. 1961. Reprint. New York: Collier Books, 1979.

Bridston, Keith, and Samuel Nafzger. "Lutheran-Reformed Dialogue." In *Lutherans in Ecumenical Dialogue: A Reappraisal*, ed. Joseph A. Burgess, 33–45. Minneapolis: Augsburg Fortress, 1990.

Broadie, Alexander. *The Shadow of Scotus: Philosophy and Faith in Pre-Reformation Scotland*. Edinburgh: T & T Clark, 1995.

——. *The Tradition of Scottish Philosophy: A New Perspective on the Enlightenment*. Edinburgh: Polygon, 1990.

Brooke, John Hedley. *Science and Religion: Some Historical Perspectives*. Cambridge: Cambridge University Press, 1991.

——. "Science and Theology in the Enlightenment." In *Religion and Science: History, Method, Dialogue*, ed. W. Mark Richardson and Wesley J. Wildman, 7–27. New York: Routledge, 1996.

Brümmer, Vincent. "Does God Really Exist?" In *Philosophical Studies in Religion, Metaphysics, and Ethics: Essays in Honour of Heikki Kirjavainen*, ed. Timo Koistinen and Tommi Lehtonen, 17–34. Schriften der Luther-Agricola-Gesellschaft 38. Helsinki: Luther-Agricola-Society, 1997.

Buckley, Michael. *At the Origins of Modern Atheism*. New Haven, Conn.: Yale University Press, 1987.

Bultmann, Rudolf. *Existence and Faith: Shorter Writings of Rudolf Bultmann*, ed. Schubert M. Ogden. London: Collins, 1964.

Bunge, Mario. *Intuition and Science*. 1962. Reprint. Westport, Conn.: Greenwood Press, 1975.

Burtt, Edwin Arthur. *The Metaphysical Foundations of Modern Physical Science*. 2d ed., 1932. Reprint. London: Routledge and Kegan Paul, 1980.

Byrne, James. *Glory, Jest and Riddle: Religious Thought in the Enlightenment*. London: SCM Press, 1996.

Cahalan, John C. *Causal Realism: An Essay on Philosophical Method and the Foundations of Knowledge*. Sources in Semiotics Series, vol. 2. Lanham, Md.: University Press of America, 1985.

Calvin, Jean. *Institutio Christianae Religionis*. Ioannis Calvini opera omnia, vol. 2, ed. Guilielmus Baum, Eduardus Cunitz, and Eduardus Reuss. Brunsvig: C. A. Schwetschke et Filium, 1864.

Čapek, Milič. *The Philosophical Impact of Contemporary Physics*. Princeton, N.J.: Van Nostrand, 1961.

——. "What Survives from the Classical Concept of Absolute Time." In *Newton's Scientific and Philosophical Legacy*, ed. P. B. Scheurer and G. Debrock, 309–19. International Archives of the History of Ideas 123. Dodrecht: Kluwer Academic Publishers, 1988.

Case-Winters, Anna. "The Question of God in an Age of Science: Constructions of Reality and Ultimate Reality in Theology and Science." *Zygon* 32, no. 3 (Sept. 1997): 351–75.

Chadwick, Henry. *The Early Church*. Penguin History of the Church, vol. 1. 1972. Reprint, London: Penguin Books, 1993.

Chadwick, Owen. *The Reformation*. Penguin History of the Church, vol. 3. London: Penguin Books, 1990.

Chaitin, Gregory J. "Number and Randomness: Algorithmic Information Theory." In *Mathematical Undecidability, Quantum Nonlocality and the Question of the Existence of God*, ed. Alfred Driessen and Antoine Suarez, 15–25. Dordrecht: Kluwer Academic Publishers, 1997.

Chalmers, A. F. *What Is This Thing Called Science? An Assessment of the Nature and Status of Science and Its Methods*. 1982. Reprint of 2d ed. Milton Keynes: Open University Press, 1988.

Chapman, Mark E. "Book Review: The Trinitarian Faith." *Pro Ecclesia: A Journal of Catholic and Evangelical Theology* 2, no. 4 (Fall 1993): 499–502.

Christie-Murray, David. *A History of Heresy*. Oxford: Oxford University Press, 1989.

Christianson, Gale E. *In the Presence of the Creator: Isaac Newton and His Times*. New York: Free Press, 1984.

Clarke, W. Norris, S.J. "Is a Natural Theology Still Possible Today?" In *Physics, Philosophy, and Theology: A Common Quest for Understanding*, ed. Robert J. Russell, William R. Stoeger, S.J., and George V. Coyne, S.J., 103–23. Vatican: Vatican Observatory, 1988.

Clayton, Philip. *Explanation from Physics to Theology: An Essay in Rationality and Religion*. New Haven, Conn.: Yale University Press, 1989.

Clifford, Anne M. "Darwin's Revolution in *The Origin of Species*: A Hermeneutical Study of the Movement from Natural Theology to Natural Selection." In *Evolutionary and Molecular Biology: Scientific Perspectives on Divine Action*, ed. Robert John Russell, William R. Stoeger, S.J., and Francisco J. Ayala, 281–302. Vatican: Vatican Observatory Publications, 1998.

Cohen, H. Floris. *The Scientific Revolution: A Historiographical Inquiry*. Chicago: University Press of Chicago, 1994.

Colyer, Elmer Monroe. "The Nature of Doctrine in the Theology of T. F. Torrance." Ph.D. diss., Boston College, 1992.

Confession. Confession of Faith, the Subordinate Standards and Other Authoritative Documents of the Free Church of Scotland. Edinburgh: Free Church of Scotland, 1973.

Copleston, Frederick. *A History of Philosophy*. 9 vols. Garden City, N.Y.: Image Books, 1962–77.

Craig, Edward. *The Mind of God and the Works of Man*. Oxford: Clarendon Press, 1987.

Craig, William Lane, and Quentin Smith. *Theism, Atheism, and Big Bang Cosmology*. Oxford: Clarendon Press, 1995.

Dahlfert, Ingolf. *Theology and Philosophy*. Oxford: Basil Blackwell, 1988.

Dancy, Jonathan. *An Introduction to Contemporary Epistemology*. Oxford: Basil Blackwell, 1985.

Darwin, Charles. *The Origins of Species*. Ware, Hertfordshire: Wordsworth Editions Limited, 1998.

Davies, Paul. *About Time: Einstein's Unfinished Revolution*. London: Penguin Books, 1995.

——. *God and the New Physics*. 1983. Reprint. London: Penguin Books, 1990.

——. *The Mind of God: Science and the Search for Ultimate Meaning.* London: Penguin Books, 1993.

Dawkins, Richard. *The Blind Watchmaker.* Harlow: Longman, 1986.

Dawson, Gerrit, and Jock Stein, eds. *A Passion for Christ: The Vision That Ignites Ministry.* Edinburgh: Handsel Press, 1999.

D'Costa, Gavin. *The Meeting of Religions and the Trinity.* Edinburgh: T & T Clark, 2000.

Deason, G. B. "The Protestant Reformation and the Rise of Modern Science." *Scottish Journal of Theology* 38 (1985): 221–40.

——. "Reformation Theology and the Mechanistic Conception of Nature." In *God and Nature: Historical Essays on the Encounter between Christianity and Science*, ed. David C. Lindberg and Ronald L. Numbers, 167–91. Berkeley: University of California Press, 1986.

Descartes, René. *Essential Descartes*, ed. Margaret D. Wilson. New York: New American Library, 1983.

Dobbs, Betty Jo Teeter, and Margaret C. Jacob. *Newton and the Culture of Newtonianism.* The Control of Nature Series. New Jersey: Humanities Press, 1995.

Drees, Willem B. *Beyond the Big Bang: Quantum Cosmologies and God.* La Salle, Ill.: Open Court, 1990.

Driessen, Alfred. "The Question of the Existence of God in the Book of Stephen Hawking: A Brief History of Time." In *Mathematical Undecidability, Quantum Nonlocality and the Question of the Existence of God*, ed. Alfred Driessen and Antoine Suarez, 203–13. Dordrecht: Kluwer Academic Publishers, 1997.

Duhem, Pierre. *Medieval Cosmology: Theories of Infinity, Place, Time, Void, and the Plurality of Worlds*, ed. Roger Ariew. Chicago: University of Chicago Press, 1985.

Dupree, A. Hunter. "Christianity and the Scientific Community in the Age of Darwin." In *God and Nature: Historical Essays on the Encounter between Christianity and Science*, ed. David C. Lindberg and Ronald L. Numbers, 351–68. Berkeley: University of California Press, 1986.

Earman, John. *World Enough and Space-Time: Absolute versus Relational Theories of Space and Time.* Cambridge, Mass.: MIT Press, 1989.

Einstein, Albert. *Ideas and Opinions.* 1954. Reprint. New York: Three Rivers Press, 1982.

——. *Mein Weltbild.* Zürich: Europa Verlag, 1953.

Emerton, Norma. "Arguments for the Existence of God from Nature and Science." In *Science and Theology: Questions at the Interface*, ed. Murray Rae, Hilary Regan, and John Stenhouse, 72–86. Edinburgh: T & T Clark, 1994.

Fergusson, David A. S. "Predestination: A Scottish Perspective." *Scottish Journal of Theology* 46 (1993): 457–78.

Feuerbach, Ludwig. *The Essence of Christianity. With Introductory Essay by Karl Barth and Foreword by H. Richard Niebuhr.* New York: Harper Torchbooks, 1957.

Fine, Arthur. *The Shaky Game: Einstein, Realism and the Quantum Theory.* Science and Its Conceptual Foundations Series. Chicago: University of Chicago Press, 1986.

Fleming, Richard. "Einstein and the Limits of Reason." In *Einstein and the Humanities*, ed. Dennis P. Ryan. Contributions in Philosophy no. 32, 81–86. New York: Greenwood Press, 1987.

Fontaine, P. F. M. *The Light and the Dark: A Cultural History of Dualism.* 14 vols. Amsterdam: J. C. Gieben, 1986–99.

Force, James E. *Essays on the Context, Nature, and Influence of Isaac Newton's Theology* (& Richard H. Popkin). International Archives of the History of Ideas 129. Dordrecht: Kluwer Academic Publishers, 1990.

Gigerenzer, Gerd, Zeno Swijtink, Theodore Porter, Lorraine Daston, John Beatty, and Lorenz Krüger. *The Empire of Chance: How Probability Changed Science and Everyday Life.* Ideas in Context Series. Cambridge: Cambridge University Press, 1989.

Gingerich, Owen. "Is There a Role for Natural Theology Today?" In *Science and Theology: Questions at the Interface*, ed. Murray Rae, Hilary Regan, and John Stenhouse, 29–48. Edinburgh: T & T Clark, 1994.

Gjertsen, Derek. *The Newton Handbook*. London: Routledge & Kegan Paul, 1986.

Gloege, G. "Dualismus–Theologisch." In *Die Religion in Geschichte und Gegenwart: Handwörterbuch für Theologie und Religionswissenschaft*, vol. 2, cols. 274–76. 3d ed. Tübingen: Mohr, 1958.

Grant, Edward. "The Effect of the Condemnation of 1277." In *The Cambridge History of Later Medieval Philosophy: From the Rediscovery of Aristotle to the Disintegration of Scholasticism 1100–1600*, ed. Norman Kretzmann, Anthony Kenny, and Jan Pinborg, 537–39. Cambridge: Cambridge University Press, 1982.

——. *The Foundations of Modern Science in the Middle Ages: Their Religious, Institutional, and Intellectual Contexts*. Cambridge: Cambridge University Press, 1996.

——. "Science and Theology in the Middle Ages." In *God and Nature: Historical Essays on the Encounter between Christianity and Science*, ed. David C. Lindberg and Ronald L. Numbers, 49–75. Berkeley: University of California Press, 1986.

Gray, Bryan J. A. "Theology as Science: An Examination of the Theological Methodology of Thomas F. Torrance." Sacred theology doctorate diss. Katholieke Universiteit te Leuven, 1975.

Gregersen, Nils Henrik, Willem B. Drees, and Ulf Görman, eds. *The Human Person in Science and Theology*. Issues in Science and Theology Series. Edinburgh: T & T Clark, 2000.

Gregory, Frederick. "The Impact of Darwinian Evolution on the Protestant Theology in the Nineteenth Century." In *God and Nature: Historical Essays on the Encounter between Christianity and Science*, ed. David C. Lindberg and Ronald L. Numbers, 369–90. Berkeley: University of California Press, 1986.

——. *Nature Lost? Natural Science and the German Theological Traditions of the Nineteenth Century*. Cambridge, Mass.: Harvard University Press, 1992.

Gunton, Colin E. *The Promise of Trinitarian Theology*. 2d ed. Edinburgh: T & T Clark, 1997.

——. "Relation and Relativity: The Trinity and the Created World." In *Trinitarian Theology Theology Today: Essays on Divine Being and Act*, ed. Christoph Schwöbel, 92–112. Edinburgh: T & T Clark, 1995.

Gustafsson, Bengt. "The Current Scientific World View." In *The New Science-Faith Debate: Probing Cosmology, Technology, and Theology*, ed. John M. Mangum, 1–9. Minneapolis: Fortress Press, 1989.

Guthridge, Joannes, S.J. *The Christology of T. F. Torrance: Revelation and Reconciliation in Christ*. Excerpta ex diss. ad Lauream. Melbourne: Pontificia Universitas Gregoriana, 1967.

Hahn, Roger. "Laplace and the Mechanistic Universe." In *God and Nature: Historical Essays on the Encounter between Christianity and Science*, eds. David C. Lindberg and Ronald L. Numbers, 256–76. Berkeley: University of California Press, 1986.

Hall, Roland. "Monism and Pluralism." In *Encyclopedia of Philosophy*, vol. 5, 363–65. New York: Macmillan, 1967.

Hallamaa, Jaana. "The Concept of Person and God as Trinity of Persons." In *Philosophical Studies in Religion, Metaphysics, and Ethics: Essays in Honour of Heikki Kirjavainen*, ed. Timo Koistinen and Tommi Lehtonen, 140–59. Schriften der Luther-Agricola-Gesellschaft 38. Helsinki: Luther-Agricola-Society, 1997.

Hardy, Daniel W. "Theology through Philosophy." In *The Modern Theologians: An Introduction to Christian Theology in the Twentieth Century*, ed. David F. Ford, 252–85. 2d ed. Oxford: Blackwell Publishers, 1997.

Hartwell, Herbert. *The Theology of Karl Barth: An Introduction*. Philadelphia: Westminster Press, 1964.

Hawking, Stephen W. *A Brief History of Time: From the Big Bang to Black Holes*. London: Bantam Press, 1988.

Heick, Otto W. *A History of Christian Thought*. 2 vols. Philadelphia: Fortress Press, 1965–66.

Heisenberg, Werner. *Physics and Philosophy: The Revolution in Modern Science*. London: George Allen & Unwin, 1959.

Heller, Michael. "Scientific Rationality and Christian Logos." In *Physics, Philosophy, and Theology: A Common Quest for Understanding*, ed. Robert J. Russell, William R. Stoeger, S.J., and George V. Coyne, S.J., 141–50. Vatican: Vatican Observatory, 1988.

Hempel, Carl G. *Philosophy of Natural Science*. Foundations of Philosophy Series. Englewood Cliffs, N.J.: Prentice-Hall, 1966.

Hendry, John. *James Clerk Maxwell and the Theory of the Electromagnetic Field*. Bristol: Hilger, 1986.

Heppe, Heinrich. *Reformed Dogmatics: Set Out and Illustrated from the Sources*. 1950. Reprint. Grand Rapids, Mich.: Baker Book House, 1984.

Herrman, Eberhard. "Realism, Semantics and Religion." In *Philosophical Studies in Religion, Metaphysics, and Ethics: Essays in Honour of Heikki Kirjavainen*, ed. Timo Koistinen and Tommi Lehtonen, 77–94. Schriften der Luther-Agricola-Gesellschaft 38. Helsinki: Luther-Agricola-Society, 1997.

Hesse, Mary B. "Physics, Philosophy, and Myth." In *Physics, Philosophy, and Theology: A Common Quest for Understanding*, ed. Robert J. Russell, William R. Stoeger, S.J., and George V. Coyne, S.J., 185–202. Vatican: Vatican Observatory, 1988.

Hick, John. *Evil and the God of Love*. 2d ed. London: Macmillan, 1977.

——. "Religious Realism and Non-Realism: Defining the Issue." In *Is God Real?*, ed. Joseph Runzo, 3–16. New York: St. Martin's Press, 1993.

Hiebert, Erwin N. "Modern Physics and Christian Faith." In *God and Nature: Historical Essays on the Encounter between Christianity and Science*, ed. David C. Lindberg and Ronald L. Numbers, 424–47. Berkeley: University of California Press, 1986.

Hilgenfeld, Hartmut. *Mittelalterlich-traditionelle Elemente in Luthers Abendmahlsschriften*. Studien zur Dogmengeschichte und systematischen Theologie, no. 29. Zürich: Theologischer Verlag, 1971.

Hooker, C. A. *A Realistic Theory of Science*. Albany: State University of New York Press, 1987.

Hooykaas, R. *Religion and the Rise of Modern Science*. 1973. Reprint. Edinburgh: Scottish Academic Press, 1984.

Howard, Don, and John Stachel, eds. *Einstein and the History of General Relativity*. Einstein Studies, vol. 1. Boston: Birkhäuser, 1989.

Howe, Günter. "Parallelen zwischen der theologie Karl Barths und der heutigen Physik." In *Antwort: Karl Barth zum siebzigsten Geburtstag am 10. Mai 1956*, ed. Rudolf Frey, 409–22. Zollikon-Zürich: Evangelischer Verlag, 1956.

Hume, David. *Dialogues Concerning Natural Religion and the Natural History of Religion*, ed. J. C. A. Gaskin. Oxford: Oxford University Press, 1993.

Hunsinger, George. *How to Read Karl Barth: The Shape of His Theology*. New York: Oxford University Press, 1991.

Hunter, Michael. *Science and the Shape of Orthodoxy: Intellectual Change in Late Seventeenth-Century Britain*. Woodbridge, Suffolk: Boydell Press, 1995.

Isham, C. J. "Creation of the Universe as a Quantum Process." In *Physics, Philosophy, and Theology: A Common Quest for Understanding*, ed. Robert J. Russell, William R. Stoeger, S.J., and George V. Coyne, S.J., 375–408. Vatican: Vatican Observatory, 1988.

Jacob, Margaret C. "Christianity and the Newtonian Worldview." In *God and Nature: Historical Essays on the Encounter between Christianity and Science*, ed. David C. Lindberg and Ronald L. Numbers, 238–55. Berkeley: University of California Press, 1986.

Jaki, Stanley L. "The Absolute Beneath the Relative: Reflections on Einstein's Theories." In *Einstein and the Humanities*, ed. Dennis P. Ryan. Contributions in Philosophy no. 32, 5–18. New York: Greenwood Press, 1987.

——. "Theological Aspects of Creative Science." In *Creation, Christ and Culture: Studies in Honour of T. F. Torrance*, ed. Richard W. A McKinney, 149–66. Edinburgh: T & T Clark, 1976.

——. *The Relevance of Physics*. 2d ed. Edinburgh: Scottish Academic Press, 1992.

——. *Science and Creation: From Eternal Cycles to an Oscillating Universe*. Edinburgh: Scottish Academic Press, 1986.

Jammer, Max. "The Concept of Space in Antiquity." In *Problems of Space and Time*, ed. J. J. C. Smart, 27–42. 1964. Reprint. New York: Macmillan, 1979.

——. "Einstein and Quantum Physics." In *Albert Einstein: Historical and Cultural Perspectives*, ed. Gerald Holton and Yehuda Elkana, 59–76. Princeton, N.J.: Princeton University Press, 1982.

——. *The Philosophy of Quantum Mechanics: The Interpretations of Quantum Mechanics in Historical Perspective*. New York: John Wiley & Sons, 1974.

Jenson, Robert W. "What Is the Point of Trinitarian Theology?" In *Trinitarian Theology Today: Essays on Divine Being and Act*, ed. Christoph Schwöbel, 1–30. Edinburgh: T & T Clark, 1995.

——. "Karl Barth." In *The Modern Theologians: An Introduction to Christian Theology in the Twentieth Century*, ed. David F. Ford, 21–36. 2d ed. Oxford: Blackwell Publishers, 1997.

Jüngel, Eberhard. *Karl Barth, a Theological Legacy*. Philadelphia: Westminster Press, 1986.

Kaiser, Christopher B. *Creation and the History of Science*. History of Christian Theology Series, vol 3. London: Marshall Pickering, 1991.

——. "Quantum Complementarity and Christological Dialectic." In *Religion and Science: History, Method, Dialogue*, ed. W. Mark Richardson and Wesley J. Wildman, 291–98. New York: Routledge, 1996.

Kang, Phee Seng. "The Concept of the Vicarious Humanity of Christ in the Theology of Thomas Forsyth Torrance." Ph.D. diss., University of Aberdeen, 1983.

——. "The Epistemological Significance of ὁμοούσιον in the Theology of Thomas F. Torrance." *Scottish Journal of Theology* 45 (1992): 341–66.

Kannengiesser, Charles. *Arius and Athanasius: Two Alexandrian Theologians*. Collected Studies Series. Hampshire: Variorum, 1991.

Kant, Immanuel. *Critique of Pure Reason*, ed. Vasilis Politis. London: J. M. Dent, 1993.

Kelly, J. N. D. *Early Christian Doctrines*. 5th rev. ed. 1977. Reprint. London: A & C Black, 1989.

Kettler, Christian Delvaux. "The Vicarious Humanity of Christ and the Reality of Salvation." Ph.D. diss., Fuller Theological Seminary, 1986.

Knight, David. *The Age of Science. The Scientific World-view in the Nineteenth Century*. Oxford: Basil Blackwell, 1986.

Knox, Crawford. *Changing Christian Paradigms and Their Implications for Modern Thought*. Studies in the History of Religions Series, no. 57. Leiden: E. J. Brill, 1993.

Knuuttila, Simo. *Modalities in Medieval Philosophy*. London: Routledge, 1993.

——. "Nomic Necessities in Late Medieval Thought." In *Knowledge and the Sciences in Medieval Philosophy: Proceedings of the 8th International Congress of Medieval Philosophy*. Vol. 2, ed. Simo Knuuttila, Reijo Työrinoja, and Sten Ebbesen, 222–30. Publications of Luther-Agricola-Society Series B 19. Helsinki: Luther-Agricola-Society, 1990.

Knuuttila, Simo, and Risto Saarinen. "Innertrinitarische Theologie in der Scholastik und bei Luther." In *Caritas Dei: Beiträge zum Verständnis Luthers und der gegenwärtigen Ökumene. Festschrift für Tuomo Mannermaa zum 60. Geburtstag*, ed. Oswald Bayer, Robert W. Jenson, and Simo Knuuttila, 243–64. Schriften der Luther-Agricola-Gesellschaft 39. Helsinki: Luther-Agricola-Gesellschaft, 1997.

Koyré, Alexandre. *From the Closed World to the Infinite Universe*. New York: Harper Torchbooks, 1958.

Kruger, C. Baxter. "The Doctrine of the Knowledge of God in the Theology of T. F. Torrance: Sharing in the Son's Communion with the Father in the Spirit." *Scottish Journal of Theology* 43 (1990): 366–89.

Kuhn, Thomas S. *The Structure of Scientific Revolutions*. 2d ed. Chicago: University Press of Chicago, 1970.

Küng, Hans. "Karl Barth and the Postmodern Paradigm." *The Princeton Seminary Bulletin* 9, no. 1 (1988): 8-31.

——. *On Being a Christian*. Glasgow: Collins, 1978.

Laeuffer, Jacques. "Scientism and Scientific Knowledge of Things and God." In *Mathematical Undecidability, Quantum Nonlocality and the Question of the Existence of God*, ed. Alfred Driessen and Antoine Suarez, 185–91. Dordrecht: Kluwer Academic Publishers, 1997.

Lakatos, Imre. "Falsification and the Methodology of Scientific Research Programmes." In *Criticism and the Growth of Knowledge*, ed. Imre Lakatos and Alan Musgrave. Proceedings of the International Colloqium in the Philosophy of Science, London, 1965. Vol. 4. 1970. Reprint. London: Cambridge University Press, 1977.

Lamont, Daniel. *Christ and the World of Thought*. Edinburgh: T & T Clark, 1935.

Lanczkowski, Günter. "Dualismus." In *Theologische Realenzyklopädie*, vol. 9, 199–202. Berlin: de Gruyter, 1982.

Laurikainen, K. V. *Message of Atoms: Essays on Wolfgang Pauli and the Unspeakable*. Berlin: Springer Verlag, 1997.

——. "Quantum Physics, Philosophy, and the Image of God: Insights from Wolfgang Pauli." *Zygon* 25, no. 4 (Dec. 1990): 391–404.

Lawhead, William F. "Religion, Relativity, and Common Sense: Einstein and the Religious Imagination." In *Einstein and the Humanities*, ed. Dennis P. Ryan, 37–45. Contributions in Philosophy no. 32. New York, Greenwood Press, 1987.

Leibniz, Gottfried. "The Relational Theory of Space and Time." In *Problems of Space and Time*, ed. J. J. C. Smart, 89–98. 1964. Reprint. New York: Macmillan, 1979.

Levinas, Emmanuel. *Time and the Other [and additional essays]*. Pittsburgh: Duquesne University Press, 1987.

Lindbeck, George A. *The Nature of Doctrine: Religion and Theology in a Postliberal Age*. London: SPCK, 1984.

Lindberg, David C. *The Beginnings of Western Science: The European Scientific Tradition in Philosophical, Religious, and Institutional Context, 600 B.C. to A.D. 1450*. Chicago: University of Chicago Press, 1992.

——. "Science and the Early Church." In *God and Nature: Historical Essays on the Encounter between Christianity and Science*, ed. David C. Lindberg and Ronald L. Numbers, 19–48. Berkeley: University of California Press, 1986.

Lindberg, David C., and Ronald L. Numbers, eds. *God and Nature: Historical Essays on the Encounter between Christianity and Science*. Berkeley: University of California Press, 1986.

Lindroth, Hjalmar. *Kyrklig Dogmatik*. 3 vols. Studia Doctrinae Christianae Upsalensia, 12:1–3. Uppsala: Almqvist & Wiksell, 1975.

Locke, John. *An Essay Concerning Human Understanding*, ed. John W. Yolton. 1976. Reprint. London: J. M. Dent, 1992.

Loder, James E., and W. Jim Neidhardt. "Barth, Bohr, and Dialectic." In *Religion and Science: History, Method, Dialogue*, ed. W. Mark Richardson and Wesley J. Wildman, 271–89. New York: Routledge, 1996.

Lonergan, Bernard J. F. *The Way to Nicea: The Dialectical Development of Trinitarian Theology*. London: Darton, Longmann & Todd, 1976.

Luther, Martin. "Vom Abendmahl Christi." In *D. Martin Luthers Werke. Kritische Gesamtausgabe*, vol. 26, 241–509. Weimar: Hermann Böhlaus Nachfolger, 1909.

Lyman, J. Rebecca. *Christology and Cosmology: Models of Divine Activity in Origen, Eusebius, and Athanasius*. Oxford: Clarendon Press, 1993.

McGrath, Alister E. *Christian Theology: An Introduction*. Oxford: Blackwell, 1994.

——. *The Foundations of Dialogue in Science and Religion*. Malden, Mass.: Blackwell, 1998.

——. *A Life of John Calvin: A Study in the Shaping of Western Culture*. Oxford: Blackwell, 1990.

——. *Thomas F. Torrance: An Intellectual Biography*. Edinburgh: T & T Clark, 1999.

McGuire, J. E. *Tradition and Innovation: Newton's Metaphysics of Nature*. University of Western Ontario Series in Philosophy of Science, vol. 56. Dordrecht: Kluwer Academic Publishers, 1995.

McGuire, J. E., and Martin Tamny. *Certain Philosophical Questions: Newton's Trinity Notebook*. Cambridge: Cambridge University Press, 1983.

McInerny, R. M. "Dualism." In *New Catholic Encyclopedia*, 4:1073–75. New York: McGraw-Hill, 1967.

MacKinnon, D. M. "The Relation of the Doctrines of the Incarnation and the Trinity." In *Creation, Christ and Culture: Studies in Honour of T. F. Torrance*, ed. Richard W. A. McKinney, 92–107. Edinburgh: T & T Clark, 1976.

MacKinnon, Edward. "Complementarity." In *Religion and Science: History, Method, Dialogue*, ed. W. Mark Richardson and Wesley J. Wildman, 255–70. New York: Routledge, 1996.

McMullin, Ernan. "How Should Cosmology Relate to Theology?" In *The Sciences and Theology in the Twentieth Century*, ed. A. R. Peacocke, 17–57. Stocksfield, Northumberland: Oriel Press, 1981.

——. "Natural Science and Belief in a Creator: Historical Notes." In *Physics, Philosophy, and Theology: A Common Quest for Understanding*, ed. Robert J. Russell, William R. Stoeger, S.J., and George V. Coyne, S.J., 49–79. Vatican: Vatican Observatory, 1988.

Macquarrie, John. "Laberthonnière, Lucien." In *The Encyclopedia of Philosophy*, 3:372. 1967. Reprint. New York: Macmillan, 1972.

Mannermaa, Tuomo. *Von Preussen nach Leuenberg: Hintergrund und Entwicklung der theologischen Methode der Leuenberger Konkordie*. Hamburg: Lutherisches Verlagshaus, 1981.

Manuel, Frank E. *The Religion of Isaac Newton*. Oxford: Clarendon Press, 1974.

Marley, Alan G. *T. F. Torrance in a Nutshell*. Nutshell Series, no. 4. Edinburgh: Handsel Press, 1992.

Martikainen, Eeva. "Kommunion der Liebe oder Relationen der Personen?" In *Caritas Dei: Beiträge zum Verständnis Luthers und der gegenwärtigen Ökumene. Festschrift für Tuomo Mannermaa zum 60. Geburtstag*, ed. Oswald Bayer, Robert W. Jenson, and Simo Knuuttila, 287–314. Schriften der Luther-Agricola-Gesellschaft 39. Helsinki: Luther-Agricola-Gesellschaft, 1997.

Maxwell, James Clerk. *A Dynamical Theory of the Electromagnetic Field*, ed. Thomas F. Torrance. Edinburgh: Scottish Academic Press, 1982.

Meardon, Brian H. "A Study of Space, Time and Incarnation in the Work of Thomas F. Torrance." M.Ph. diss., Oak Hill College, London, 1984.

Meijering, E. P. *God Being History: Studies in Patristic Philosophy*. Amsterdam: North-Holland Publishing Company, 1975.

——. *Orthodoxy and Platonism in Athanasius: Synthesis or Antithesis?* 1968. Reprint. Leiden: E. J. Brill, 1974.

Mensching, G. "Dualismus–Religionsgeschichtlich." In *Die Religion in Geschichte und Gegenwart: Handwörterbuch für Theologie und Religionswissenschaft*, vol 2, columns 272–74. 3rd edition. Tübingen: Mohr, 1958.

Merton, Robert K. *Social Theory and Social Structure*. Enlarged ed. New York: The Free Press, 1968.

Metzke, Erwin. *Coinicidentia oppositorum: Gesammelte Studien zur Philosophiegeschichte*, ed. Karlfried Gründer. Forschungen und Berichte der Evangelischen Studiengemeinschaft, vol. 19. Witten: Luther-Verlag, 1961.

Midgley, Mary. *Science as Salvation: A Modern Myth and Its Meaning.* London: Routledge, 1992.

Miller, James Bradley. "Beyond Dualism: Cosmological Issues for Christian Theology in a Postmodern, Post-critical Cultural Context." Ph.D. diss., Marquette University (Milwaukee, Wis.), 1986.

Minkowski, Hermann. "Space and Time." In *Problems of Space and Time*, ed. J. J. C. Smart, 297–312. 1964. Reprint. New York: MacMillan Publishing Company, 1979.

Monod, Jacques. *Chance and Necessity: An Essay on the Natural Philosophy of Modern Biology.* New York: Vintage Books, 1972.

Moore, James R. "Geologists and Interpreters of Genesis in the Nineteenth Century." In *God and Nature: Historical Essays on the Encounter between Christianity and Science*, eds. David C. Lindberg and Ronald L. Numbers, 322–50. Berkeley: University of California Press, 1986.

——. *The Post-Darwinian Controversies: A Study of the Protestant Struggle to Come to Terms with Darwin in Great Britain and America*, 1870–1900. Cambridge: Cambridge University Press, 1979.

Morrison, John Douglas. "The Self-Given Knowledge of God as He Is in the World: Thomas Forsyth Torrance's Christological-Trinitarian Assertion of Scientific Theo-Logical Realism." Ph.D. diss. University of Virginia, 1993.

Morrison, Roy D., II. "Einstein on Kant, Religion, Science, and Methodological Unity." In *Einstein and the Humanities*, ed. Dennis P. Ryan, 47–57. Contributions in Philosophy no. 32. New York: Greenwood Press, 1987.

——. *Science, Theology and the Transcendental Horizon: Einstein, Kant and Tillich.* American Academy of Religion Studies in Religion, no. 67. Atlanta: Scholars Press, 1994.

Muller, Richard A. "The Barth Legacy: New Athanasius or Origen Redivivus? A Response to T. F. Torrance." *Thomist* 54, no. 4 (1990): 673–704.

Murphy, Nancey. "Postmodern Apologetics, or Why Theologians *Must* Pay Attention to Science." In *Religion and Science: History, Method, Dialogue*, ed. W. Mark Richardson and Wesley J. Wildman, 105–20. New York: Routledge, 1996.

——. "What Has Theology to Learn from Scientific Methodology?" In *Science and Theology: Questions at the Interface*, ed. Murray Rae, Hilary Regan, and John Stenhouse, 101–26. Edinburgh: T & T Clark, 1994.

Nebelsick, Harold. P. *Circles of God: Theology and Science from the Greeks to Copernicus.* Theology and Science at the Frontiers of Knowledge, no. 2. Edinburgh: Scottish Academic Press, 1985.

Neher, André. *Jewish Thought and the Scientific Revolution of the Sixteenth Century: David Gans (1541–1613) and His Times.* Oxford: Oxford University Press, 1986.

Neidhardt, W. Jim. "Introduction." In *The Christian Frame of Mind: Reason, Order, and Openness in Theology and Natural Science*, xi–xliv. Colorado Springs, Colo.: Helmers & Howard, 1989.

Newbigin, Lesslie. *The Gospel in a Pluralist Society.* London: SPCK, 1989.

——. *The Other Side of 1984: Questions for the Churches.* Risk Book Series. Geneva: WCC Publishing, 1983.

Newton, Isaac. "God and Gravity: Four Letters to Richard Bentley." In *Newton's Philosophy of Nature: Selections from His Writings*, ed. H. S. Thayer, 46–58. Hafner Library of Classics, no. 16. New York: Hafner Publishing, 1965.

——. *The Mathematical Principles of Natural Philosophy.* Introduction by Alfred Del Vecchio. Science Classics Library. New York: Philosophical Library, 1964.

——. *Opticks: Or, a Treatise of the Reflections, Refractions, Inflections and Colours of Light.* 1730. Reprint. London: G. Bell & Sons, 1931.

Norris, Frederick W. "Mathematics, Physics and Religion: A Need for Candor and Rigor." *Scottish Journal of Theology* 37 (1984): 457–70.

Northrop, F. S. C. "Causation, Determinism, and the 'Good.'" In *Determinism and Freedom in the Age of Modern Science*, ed. Sidney Hook, 201–11. 1961. Reprint. New York: Collier Books, 1979.

Numbers, Ronald L. "The Creationists." In *God and Nature: Historical Essays on the Encounter between Christianity and Science*, ed. David C. Lindberg and Ronald L. Numbers, 391–423. Berkeley: University of California Press, 1986.

Nussbaum, Martha C. "Aristotelian Dualism: Reply to Howard Robinson." In *Oxford Studies in Ancient Philosophi*, vol. 2, ed. Julia Annas, 197–207. Oxford: Clarendon Press, 1984.

Olding, Alan. *Modern Biology and Natural Theology*. London: Routledge, 1991.

Owens, Joseph. "Faith, Ideas, Illumination, and Experience." In *The Cambridge History of Later Medieval Philosophy: From the Rediscovery of Aristotle to the Disintegration of Scholasticism, 1100–1600*, ed. Norman Kretzmann, Anthony Kenny, and Jan Pinborg, 440–59. Cambridge: Cambridge University Press, 1982.

Pais, Abraham. *"Subtle Is the Lord . . .": The Science and the Life of Albert Einstein*. Oxford: Clarendon Press, 1982.

Palma, Robert J. "Thomas F. Torrance's Reformed Theology." *Reformed Review* 38, no. 1 (Autumn 1984): 2–46.

——. "Torrance, Thomas F." *Encyclopedia of the Reformed Faith*. Louisville, Ky.: Westminster/John Knox Press, 1992.

Pannenberg, Wolfhart. *Basic Questions in Theology: Collected Essays*. 2 vols. Philadelphia: Westminster Press, 1983.

——. *Systematic Theology*. 2 vols. Edinburgh: T & T Clark, 1991–94.

——. *Theology and the Philosophy of Science*. London: Darton, Longman & Todd, 1976.

Paul, Iain. *Knowledge of God: Calvin, Einstein, and Polanyi*. Edinburgh: Scottish Academic Press, 1987.

——. *Science and Theology in Einstein's Perspective*. Theology and Science at the Frontiers of Knowledge, no. 7. Edinburgh: Scottish Academic Press, 1986.

Peacocke, Arthur. *Creation and the World of Science*. Bampton Lectures, 1978. Oxford: Clarendon Press, 1979.

——. "The Incarnation of the Informing Self-Expressive Word of God." In *Religion and Science: History, Method, Dialogue*, ed. W. Mark Richardson and Wesley J. Wildman, 321–39. New York: Routledge, 1996.

——. *Theology for a Scientific Age: Being and Becoming–Natural and Divine*. Signposts in Theology Series. Oxford: Basil Blackwell, 1990.

Peat, F. David. *Superstrings and the Search for the Theory of Everything*. 1988. Reprint. London: Abacus Books, 1992.

Peters, Ted. "On Creating the Cosmos." In *Physics, Philosophy and Theology: A Common Quest for Understanding*, ed. Robert John Russell, William R. Stoeger, S.J., and George V. Coyne, S.J., 273–96. Vatican: Vatican Observatory, 1988.

——. "Theology and Science: Where Are We?" *Zygon* 31, no. 2 (June 1996): 323–43.

Pettersen, Alvyn. *Athanasius*. Outsanding Christian Thinkers Series. London: Geoffrey Chapman, 1995.

Polanyi, Michael. *Personal Knowledge: Towards a Post-Critical Philosophy*. Chicago: University of Chicago Press, 1958.

Polkinghorne, John. *One World: The Interaction of Science and Theology*. London: SPCK, 1986.

——. "The Quantum World." In *Physics, Philosophy and Theology: A Common Quest for Understanding*, ed. Robert John Russell, William R. Stoeger, S.J., and George V. Coyne, S.J., 333–42. Vatican: Vatican Observatory, 1988.

——. *Reason and Reality: The Relationship between Science and Theology.* London: SPCK, 1991.

——. *Science and Christian Belief: Theological Reflections of a Bottom-up Thinker.* Gifford Lectures, 1993–94. London: SPCK, 1994.

——. *Science and Creation: The Search for Understanding.* London: SPCK, 1988.

——. *Science and Providence: God's Interaction with the World.* London: SPCK, 1989.

——. *Science and Theology: An Introduction.* London: SPCK, 1998.

Popkin, Richard H. *Essays on the Context, Nature, and Influence of Isaac Newton's Theology* (& James E. Force). International Archives of the History of Ideas 129. Dordrecht: Kluwer Academic Publishers, 1990.

Popper, Karl. *Conjectures and Refutations: The Growth of Scientific Knowledge.* London: Routledge and Kegan Paul, 1963.

——. *Logic of Scientific Discovery.* 1959. Reprint. London: Hutchinson, 1975.

——. *Quantum Theory and the Schism in Physics.* From the *Postscript to The Logic of Scientific Discovery*, ed. W. W. Bartley III. London: Hutchinson, 1982.

——. *Realism and the Aim of Science.* From the *Postscript to The Logic of Scientific Discovery*, ed. W. W. Bartley III. Totowa, N.J.: Rowman and Littlefield, 1983.

Rahner, Karl. *The Theology of the Spiritual Life.* Theological Investigations, vol. 3. Baltimore, Md.: Helicon Press, 1967.

——. "Weltall–Erde–Mensch. Theologische Perspektiven zum Dialog mit den Naturwissenschaften." In *Christilicher Glaube in moderner Gesellschaft.* Enzyklopädische Bibliothek, vol. 3, ed. Franz Böckle, Franz-Xaver Kaufmann, Karl Rahner, and Bernhard Welte, 34–85. Freiburg: Herder, 1981.

Reichenbach, Hans. *The Theory of Relativity and A Priori Knowledge.* Berkeley: University of California Press, 1965.

Reid, Thomas. *An Inquiry Concerning into the Human Mind, on the Principles of Common Sense.* 5th ed. Edinburgh: Bell & Bradfute, 1801.

Richardson, Kurt Anders. "Trinitarian Reality: The Interrelation of Uncreated and Created Being in the Thought of Thomas F. Torrance." Ph.D. diss., University of Basel, 1993.

Richardson, W. Mark, and Wesley J. Wildman, eds. *Religion and Science: History, Method, Dialogue.* New York: Routledge, 1996.

Robinson, Howard. "Aristotelian Dualism." In *Oxford Studies in Ancient Philosophy*, vol. 1, ed. Julia Annas, 123–44. Oxford: Clarendon Press, 1983.

Robinson, N. H. G. "Barth or Bultmann?" *Religious Studies* 14 (1978): 275–90.

Rolston, Holmes, III. "Science, Religion, and the Future." In *Religion and Science: History, Method, Dialogue*, ed. W. Mark Richardson and Wesley J. Wildman, 61–82. New York: Routledge, 1996.

Runzo, Joseph, ed. *Is God Real?* New York: St. Martin's Press, 1993.

Russell, Robert John. "Quantum Physics in Philosophical and Theological Perspective." In *Physics, Philosophy, and Theology: A Common Quest for Understanding*, ed. Robert J. Russell, William R. Stoeger, S.J., and George V. Coyne, S.J., 343–74. Vatican: Vatican Observatory, 1988.

Russell, Robert John, William R. Stoeger, S.J. and Francisco J. Ayala, eds. *Evolutionary and Molecular Biology: Scientific Perspectives on Divine Action.* Vatican: Vatican Observatory Publications, 1998.

Scheurer, P. B., and G. Debrock, eds. *Newton's Scientific and Philosophical Legacy.* International Archives of the History of Ideas 123. Dodrecht: Kluwer Academic Publishers, 1988.

Schins, Juleon M. "Mathematics: A Pointer to an Independent Reality. Penrose's Interpretation of the Gödel and Turing Theorems." In *Mathematical Undecidability, Quantum Nonlocality and the Question of the Existence of God*, ed. Alfred Driessen and Antoine Suarez, 49–56. Dordrecht: Kluwer Academic Publishers, 1997.

Schleiermacher, Friedrich. *The Christian Faith*. 2 vols. New York: Harper Torchbooks, 1963.

Schwab, Wolfgang. *Entwicklung und Gestalt der Sakramententheologie bei Martin Luther*. European University Papers, series 23, vol. 79. Frankfurt: Peter Lang, 1977.

Schwöbel, Christoph. "Christology and Trinitarian Thought." In *Trinitarian Theology Today: Essays on Divine Being and Act*, ed. Christoph Schwöbel, 1-30. Edinburgh: T & T Clark, 1995.

——. "Introduction–The Renaissance of Trinitarian Theology: Reasons, Problems and Tasks." In *Trinitarian Theology Today: Essays on Divine Being and Act*, ed. Christoph Schwöbel, 1-30. Edinburgh: T & T Clark, 1995.

Scotsman. "Moderator Tells Kirk That Virgin Birth Was 'Symbolic.'" *The Scotsman* (3 Jan. 1994): 1.

Seitz, F. *The Science Matrix: The Journey, Travails, Triumphs*. New York: Springer Verlag, 1992.

Sharpe, Kevin J. *From Science to an Adequate Mythology*. Science, Religion and Society Series, no. 1. Auckland: Interface Press, 1984.

Shea, William R. "Galileo and the Church." In *God and Nature: Historical Essays on the Encounter between Christianity and Science*, ed. David C. Lindberg and Ronald L. Numbers, 114-35. Berkeley: University of California Press, 1986.

Sheldon-Williams, I. P. "The Greek Christian Platonist Tradition from the Cappadocians to Maximus and Eriugena." In *The Cambridge History of Later Greek and Early Medieval Philosophy*, ed. A. H. Armstrong, 425-536. Cambridge: Cambridge University Press, 1967.

Shinn, Roger L., and Paul Abrecht, eds. *Faith and Science in an Unjust World: Report of the World Council of Churches' Conference on Faith, Science and the Future*. 2 vols. Geneva: World Council of Churches, 1980.

Sigurdson, Ola. *Karl Barth som den andre: En studie i den svenska teologins Barth-reception*. Stockholm: Brutus Östlings Bokförlag Symposion, 1996.

Simmons, Stephen Arthur. "The Semantics of God and Nature in the Writings of Thomas F. Torrance." Ph.D. diss., University of Chicago, 1995.

Simons, Eberhard. "Dualism." In *Encyclopedia of Theology: A Concise Sacramentum Mundi*, ed. Karl Rahner, 370-74. London: Burns & Oates, 1977.

Sklar, Lawrence. *Philosophy and Spacetime Physics*. Berkeley: University of California Press, 1985.

——. *Space, Time, and Spacetime*. Berkeley: University of California Press, 1974.

Smart, J. J. C., ed. *Problems of Space and Time*. 1964. Reprint. New York: Macmillan, 1979.

Smith, G. "Natural Theology." In *New Catholic Encyclopedia*, 14:61-62. New York: McGraw-Hill, 1967.

Sorabji, Richard. *Time, Creation and the Continuum: Theories in Antiquity and the Early Middle Ages*. London: Gerald Duckworth, 1983.

Soskice, Janet. "Knowledge and Experience in Science and Religion: Can We Be Realists?" In *Physics, Philosophy, and Theology: A Common Quest for Understanding*, ed. Robert J. Russell, William R. Stoeger, S.J., and George V. Coyne, S.J., 173-84. Vatican: Vatican Observatory, 1988.

Spjuth, Roland. *Creation, Contingency and Divine Presence in the Theologies of Thomas F. Torrance and Eberhard Jüngel*. Studia Theologica Lundensia, no. 51. Lund: Lund University Press, 1995.

Sprigge, Timothy L. S. "Relation–III: Internal Relations." In *Handbook of Metaphysics and Ontology*, vol. 2, ed. Hans Burkhardt and Barry Smith, 781-83. München: Philosophia Verlag, 1991.

Stamps, Robert Julian. "The Sacrament of the Word Made Flesh: The Eucharistic Theology of Thomas F. Torrance." Ph.D. diss. St. John's College (Nottingham), 1986.

Stead, Christopher. *Divine Substance*. Oxford: Clarendon Press, 1977.

——. "Greek Influence on Christian Thought." In *Early Christianity: Origins and Evolution to AD 600*, ed. Ian Hazlett, 175-85. 1991. Reprint. London: SPCK, 1993.

Stoeger, William R., S.J. "Key Developments in Physics Challenging Philosophy and Theology." In *Religion and Science: History, Method, Dialogue*, ed. W. Mark Richardson and Wesley J. Wildman, 183-200. New York: Routledge, 1996.

Swinburne, Richard. *The Existence of God*. Oxford: Clarendon Press, 1979.

——. *Faith and Reason*. Oxford: Clarendon Press, 1981.

——. *Revelation: From Metaphor to Analogy*. Oxford: Clarendon Press, 1992.

Tal, Uriel. "Jewish and Universal Social Ethics in the Life and Thought of Albert Einstein." In *Albert Einstein: Historical and Cultural Perspectives*, ed. Gerald Holton and Yehuda Elkana, 297–318. Princeton, N.J.: Princeton University Press, 1982.

Thiselton, Anthony C. *The Two Horizons: New Testament Hermeneutics and Philosophical Description with Special Reference to Heidegger, Bultmann, Gadamer, and Wittgenstein*. Exeter: Paternoster Press, 1980.

Tillich, Paul. *Systematic Theology*. 3 vols. London: SCM Press, 1978.

Tolstoy, Ivan. *James Clerk Maxwell: A Biography*. Edinburgh: Canongate, 1981.

Torrance, Thomas F., ed. *Belief in Science and in Christian Life: The Relevance of Michael Polanyi's Thought for Christian Faith and Life*. Edinburgh: Handsel Press, 1980.

——. *Theological Dialogue between Orthodox and Reformed Churches*. 2 vols. Edinburgh: Scottish Academic Press, 1985–93.

Trook, Douglas Alan. "The Unified Christocentric Field: Toward a Time-Eternity Relativity Model for Theological Hermeneutics in the Onto-Relational Theology of Thomas F. Torrance." Ph.D. diss., Drew University (Madison, N.J.), 1986.

Tuttle, George M. *So Rich a Soil: John McLeod Campbell on Christian Atonement*. Edinburgh: Handsel Press, 1986.

Työrinoja, Reijo. "Nova vocabula et nova lingua." In *Thesaurus Lutheri: Auf der Suche nach neuen Paradigmen der Luther-Forschung*, ed. Tuomo Mannermaa, Anja Ghiselli, and Simo Peura, 221–36. Helsinki: Die Finnishe theologische Literaturgesellschaft, 1987.

Voorhees, Burton H., and Joseph R. Royce. "Einstein and Epistemology." In *Einstein and the Humanities*, ed. Dennis P. Ryan, 25–33. Contributions in Philosophy no. 32. New York: Greenwood Press, 1987.

Vos Jaczn., A., H. Veldhuis, A. H. Looman-Graaskamp, E. Dekker, and N. W. den Bok. "Introduction." In *Contingency and Freedom*, by John Duns Scotus, ed. Vos Jaczn., Veldhuis, Looman-Graaskamp, Dekker, and den Bok, 1–41. Texts and Studies in the History of Philosophy, vol. 42. Dordrecht: Kluwer Academic Press, 1994.

Weber, Max. *The Protestant Ethic and the Spirit of Capitalism*. London: Unwin, 1970.

Weinandy, Thomas G. *The Father's Spirit of Sonship: Reconceiving the Trinity*. Edinburgh: T & T Clark, 1995.

Westfall, Richard S. *Science and Religion in Seventeenth-Century England*. New Haven, Conn.: Yale University Press, 1958.

——. "The Rise of Science and the Decline of Orthodox Christianity: A Study of Kepler, Descartes, and Newton." In *God and Nature: Historical Essays on the Encounter between Christianity and Science*, ed. David C. Lindberg and Ronald L. Numbers, 218–37. Berkeley: University of California Press, 1986.

Westman, Robert S. "The Copernicans and the Churches." In *God and Nature: Historical Essays on the Encounter between Christianity and Science*, ed. David C. Lindberg and Ronald L. Numbers, 76–113. Berkeley: University of California Press, 1986.

White, Andrew Dickson. *A History of the Warfare of Science with Theology in Christendom*. 2 vols. New York: D. Appleton, 1898.

White, Michael. *Isaac Newton: The Last Sorcerer*. London: Fourth Estate, 1998.

Wildman, Wesley J. "The Quest for Harmony: An Interpretation of Contemporary Theology and Science." In *Religion and Science: History, Method, Dialogue*, ed. W. Mark Richardson and Wesley J. Wildman, 41–60. New York: Routledge, 1996.

Williams, Rowan. *Arius: Heresy and Tradition*. London: Darton, Longman and Todd, 1987.

Yandell, Keith E. "Protestant Theology and Natural Science in the Twentieth Century." In *God and Nature: Historical Essays on the Encounter between Christianity and Science*, ed. David C. Lindberg and Ronald L. Numbers, 448–71. Berkeley: University of California Press, 1986.

Yeung, Hing Kau. "Being and Knowing: An Examination of T. F. Torrance's Christological Science." Ph.D. diss., London University, 1993.

Yu, Carver T. *Being and Relation: A Theological Critique of Western Dualism and Individualism*. Theology and Science at the Frontiers of Knowledge, no. 8. Edinburgh: Scottish Academic Press, 1987.

———. "The Principle of Relativity as a Conceptual Tool in Theology." In *Science and Theology: Questions at the Interface*, ed. Murray Rae, Hilary Regan, and John Stenhouse, 180–210. Edinburgh: T & T Clark, 1994.

Zizioulas, John D. *Being as Communion: Studies in Personhood and the Church*. Contemporary Greek Theologians, no. 4. Crestwood, N.Y.: St. Vladimir's Seminary Press, 1993.

Index